Rudolf Seising

The Fuzzification of Systems

Studies in Fuzziness and Soft Computing, Volume 216

Editor-in-chief
Prof. Janusz Kacprzyk
Systems Research Institute
Polish Academy of Sciences
ul. Newelska 6
01-447 Warsaw
Poland
E-mail: kacprzyk@ibspan.waw.pl

Further volumes of this series can be found on our homepage: springer.com

Vol. 202. Patrick Doherty, Witold Łukaszewicz, Andrzej Skowron, Andrzej Szalas
Knowledge Representation Techniques: A Rough Set Approach, 2006
ISBN 978-3-540-33518-4

Vol. 203. Gloria Bordogna, Giuseppe Psaila (Eds.)
Flexible Databases Supporting Imprecision and Uncertainty, 2006
ISBN 978-3-540-33288-6

Vol. 204. Zongmin Ma (Ed.)
Soft Computing in Ontologies and Semantic Web, 2006
ISBN 978-3-540-33472-9

Vol. 205. Mika Sato-Ilic, Lakhmi C. Jain
Innovations in Fuzzy Clustering, 2006
ISBN 978-3-540-34356-1

Vol. 206. A. Sengupta (Ed.)
Chaos, Nonlinearity, Complexity, 2006
ISBN 978-3-540-31756-2

Vol. 207. Isabelle Guyon, Steve Gunn, Masoud Nikravesh, Lotfi A. Zadeh (Eds.)
Feature Extraction, 2006
ISBN 978-3-540-35487-1

Vol. 208. Oscar Castillo, Patricia Melin, Janusz Kacprzyk, Witold Pedrycz (Eds.)
Hybrid Intelligent Systems, 2007
ISBN 978-3-540-37419-0

Vol. 209. Alexander Mehler, Reinhard Köhler
Aspects of Automatic Text Analysis, 2007
ISBN 978-3-540-37520-3

Vol. 210. Mike Nachtegael, Dietrich Van der Weken, Etienne E. Kerre, Wilfried Philips (Eds.)
Soft Computing in Image Processing, 2007
ISBN 978-3-540-38232-4

Vol. 211. Alexander Gegov
Complexity Management in Fuzzy Systems, 2007
ISBN 978-3-540-38883-8

Vol. 212. Elisabeth Rakus-Andersson
Fuzzy and Rough Techniques in Medical Diagnosis and Medication, 2007
ISBN 978-3-540-49707-3

Vol. 213. Peter Lucas, José A. Gámez, Antonio Salmerón (Eds.)
Advances in Probabilistic Graphical Models, 2007
ISBN 978-3-540-68994-2

Vol. 214. Irina Georgescu
Fuzzy Choice Functions, 2007
ISBN 978-3-540-68997-3

Vol. 215. Paul P. Wang, Da Ruan, Etienne E. Kerre (Eds.)
Fuzzy Logic, 2007
ISBN 978-3-540-71257-2

Vol. 216. Rudolf Seising
The Fuzzification of Systems, 2007
ISBN 978-3-540-71794-2

Rudolf Seising

The Fuzzification of Systems

The Genesis of Fuzzy Set Theory and its Initial Applications – Developments up to the 1970s

With 139 Figures

Dr. Rudolf Seising
Medical University of Vienna
Core Unit for Medical Statistics
and Informatics
Spitalgasse 23
1090 Vienna
Austria
Rudolf.Seising@meduniwien.ac.at

Originally published in German "Die Fuzzifizierung der Systeme. Die Entstehung der Fuzzy Set Theorie und ihrer ersten Anwendungen - Ihre Entwicklung bis in die 70er Jahre des 20. Jahrhunderts", Boethius, Band 54 ISBN: 9783515087681, Steiner Franz Verlag, 2005.

The translation of this book was undertaken by Gavin Bruce (Lawrence, Kansas, USA).

The printing of the book was funded by a grant from the Kurt Vogel Foundation (Munich).

Marek Reformat (Edmonton, Canada) kindly provided the photograph used in the front matter.

Library of Congress Control Number: 2007926120

ISSN print edition: 1434-9922
ISSN electronic edition: 1860-0808
ISBN   978-3-540-71794-2 Springer Berlin Heidelberg New York

This work is subject to copyright. All rights are reserved, whether the whole or part of the material is concerned, specifically the rights of translation, reprinting, reuse of illustrations, recitation, broadcasting, reproduction on microfilm or in any other way, and storage in data banks. Duplication of this publication or parts thereof is permitted only under the provisions of the German Copyright Law of September 9, 1965, in its current version, and permission for use must always be obtained from Springer. Violations are liable for prosecution under the German Copyright Law.

Springer is a part of Springer Science+Business Media
springer.com
© Springer-Verlag Berlin Heidelberg 2007

The use of general descriptive names, registered names, trademarks, etc. in this publication does not imply, even in the absence of a specific statement, that such names are exempt from the relevant protective laws and regulations and therefore free for general use.

Typesetting: Integra Software Services Pvt. Ltd., India
Cover design: WMX Design, Heidelberg

Printed on acid-free paper   SPIN: 11898474   42/3100/Integra   5 4 3 2 1 0

For Julia

Foreword by Rudolf Kruse

Fuzzy Set Theory is turning 40 this year. I am very pleased that the book which emerged from Rudolf Seising's Habilitationsschrift (professorial dissertation) is being published in time for this anniversary. I have been working in the field of fuzzy systems for more than 25 years myself, and during this time I have been able to witness the development of the Fuzzy Set Theory as an insider.

One area in which fuzzy methods are typically applied is control engineering. Control systems[1] are most often described and developed by means of mathematical models in the form of systems of differential equations, which normally adhere to physical laws. A common difficulty in using models to create descriptions is the fact that significant idealizations are often necessary in order to move from a concrete problem to a suitable mathematical model. The reasons for this can be attributed to the generally high complexity of the models or to an incomplete understanding of the problem to be described. Lotfi A. Zadeh's idea in the mid-1960s was to develop methods that are frequently more tolerant than traditional exact methods when it comes to phenomena such as uncertainty, fuzziness, incomplete information and high complexity. The great popularity of fuzzy forms of description was sparked by the amazing success of fuzzy logic in cybernetics.

The subject of the development of the Fuzzy Set Theory is particularly exciting to a science historian, especially since most of those involved are still alive and can be consulted. It is for this purpose that Rudolf Seising has traveled often to the University of California in Berkeley in recent years to conduct numerous interviews with Lotfi Zadeh. Zadeh's little office really is a treasure trove of science history; it can be difficult just to find one's way through the mounds of papers to his desk. Seising was granted access to previously unknown original sources and reports that cast more light on the development of the field of fuzzy systems. Moreover, Dr. Seising has utilized a

---

[1] [400] Michels et al. (2006).

huge base of sources in the form of original articles and literature in the field of science history.

There has been a whole slew of popular scientific publications about the development of the Fuzzy Set Theory or about Zadeh himself. Examples include the populist book by Daniel McNeill and Paul Freiberger: *Fuzzy Logic: The Discovery of a Revolutionary Computer Technology – and How it is Changing Our World*[2] and the adorable book by Lotfi's wife Fay Zadeh: *My Life and Travels with the Father of Fuzzy Logic*[3]. To my knowledge, however, the present work is the first treatment of the history of the Fuzzy Set Theory from a science history perspective.

Particularly interesting is Seising's incorporation of the history of fuzzy sets into two developments in the history of sciences and technology. He demonstrates that electrical engineering, with its focus on information and communication theory, as well as multi-valued logic were very influential in the development of the Fuzzy Set Theory.

Until 1965, most scientists subscribed to the notion expressed, for example, by Gottlob Frege in the book *The Basic Laws of Arithmetic* 2, published by Verlag von Hermann Pohle in Jena, 1903: "... A concept that is not sharply limited would be akin to an area that did not have a consistently sharp boundary line, but that in some places was completely blurred and crossed over into its surroundings. That would not actually be an area at all; and thus an unsharply defined concept is incorrectly called a concept. Logic cannot accept concept-like formations such as this; it is impossible to establish precise laws from them. After all, the law of the excluded middle is actually just another form of the requirement that the concept be sharply limited. Any given object x either falls under the concept y or it does not: tertium non datur...".

It is clear that the rejection of this principle by Berkeley's Zadeh, already internationally renowned as an electrical engineer at that time, would provoke passionate reactions in professional circles. Zadeh's paper *Fuzzy Sets*, which appeared in the journal *Information and Control*[4] in 1965, has accordingly also become an extremely frequently cited article. In my (subjective) view, this paper was accepted because it was submitted by a professor of electrical engineering at the university in Berkeley who also happened to be a member of the editorial board of the journal *Information and Control*; additionally, the title *Fuzzy Sets* was very unusual. The paper contains many concepts used in later applications, such as fuzzy operators, fuzzy intersections, fuzzy relations and the combinations of these, as well as convex fuzzy sets. These concepts are still discussed today in every lecture on the subject of fuzzy systems.

In this context, it is interesting to see that earlier related approaches to model imprecision were not accorded the same level of attention. Let me just

---

[2] [370] McNeil et al. Fuzzy (1993).
[3] [574] Zadeh F., Life (1998).
[4] [612] Zadeh, Fuzzy (1965).

mention the paper by the philosopher Max Black, who wrote his article on *Vagueness*[5] in 1937, and the paper by Łukasiewics, who studied many-valued systems of propositional logic in 1930.[6] In his book, Seising defines Zadeh's discrete perception of the theory of fuzzy systems quite conclusively as a general system theory by means of which so-called input-output analyses of systems can be carried out.

During the period Seising describes, the exact mathematical foundations of the methods used in fuzzy systems (such as the precise semantics of degrees of membership) were seldom questioned by users in the industry. This necessarily led to conflicts with proponents of other schools of science. In particular, there were fervent discussions with researchers from the fields of (subjective) probability theory and logic, but even classically trained control engineers were extremely skeptical. It was particularly the mathematical simplicity and tolerance of suboptimal solutions to real problems, qualities that are appreciated today, that were a thorn in the side of many researchers at the time.

Seising reports in detail on these (distribution) struggles and, in his discussion of fuzzy automata, fuzzy algorithms and fuzzy Turing machines, he introduces three examples of areas in which research was being conducted during this early stage. Researchers in this field used the principle of fuzzification, by which any theory involving the concept of sets can be expanded into a corresponding theory with fuzzy sets: The extension principle as well as the representation theorem are applied here, and this sometimes results in interesting generalizations. In 1983 I used this method myself to generalize concepts from statistics – although at the time, the publisher at Reidel Verlag urged me, for various reasons, to publish the results in a book entitled *Statistics With Vague Data*[7] and thus at least avoid using the word "Fuzzy" in the title. As Seising describes, the concept of the first fuzzy logic controller (a simple characteristic diagram controller) was developed by Mamdani and his student Assilian over the course of one weekend with the help of a fuzzification process. This methodology is still routinely used for many simple feedback control problems today, and even the AG4 automatic transmission in the Volkswagen New Beetle was developed using this design methodology.

Today fuzzy techniques, especially those logical combinations of different methods that have been developed in the last few years (such as neuro-fuzzy systems), are being applied in numerous other technical and non-technical fields and have meanwhile also become part of a standard education at universities.

This account of the history of the Fuzzy Set Theory is very well done, in my opinion. I consider it technically and factually correct.

---

[5] [83] Black, Vagueness (1937).
[6] [331] Łukasiewics, Remarks (1930).
[7] [300] Kruse et al., Statistics (1987).

I hope the book enjoys broad distribution, and not just among historians of science and technology but also among all scientists and engineers who use fuzzy methods and have ever asked themselves how this theory of fuzzy sets came about.

Magdeburg, Germany *Rudolf Kruse*
June 2006

Preface to the English Edition

The German edition of this book was published in the summer of 2005. It grew out of my habilitation thesis on the subject of the history of the natural sciences, which I submitted at Ludwig-Maximilians University in Munich in the summer of 2003 and which was accepted a year later.

Since the book was originally written for readers in the German-speaking world, a significant portion of the cited literature (particularly in chapter 2) refers to works written in German. I also frequently used German translations of English literature. In some instances, I was not able to track down the exact word-for-word quotations in the English-language sources. I ask for understanding in these cases.

The publication of the English edition of this book just two years after the German original is due to the support of many people, whom I would like to mention here:

Gavin Bruce (Lawrence, Kansas) translated the German text into what I consider to be excellent English. For this reason, I would also like to thank Sabine Weisheit (Norderstedt) and Devika Seupaul (Berlin) for their assistance in the search for the best translator for this book: They found him!

Ecaterina Dodu (Vienna) took on the arduous task – as she did for the German edition – of importing the text into LaTeX, repeatedly carrying out any necessary corrections and converting the original German indices of names and subjects into English-language indices.

I owe a debt of gratitude to Prof. Menso Folkerts (Munich) and the Kurt Vogel Foundation for a very generous grant to cover printing costs, which allowed me to realize this project.

Prof. Lotfi Zadeh (Berkeley) has encouraged and supported me with constant interest and assistance over the many years of this project. Prof. Rudolf Kruse (Magdeburg), Prof. Janusz Kacprzyk (Warsaw) and Prof. Witold Pedrycz (Edmonton) have repeatedly urged me to bring out an English version since the German edition was published.

I thank Prof. Reinhard Siegmund-Schultze (Kristiansand) for valuable suggestions which resulted in the correction of a few minor errors in the original edition. Other such corrections in this edition are the work of Julia Limberg (Vienna), who proofread large portions of the nearly completed manuscript over the final days of this project and assisted me with tips on LaTeX and Photoshop. Prof. Martin Posch (Vienna) and (once again) Reinhard Pitsch (Vienna) also helped me many times with their knowledge of LaTeX.

The Franz Steiner Verlag (Stuttgart), which published the German edition of the book, granted me permission to produce an English version. I am very grateful to the Springer Verlag (Heidelberg) and in particular to Dr. Thomas Ditzinger (Heidelberg) for helping this edition find its way onto the publisher's list; likewise to Prof. Janusz Kacprzyk (Warsaw), who accepted the book into the series Studies in Fuzziness and Soft Computing.

Finally, I would like to thank Dr. Marek Reformat (Edmonton) for the photograph which he provided to the publisher and which graces the cover of this book. The picture was taken at the celebration of "40 Years of Fuzzy Sets and 20 Years of IFSA" during the 11th International Fuzzy Systems Association World Congress (IFSA2005) that took place on July 28-31, 2005, at Tsinghua University, Beijing.

In conclusion, to all of the people and institutions listed in the foreword to the German edition I gladly and emphatically reiterate my thanks.

Vienna, Austria *Rudolf Seising*
January 2007

Preface

It will be the summer of 2005 when this book is published, thus halfway through the "Einstein year" and possibly a suitable time to mention another event, albeit one not quite as momentous and also not quite 100 – not even 50 – years ago: The Fuzzy Set Theory will be 40 years old this summer.

The anniversaries of the deaths of Carl Friedrich Gauß and Albert Einstein will have been commemorated midway through 2005. Gauß passed away on February 23, 1855; Einstein on April 18, 1955. For all of us who have always moved much more slowly than the speed of light (and will also almost certainly continue to do so for the rest of our lives), this means that Gauss died 150 years ago and Einstein 50 years ago and that, according to our time scale, their deaths occurred approximately 100 years apart. In other systems moving relative to us, this may not be the case!

We humans are far from being able to travel at light speed, and this limitation highlights the important opportunity to celebrate the year 2005 as an Einstein year: One hundred years ago, Einstein devised the Special Theory of Relativity. It states, among other things, that time passes more slowly in systems which, compared to us, move at nearly the speed of light. If there were intelligent life forms in these systems, then they would at the very least not celebrate the same anniversaries.

The year 1905 is referred to as Einstein's "annus mirabilis". The five works he published that year seem "miraculous", and with them he revolutionized 20th century physics and left a lasting impression on the history of science in the 20th century. In these papers, Einstein established the Special Theory of Relativity[8], he analyzed Brownian motion[9] and he explained the photoelectric effect on the basis of the assumption borrowed from Max Planck that light is quantized.[10] According to the Special Theory of Relativity, time passes relative to its frame of reference. Time, like the coordinates of a space, is

---

[8] [157] Einstein, Elektrodynamik (1905).
[9] [158] Einstein, Wärme (1905).
[10] [159] Einstein, Licht (1905).

accordingly not an absolute and thus not a universally observable quantity, as it had been considered until that time. It would henceforth be identified as a relative quantity!

In his work on Brownian molecular motion, Einstein distanced himself from the hypothesis that the velocity of small particles (molecules) could be observed. Instead, he introduced into this theory of Brownian motion the concept of "mean displacement" as a very well observable quantity.

Einstein's hypothesis of light quanta turned out to be a building block for the subsequent fundamental physical theory of quantum mechanics, though Einstein never agreed with the Copenhagen interpretation. That interpretation of quantum theory is based largely on Werner Heisenberg's uncertainty relations. The young Werner Heisenberg conducted one of his "conversations in the field of atomic physics" with Albert Einstein in the spring of 1926. In a paper which preceded this conversation, Heisenberg had distanced himself from the assumption that the electrons in an atom move on paths. Instead of such hypothetical electron paths, he considered the oscillation frequencies and amplitudes of the observable radiation that is emitted by electrons during the discharge process. In contrast to the unobservable electron paths, he explained to Einstein, who had invited him to his home, these observable quantities were more plausible quantities for the new theory. A discussion then developed between the two scientists, which Heisenberg later recalled as follows:

> "But you don't actually believe," countered Einstein, "that it is possible in a physical theory to record only observable quantities." "I thought", I asked astounded, "that you of all people would have made this idea a basis for your Theory of Relativity? After all, you stressed that we cannot speak of absolute time, since it is not possible to observe this absolute time. Only what clocks tell us, be it in a moving or stationary frame of reference, is authoritative in determining time". "Maybe I did use a philosophy like this", Einstein answered, "but it is nonsense nonetheless. Or I can state more cautiously that it may be heuristically valuable to remember what one is observing. However, from a fundamental standpoint, it is entirely false to want to base a theory on observable quantities alone. For in reality it is just the opposite. It is the theory that determines what one can observe".[11]

---

[11] „„Aber Sie glauben doch nicht im Ernst" entgegnete Einstein, „dass man in eine physikalische Theorie nur beobachtbare Größen aufnehmen kann." „Ich dachte", fragte ich erstaunt, „dass gerade Sie diesen Gedanken zur Grundlage Ihrer Relativitätstheorie gemacht hätten? Sie hatten doch betont, dass man nicht von absoluter Zeit reden dürfe, da man diese absolute Zeit nicht beobachten kann. Nur die Angaben der Uhren, sei es im bewegten oder im ruhenden Bezugssystem, sind für die Bestimmung der Zeit maßgebend." „Vielleicht habe ich diese Art von Philosophie benützt" antwortete Einstein, „aber sie ist trotzdem Unsinn. Oder ich kann vorsichtiger sagen, es mag heuristisch von Wert sein, sich daran zu erinnern,

Differentiating between observable and unobservable quantities is eminently important in Einstein's theoretical deliberations. However, with regard to his "cautious" assertions that, "from a fundamental standpoint", the theory determines what is observed, it must be said that it is not possible to adhere to this "fundamental standpoint" for the experimental side of the natural sciences, since real things behave differently and thus result in observations that diverge from what their ideal counterparts would lead one to expect in theory. Although the theory may determine what can be observed, the precise values of the observed quantities that are predicted by the theory do not occur as precisely in reality. Einstein had already used this discrepancy between theory and practice in the natural sciences as the subject of a public lecture delivered on January 27, 1921 on the occasion of the 209th birthday of Kaiser Frederick II. The title of his speech was *Geometry and Experience*, and its text begins as follows:

> Mathematics enjoys particular prestige, above all other sciences, for one reason; its laws are absolutely certain and indisputable, while those of all other sciences are debatable to some degree and are in constant danger of being overturned by newly discovered facts. In spite of this, a researcher in another field would not need to envy the mathematician if his laws were based not on objects of reality but only on objects of our mere imagination. For it cannot come as a surprise that people arrive at the same logical conclusions when they have agreed upon the fundamental laws (axioms) as well as the methods by which other laws are to be deduced from these fundamental laws. On the other hand, however, the great prestige of mathematics is based on the fact that it is also mathematics that affords the exact natural sciences a certain measure of security, which they could not attain without mathematics. At this point, though, a mystery arises which has troubled researchers so much throughout the ages. How is it possible that mathematics, which is after all a product of human thought and is independent of experience, is so beautifully suited to the objects of reality? Can it be that human reason is able to fathom the properties of real things merely by thinking, without experience? In my opinion, the brief answer to this question is this: Insofar as the laws of mathematics relate to reality, they are not certain; and insofar as they are certain, they do not refer to reality.[12].

---

was man wirklich beobachtet. Aber vom prinzipiellen Standpunkt aus ist es ganz falsch, eine Theorie nur auf beobachtbare Größen gründen zu wollen. Denn es ist ja in Wirklichkeit genau umgekehrt. Erst die Theorie entscheidet darüber, was man beobachten kann. " "[238] Heisenberg, Teil (1969), p. 79f.

[12] „Die Mathematik genießt vor allen anderen Wissenschaften aus einem Grunde ein besonderes Ansehen; ihre Sätze sind absolut sicher und unbestreitbar, während die aller anderer Wissenschaften bis zu einem gewissen Grad umstritten und stets in Gefahr sind, durch neu entdeckte Tatsachen umgestoßen zu werden. Trotz-

There exists a rift of uncertainty between mathematical theory on the one hand and actual objects and phenomena on the other. The accordance between experimental results and the theoretically predictable values of observable quantities is exact only in the rarest of instances and, in most cases, is only approximate. There is an area of "uncertainty" or "imprecision" or "fuzzines" here. For a long time, probability theory and statistics were the only tools with which an attempt was made to deal with this area. By developing the Fuzzy Set Theory in the mid-1960s, the electrical engineer Lotfi Zadeh provided a further scientific tool to manage the "incompatibility" between theory and practice in science and technology. It was with this in mind that in 1973 he formulated the "principle of incompatibility" that has been cited so often since:

> The closer one looks at a "real world" problem,
> the fuzzier becomes its solution.

Zadeh's corresponding explanations also show that the relationship, which he chose to characterize with the word "fuzzy", between mathematical-logical statements and the real things to which said statements refer represents the uncertainty between mathematics and reality to which Einstein had alluded over 50 years earlier:

> Stated informally, the essence of this principle is that as the complexity of a system increases, our ability to make precise and yet significant statements about its behaviour diminishes until a threshold is reached beyond which precision and significance (or relevance) become almost mutually exclusive characteristics.

This book will describe and interpret the history of the Fuzzy Set Theory and its initial applications. In so doing, the view will be taken and demonstrated that the observable rift in electrical engineering between mathematical

---

dem brauchte der auf einem anderen Gebiete Forschende den Mathematiker noch nicht zu beneiden, wenn sich seine Sätze nicht auf Gegenstände der Wirklichkeit, sondern nur auf solche unserer bloßen Einbildung bezögen. Denn es kann nicht wundernehmen, dass man zu übereinstimmenden logischen Folgerungen kommt, wenn man sich über die fundamentalen Sätze (Axiome) sowie über die Methoden geeinigt hat, vermittels welcher aus diesen fundamentalen Sätzen andere Sätze abgeleitet werden sollen. Aber jenes große Ansehen der Mathematik ruht andererseits darauf, daß die Mathematik es auch ist, die den exakten Naturwissenschaften ein gewisses Maß von Sicherheit gibt, das sie ohne Mathematik nicht erreichen könnten. An dieser Stelle nun taucht ein Rätsel auf, das Forscher aller Zeiten so viel beunruhigt hat. Wie ist es möglich, daß die Mathematik, die doch ein von aller Erfahrung unabhängiges Produkt des menschlichen Denkens ist, auf die Gegenstände der Wirklichkeit so vortrefflich paßt? Kann denn die menschliche Vernunft ohne Erfahrung durch bloßes Denken Eigenschaften der wirklichen Dinge ergründen? Hierauf ist nach meiner Ansicht kurz zu antworten: Insofern sich die Sätze der Mathematik auf die Wirklichkeit beziehen, sind sie nicht sicher und insofern sie sicher sind, beziehen sie sich nicht auf die Wirklichkeit."[160] Einstein, Geometrie (1921)

theory on the one hand and actual electrical circuits on the other hand is what led to the creation of fuzzy sets and to the establishment of the Fuzzy Set Theory as an "expanded" theory of imprecise amounts. It is important to mention that this theory is not a "fuzzy" mathematical theory – it is mathematically exact! It is the basic elements, the fuzzy sets, that lack sharpness, and for a particular reason, as the present book shall clarify.

My first and extraordinary thanks go to Dr. Lotfi Zadeh (Berkeley, California), who enthusiastically and encouragingly supported my project of writing a history of the Fuzzy Set Theory he created, who over the last six years has willingly sat for question and answer sessions in interviews, innumerable conversations and in meetings at the margins of conferences and workshops whenever he was able to do so or happened to be on the same continent. He and his wife Fay hosted me generously on many occasions. It is thanks to their helpfulness that I have many other materials for this book: a number of photographs taken by the avid amateur photographer Lotfi Zadeh (which are not fuzzy at all), letters, articles and other sources from his private papers.

I obtained more private photographs from Rosemary Menger Gilmore (Chicago), Elie Sanchez (Marseille) and Ebrahim Mamdani (London). The sources of the other images from archives, books and periodicals as well as reprint permissions can be found in the index of illustrations. Unfortunately, in some cases I was unable to contact the rights holders to images. I request that they contact me as appropriate.

Most sincere thanks go to Prof. Ivo Schneider (Munich) for the great deal of help and support I received from him while working on this book. As his assistant in the field of the history of science, I traveled twice to the US for several weeks and took many shorter trips within Germany to meet with further interview partners. I would also like to thank Ivo Schneider for his support in my completion of the professorial qualification process in science history with this work. I thank him as well as Prof. Menso Folkerts (Munich) and Prof. Rudolf Kruse (Magdeburg) for their expert opinions and for tips that assisted my work on this book. I thank Prof. Folkerts very much for accepting this book into the Boethius Series and Prof. Kruse for his preface.

I would like to express particular gratitude to Dr. Cathryn Carson, director of the Office for the History of Science and Technology in Berkeley, California. Cathryn Carson invited me to all three of my multi-week stays in Berkeley, and her immense help in finding my bearings in an unfamiliar environment, both professionally and organizationally, was by no means taken for granted. I have her to thank as well for my contact with David Farrell at the Bancroft Library in Berkeley, California, where we hopefully managed to lay the groundwork for a Lotfi Zadeh Collection. I'd also like to thank Diane West and Alex Wellerstein, who assisted me very often in word and deed on the Berkeley campus. The University of the Bundeswehr Munich in Neubiberg and the Medical Faculty of the University of Vienna (now the Medical University of Vienna) each provided me with financial support for one US trip; the third journey was made possible for Prof. Walburga von Zameck und Prof. Bernd Köhler (both of Neubiberg). I owe them all many thanks.

Prof. Andreas Nürnberger and Tanja Falkowski (Magdeburg), who were living in Richmond (and working at the university in Berkeley) at that time, became invaluable dialogue partners during my last two visits to Berkeley. I thank them for their friendly assistance!

Thanks to Dr. Uwe Langer (Munich) for information about a number of details pertaining to electrical engineering, which he explained to me during several friendly conversations, Dr. Michael Eckert (Munich) for helpful tips and discussions on Arnold Sommerfeld, Dr. Michael Plail (Wörthsee) for directing me to an editorial by Lotfi Zadeh, Timo Christoph (then in Neubiberg), who conducted a number of interviews with me, and Steffen Vogelreuter (then in Neubiberg) for unconventional assistance with many computer problems. I would like to thank Christoph Adl and Reinhard Pitsch (both in Vienna), who assisted me with various problems I had with LaTeX. Reinhard Pitsch was an invaluable help to me in the last few weeks before the book was printed.

I thank Michaela Fritsch and Erna Kirchner (both in Neubiberg) for transcribing the recordings of many interviews, Prof. Erich Peter Klement for several chats and tips which have contributed to the development of this book and Prof. Klaus-Peter Adlassnig for his benevolent support over the last three years I was an assistant at the Section on Medical Expert and Knowledge-Based Systems at the University of Vienna.

Sincere thanks to all of my interviewees for their willingness to remember and to answer my questions: Prof. Klaus-Peter Adlassnig (Vienna), Prof. Hans Bandemer (Halle), Dr. James C. Bezdek (Pensacola, Florida), Prof. Steffen Bocklisch (Chemnitz), Prof. Dimiter Driankov (Munich), Prof. Didier Dubois (Paris), Dr. Rudolf Felix (Dortmund), Dr. Erwin Gersthofer (Munich), Prof. Ulrich Höhle (Wuppertal), Dr. Josef Goguen (Los Angeles), Prof. Siegfried Gottwald (Leipzig), Prof. Peter Hajek (Prague), Prof. Harro Kiendl (Dortmund), Prof. Erich Peter Klement (Linz), Prof. Rudolf Kruse (Magdeburg), Dr. George Lakoff (Berkeley, California), Dr. Karl Lieven (Aachen), Prof. Ebrahim Mamdani (London), Dr. Rainer Palm (Munich), Prof. Manfred Peschel (Grossschönau), Prof. Zdzislaw Pawlak (Warsaw), Dr. Paul-Theo Pilgram (Munich), Dr. Michael Reinfrank (Munich), Dr. Enrico Ruspini (Stanford, California), Prof. Bernd Reusch (Dortmund), Prof. Kokichi Tanaka (Tokyo), Dr. Richard Weber (Aachen), Dr. Lotfi Zadeh and his wife Mrs. Fay Zadeh (Berkeley, California), Prof. Hans-Jürgen Zimmermann (Aachen).

I would like to thank David Farrell, Curator for History of Science and Technology at the Bancroft Library in Berkeley, for much assistance and his commitment to establish a "Lotfi Zadeh Collection" at the Bancroft Library. I hope it will be possible to make this plan a reality.

My thanks to Dr. Wolfgang Foit (Holzkirchen) for his consistent constructive criticism, helpfulness and friendship as well as for his generous aid in compiling the bibliography.

Thanks to Dr. Christian Schuh (Vienna) for his assistance over the last three years, which apparently always seemed to be a matter of course to him but certainly was not to me.

A very special thank you to Bärbel Rutert (Duisburg), who read the manuscript of my professorial dissertation and modified it according to the new German spelling conventions – very difficult work that cannot be praised too highly. Ms. Ecaterina Dodu (Vienna) converted the entire book manuscript from Word into LaTeX and then corrected the resulting files again and created the present layout. Furthermore, she was of inestimable assistance in compiling the index. I thank her for performing all of these time-sensitive and tedious tasks! Naturally, the fault for any remaining errors or lapses in clarity is mine.

I would like to express my gratitude to the Kurt Vogel Foundation (Munich) for a generous subsidy for printing costs, without which this book would not have been published in the beautiful form in which it now appears. This is also thanks to the Franz Steiner Verlag (Stuttgart) and in particular to Dr. Thomas Schaber and Ms. Angela Höld for their excellent cooperation.

Finally, I would like to thank Julia Marquardt (Munich) for the patience and understanding she exhibited during the creation of this book; I dedicate this book to her.

Vienna, Austria *Rudolf Seising*
May 2005

Contents

Introduction 1

1 Beginnings of Communication Technology in the 20th Century 7
   1.1 Communication Network 9
   1.2 Network Analyses 20
   1.3 Ernst A. Guillemin, Lotfi A. Zadeh, Dr. Jekyll and Mr. Hyde . 32
   1.4 Ideal and Optimum Filters 39
   1.5 Generalizations.. 45
   1.6 Let's go digital! 48
       1.6.1 A View from 1950 into the Future of 1965 55
       1.6.2 System Theory................................... 57
       1.6.3 From System Theory to the Theory of Fuzzy Systems .. 63

2 Logical Tolerance, *Ensembles Flous* and *Probabilistic Metrics* 65
   2.1 The Vienna Circle 65
   2.2 Karl Menger: "From the Criterion of Meaning to the Principle of Tolerance" 77
       2.2.1 Menger's Conventionalism 85
       2.2.2 Alfred Tarski's Support 89
       2.2.3 The Principle of Logical Tolerance 93
   2.3 Probabilistic or Statistical Metrics and *Ensembles Flous* 98

3 General Systems Theory and Cybernetics 111
   3.1 General Systems Theory................................ 113
   3.2 Cybernetics... 120
   3.3 Communication Theory 129
   3.4 "The Cybernetics Group" 137
   3.5 Von Neumann, the Computer and the Brain 148
   3.6 Pattern Recognition with the Perceptron 153
   3.7 Automata 158

4 From Circuit Theory to System Theory 165
- 4.1 Dynamic Programming 167
- 4.2 Identification Problems and a "Time Out" in Princeton 170
- 4.3 "Optimal Fetishism", Expanded Linearity Concepts and Adaptive Systems 176
  - 4.3.1 Optimality 177
  - 4.3.2 Linearity 179
  - 4.3.3 "Non-Inferiority" 180
  - 4.3.4 Adaptivity 183
- 4.4 Zadeh's System Theory 186
  - 4.4.1 The State Space Approach 188
  - 4.4.2 A Renaissance of General Systems Theory? 192
  - 4.4.3 System States as Input-Output State Relation Pairs ... 195
- 4.5 Zadeh's Decision 197

5 Fuzzy Sets and Fuzzy Systems 201
- 5.1 The Genesis of Fuzzy Set Theory 203
- 5.2 Fuzzy Sets and Fuzzy Systems 212
  - 5.2.1 Fuzzy Sets 213
  - 5.2.2 Fuzzy Systems 215
  - 5.2.3 Fuzzy Classes of Systems 216
  - 5.2.4 Optimization under Fuzzy Conditions 217
  - 5.2.5 First Papers on Fuzzy Sets 218
- 5.3 The Article *Fuzzy Sets* 219
- 5.4 An Interpretation for Unions and Intersections 224
- 5.5 System Theory and Fuzzy Systems 228

6 Fuzzifications 235
- 6.1 Reactions 235
- 6.2 Fuzzy Automata 241
- 6.3 Fuzzy Algorithms 245
- 6.4 Fuzzy Turing Machines 254
- 6.5 Fuzzifications of Elements of Mathematical Theory 254
- 6.6 Other Fuzzifications 259
- 6.7 Fuzzy Control 266
- 6.8 The First "Fuzzy Logic Controller" for Controlling a Steam Engine 270

7 The Fuzzification of Medical Diagnostics 283
- 7.1 Mechanization in Medical Diagnostics 286
- 7.2 Biomedicine and Digital Computers 288
- 7.3 Computer Systems in Medicine 303
- 7.4 Computer-Assisted Diagnosis at Vienna General Hospital 307
  - 7.4.1 CADIAG 309
  - 7.4.2 CADIAG-I 314
  - 7.4.3 From CADIAG-I to CADIAG-II 317

7.5	Fuzziness in Medicine	320
	7.5.1 "Medical Knowledge"	325
	7.5.2 CADIAG-II	331

8 Conclusion 337

Afterword by Lotfi A. Zadeh 351

References 353

List of Figures 391

Index of Names 399

Index of Subjects 405

Introduction

Fuzzy Sets – Lotfi Zadeh chose this designation in 1964 as a name for the basic elements of his new mathematical theory. It was unusual and provocative. The word *fuzzy* means *blurred, vague, indistinct*. As an attribute of *sets*, meaning *quantities*, this nomenclature seemed peculiarly contradictory. In the Set Theory devised by Georg Cantor, *collections* of the objects we perceive or think are referred to as sets and the objects belonging to said sets are called their elements. For every object, there is a clear determination as to whether or not it is an element of a particular set.

Fuzzy sets – blurred, vague, indistinct quantities – this means that sets and their surroundings, their complement within the universe of discourse, cannot be clearly separated from one another. In the usual graphic representation of sets as circles, fuzzy sets appear to be sets with indistinct edges. There are objects at the margins that are elements of these sets only to a certain degree. They are elements both of the fuzzy set and of its complement, and therefore the complement of a fuzzy set is also a fuzzy set.

Fuzzy Sets – this was the title of Zadeh's first article about his new mathematical theory, which appeared in a scientific journal in 1965. In it he defined the basic concepts and – similar to the example of the conventional Set Theory – combination operations such as intersection and union of fuzzy sets, fuzzy relations and a separation theorem with applications in the area of pattern recognition. This application aspect is a first indication of an interesting story of the theory's origins.

If popular scientific books about the Fuzzy Set Theory discuss its historical development at all, they do so only very briefly and on a very superficial level.[13] Zadeh reportedly generalized the Set Theory into the Fuzzy Set Theory after it suddenly occurred to him. At that time, he was engaged in the study of complex systems and recognized that the path of precision could

---

[13] See [370] McNeill et al., Fuzzy (1993), [296] Kosko, Thinking (1994), [147] Drösser, Fuzzy (1994), [295] Kosko, Fuzzy (2001).

not be traveled endlessly in science, since one would soon become a prisoner of strict argumentation. *High complexity and high precision are not reconcilable.*[14] Zadeh formulated a *principle of incompatibility* in 1973 by stating: *The closer one looks at a "real world" problem, the fuzzier becomes its solution.* He then explained:

> Stated informally, the essence of this principle is that as the complexity of a system increases, our ability to make precise and yet significant statements about its behaviour diminishes until a threshold is reached beyond which precision and significance (or relevance) become almost mutually exclusive characteristics.[15]

Zadeh had arrived at the conviction that we encounter classes or sets of objects in the real world that do not have precisely defined criteria of inclusion:

> For example, the class of animals clearly includes dogs, horses, birds, etc. as its members, and clearly excludes such objects as rocks, fluids, plants, etc. However, such objects as starfish, bacteria, etc. have an ambiguous status with respect to the class of animals.[16]

His examples illustrate that areas of indistinctness are produced in the linguistic concepts we use to describe sets in the real world. In contrast to the sharply defined sets in exact mathematics, it is fuzzy sets that we deal with in the empirical sciences and technology.

It is not the task of popular scientific books to address the historical development of their subjects. In these publications, it is surely sufficient to simplify and condense and largely forego historical research. The great success of the Fuzzy Set Theory in the field of technical applications since the 1980s has awakened widespread interest in explanations of the principles of this theory and its applications that can be understood by all. The books by Kosko[17], McNeill and Freiberger[18] and Drösser[19] have been able to meet this demand in various ways.

The questions as to why and how this theory came to be have heretofore not been answered. Now 40 years later, this book, which is built upon years of studies, represents a first attempt to trace the developments in scientific and technological history that led to the Fuzzy Set Theory and its initial applications.

The Fuzzy Set Theory did not arise within mathematics as a generalization of Set Theory. Zadeh was not and is not a mathematician who would have been interested in establishing an abstract theoretical construct. His interest lay in

---

[14] L. A. Zadeh in an interview with M. Chaouli in *highTech*, 11, 1990, p. 50
[15] [641] Zadeh, Outline (1973), p. 28
[16] [612] Zadeh, Fuzzy (1965), p. 338
[17] [296] Kosko, Thinking (1994), [295] Kosko, Fuzzy (2001)
[18] [370] McNeill et al., Fuzzy (1993)
[19] [147] Drösser, Fuzzy (1994)

the mathematical principles of technical applications, and he soon noticed the vast gulf that had opened up between theory and practice. When he created the Fuzzy Set Theory, it was an attempt to close this gap between pure mathematics and technical application.

The Fuzzy Set Theory was Lotfi Zadeh's work alone, and many of the developments that built upon it – "fuzzifications" – were initiated and conducted by him as well. Zadeh constantly expanded his theory over the decades, continues to do so and completes an immense program of travel every year to speak at many international congresses and workshops on fuzzy systems, " soft computing" or "computational intelligence". Today he remains the central personality in the scientific community of the Fuzzy Set Theory and its applications. The history of the Fuzzy Set Theory is so closely related to Zadeh's life story that the historical-biographical approach seemed the most appropriate, particularly as Zadeh himself and many other scientists who were involved in the development of the theory and its applications could be interviewed as contemporary witnesses. A not inconsiderable portion of the sources for this book is therefore based on interviews with Zadeh that were possible during the period from 1998 to 2002.[20] During the same period, interviews were also carried out with numerous other pioneers of the Fuzzy Set Theory and its technological applications. I was able to view all of Zadeh's early publications on the Fuzzy Set Theory, including university and research reports and almost all of his previous publications, and it was possible to include most of them. I was also granted access to several documents and writings from Zadeh's private papers. Finally, I also had the opportunity to question protagonists who were instrumental in devising the first technical applications.[21]

Although this book on the history of the Fuzzy Set Theory has a strong biographical character[22] and although the temptation was great, this did not turn into a biography. A description of Lotfi Zadeh's life to this point would have required many times the source research and text than was already necessary for this book. A biography would have had to elaborate upon a great number of events, aspects and background information that this book mentions only briefly or not at all. Such meticulousness would have come at the expense of many important developments and aspects of the creation of the Fuzzy Set Theory and its early applications. For this reason, the intention

---

[20] See the bibliography

[21] Not all of the interviews are quoted verbatim in this work. In many interviews, however, views and interpretations were expressed that it would have been impossible to locate in the literature. Moreover, there were many informal discussions that were not documented as interviews, in which advice was given that has been incorporated into this work. In this context, I should mention the tip from Prof. Erich Peter Klement, who drew my attention to the relationship between Karl Menger's "probabilistic metrics" and the Fuzzy Set Theory

[22] The science history and biographical works by Ivo Schneider served as a model for these aspects of this work: e. g. [491] Schneider, Moivre (1968), [492] Schneider, Newton (1988), [490] Schneider, Faulhaber (1993)

that has dominated from the beginning – to write a history of fuzzy sets – was followed resolutely to the end.

The Fuzzy Set Theory developed before the backdrop of the 20th century advancements in information and communication technology. When Lotfi Zadeh studied electrical engineering in the United States in the 1940s, communication technology was differentiating itself from electrical engineering. Early textbooks entitled *Communication Networks*, such as the one by Ernst A. Guillemin, under whom Zadeh studied, had been published and the professional profile of a "Communication Engineer" was beginning to take shape. Advances in the field of the theory of electrical circuits had led to the expansion of individual areas such as network analysis, network synthesis and electrical filtering. This part of the pre-history will be described in detail in the first chapter of this work. During his studies at MIT, Zadeh encountered Norbert Wiener's cybernetics and Shannon's information theory. Following from Wiener's work, Zadeh created a functional calculus for ideal filters. He then searched for a corresponding calculus for optimum filters. This work led to statistical observations. The advent of the first digital computers animated him to become involved with "Thinking Machines" and with Shannon's information theory, and, in the mid-1950s, he introduced the system concept into the electrical engineering of the United States.

The second chapter does not continue this thread of the plot. Here we start over with the scientific-philosophical direction of Logical Empiricism (or Neo-Positivism), which originated in the Vienna Circle in the early part of the 20th century. The mathematician Karl Menger is interesting in the history of the Fuzzy Set Theory for two reasons. While it was the prevailing opinion in the Vienna Circle that there could be just *one logic* and one language in science, Menger – conditioned by his involvement with Brouwer's intuitionism – espoused the view that it was perfectly reasonable to introduce and use different languages and logics in science. Rudolf Carnap was also later persuaded of this view after Alfred Tarski, at Menger's invitation, had spoken to the Vienna Circle about "metamathematics". Carnap popularized this "principle of logical tolerance" through his publications.

After his emigration to the United States, Menger had proposed a theory of probabilistic or statistical metrics, and he considered sets with elements having only one probability – possibly other than 0 and 1. In a French journal, he called these sets *ensembles flous*. The French word *flou* means blurred, indistinct, and when Menger learned of Zadeh's Fuzzy Set Theory in 1966, he immediately saw the connection to his *blurred* sets. Within the framework of his theory of statistical metrics, Menger had also introduced the term triangular norm (t-norm), which today is a permanent component of fuzzy logic.

An attempt is made in the third chapter to bring together the two plot strands traced in the first two chapters. A path leads from the unification of the sciences for which the Vienna Circle strove to the General Systems Theory proposed by biologist Ludwig von Bertalanffy. In this transdisciplinary scientific discipline, the same principles were sought in different sciences. The feedback principle, which was central to Wiener's cybernetics, constituted one of these principles. After the Second World War, a certain fusion of the General Systems Theory and cybernetics occurred, and the first works on artificial neuronal networks and automata theory arose during this phase of very fruitful interdisciplinary scientific cooperation. A typical area of activity in artificial neuronal networks was pattern recognition. With the perceptron, scientists believed they had constructed a successful artificial neuronal network in this field. Not until a decade later was it shown that the perceptron could not even perform the simple task of separating two sets by a straight line.

The fourth chapter will show how, in the course of his work on the theory of electrotechnical systems, Zadeh gradually became aware of the fact that although the descriptions of real systems using the instruments of ordinary mathematics were becoming more and more complicated, they were not appropriate to the real systems. His attempts to characterize adaptive, linear and optimum systems demonstrate his efforts to find good definitions within the framework of conventional mathematical theory. His attempts to describe real systems in a mathematically precise manner failed. During this phase, Zadeh also developed his new approach to system theory: the *state space approach*.

The fifth chapter highlights how the theory of fuzzy sets originated. There were questions about the problems of pattern recognition which Zadeh was beginning to ponder in terms of the gradual membership of elements to sets. This fact is an important detail that has not been emphasized before, if it was mentioned at all, when the genesis of the fuzzy sets concept has been discussed. Zadeh introduced the fuzzy sets in order to solve the problem of abstraction in pattern recognition in a "natural and comfortable way", in other words, to find a decision function on the basis of a random sampling. Moreover, he proved a theorem for the problem of the separability of two fuzzy sets by a hyperplane – the problem that the perceptron could not solve, as would be shown some four years later. At this point, the text will report on the main features of Zadeh's first publications about fuzzy sets and fuzzy systems. An attempt will then be made to interpret the definition of fuzzy operators for the union and intersection of fuzzy sets from Zadeh's earlier works on electrical filters. In the late 1960s, Zadeh regarded his theory of fuzzy systems as a general system theory that could be used to cope with the so-called input-output analysis of systems.

Zadeh's *fuzzification program* for mathematical concepts and theories is the subject of Chap. 6, after a few initial examples to illustrate the hostility to which the founder of the Fuzzy Set Theory was subjected. Works from the years 1969 to 1973 by Zadeh and a number of other authors will be drawn upon to show that various fuzzifications were being carried out not only for individual mathematical concepts, but also for entire theories: There were, for example, the fuzzy probability theory, fuzzy topology and fuzzy logic; Zadeh himself fuzzified automata and algorithms. Finally in 1972, he and Sheldon S. L. Chang penned an article that introduced fuzzy sets into the realm of control theory. Here the two authors proved a theorem with the statement that feedback systems could achieve a precise objective with imprecise control – fuzzy control – and that when it is possible to make an adequate observation while approaching the objective, it was also possible to determine whether said objective had been achieved well enough.

The realization of this theoretical result occurred a short time later within the scope of the dissertation by Sedrak Assilian, a doctoral student of Ebrahim Mamdani in London. Motivated by Zadeh's publications on fuzzy algorithms, these two control engineers were able to control a steam engine with a set of rules formulated in a fuzzy algorithm. The simple fuzzy IF-THEN rules were implemented as fuzzy relations. Assilian and Mamdani additionally had to interpret the composition of fuzzy relations into a new fuzzy relation as an interference rule. For this they used the max-min-composition that Zadeh had established back in his first article.

The use of the Fuzzy Set Theory in medical diagnostics, as discussed in the seventh chapter, employs the max-min-composition rule, as well. In the computer-supported diagnostic systems developed in Vienna's General Hospital, the correlations between patients' symptoms and their illnesses were represented as fuzzy relations. The description of this medical application of the Fuzzy Set Theory is preceded by a historical outline of computer-supported diagnostics in medicine. The first fuzzification of a computer-supported diagnostic system was carried out on the CADIAG-II system in Vienna. A previous version, CADIAG-I, had been endowed with a trivalent logic, since symptoms were intended to be characterized either as "present" or "not present", but also as "not examined". Illnesses can often be marked either as "certainly present" or "certainly excluded", and so a third value of "possible" was most often very useful. In the late '70s, the CADIAG-II version was placed in operation. Its hallmark is primarily the ability to represent medical knowledge in the form of fuzzy relations.

The story of the Fuzzy Set Theory and its applications presented here concludes at the end of the 1970s. A caesura in the historical development of this theory is recorded at this point. Fuzzy sets first became popular outside of science and technology in the 1980s. In Japan, fuzzy control principles were implemented in the '80s for products in the household appliance and entertainment industries. The first subway run by fuzzy methods, in the Japanese city of Sendai, was also very good publicity. A bit later, this "fuzzy boom" was also sparked in the United States and Western Europe. Somewhat before this, though barely noticed in the Western world, several states of the Soviet Union had begun implementing technological uses for the Fuzzy Set Theory. These developments are considered to be a second chapter in the history of the Fuzzy Set Theory and its applications, during which a number of journals about fuzzy sets and fuzzy systems were founded and a scientific community with its own conferences was born. Incorporating these aspects of the second era of the Fuzzy Set Theory's historical development would have exceeded the scope of the present work, which is dedicated to the initial phase of pioneering work.

1

Beginnings of Communication Technology in the 20th Century

Of the Terrible Doubt of Appearances

Of the terrible doubt of appearances,
Of the uncertainty after all, that we may be deluded,
That may-be reliance and hope are but speculations after all,
That may-be identity beyond the grave is a beautiful fable only,
May-be the things I perceive, the animals, plants, men, hills,
shining and flowing waters,
The skies of day and night, colors, densities, forms, may-be these
are (as doubtless they are) only apparitions, and the real
something has yet to be known,
(How often they dart out of themselves as if to confound me and mock me!
How often I think neither I know, nor any man knows, aught of them,)
May-be seeming to me what they are (as doubtless they indeed but
seem) as from my present point of view, and might prove (as
of course they would) nought of what they appear, or nought
anyhow, from entirely changed points of view;
To me these and the like of these are curiously answer'd by my
lovers, my dear friends,
When he whom I love travels with me or sits a long while holding me by the
hand,
When the subtle air, the impalpable, the sense that words and reason
hold not, surround us and pervade us,
Then I am charged with untold and untellable wisdom, I am silent,
I require nothing further,
I cannot answer the question of appearances or that of identity
beyond the grave,
But I walk or sit indifferent, I am satisfied,
He ahold of my hand has completely satisfied me.[1]

---

[1] [555] Walt Whitman, *Leaves of Grass* (1891), p. 101

"May-be the things I perceive, ... are only apparitions, and the real something has yet to be known."[2] The American radio engineer Lloyd Espenschied[3] quoted this line from the above poem by Walt Whitman[4], taken from his great work *Leaves of Grass*. He used it as a motto for a text in 1943 in which he illustrated his view of communication technology.

Fig. 1.1. L. Espenschied

With these words, a great poet was alluding to the state of research in his time and to its future, to new discoveries and scientific growth.[5] The impreciseness of science and technology were obviously rooted in their nature; this was clear from our perceptions of remote things by means of electricity and likewise from the use of physical systems and our interpretation of what they indicate. Electric communication technology had now achieved a broad physical basis. Espenschied refers to this in four points at the conclusion of his article entitled *Electric Communications, the Past and Present Illuminate the Future*[6]:

> Electronics, representing a new command of electricity in the form of carriers free in vacuous space capable of energizing communication systems at enormously high rates;

---

[2] [555] Walt Whitman, Doubt (1891) quoted in [167] Espenschied, Communications (1943), p. 395

[3] Lloyd Espenschied (1899–1986) came to the engineering department of American Telephone and Telegraph Company (AT&T) in 1919 and remained with the Bell Telephone Company until 1954. He was instrumental in the formation of the IRE (Institute of Radio Engineers) in 1912, became a Fellow of the IRE in 1924 and Fellow of the American Institute of Electrical Engineers (AIEE) in 1930. Espenschied worked at Bell in the field of wireless telephone and radio communication, which in 1915 was carried out using vacuum tubes. In 1940 he was awarded the IRE Medal of Honor and, in 1967, the Pioneer Award of the IEEE Aerospace and Electronic Systems Group. (*http ://www.ieee.org/organizations/history_center/legacies/espenschied.html*)

[4] Walt Whitman (1819–1892), American poet. *Leaves of Grass* is perhaps the most influential work of poetry in the history of American literature

[5] [167] Espenschied, Communications (1943), p. 395

[6] The paper bears the subtitle: *A Suggestive Interpretation*

A greatly extended frequency dimension, representing enlarged intelligence-carrying capacity and a new technique common to the radio and guided method of transmission;

Means in the forms of wave guides and directed radio for realizing transmissions of very high frequency characterized by a high degree of special definition, where-by the number of communications may be multiplied on a space-segregation basis;

A trend toward undertaking the transmission of additional forms of intelligence, in particular of the space-pattern type.[7]

Espenschied was ultimately painting a picture of a future world, which the world at the turn of the 21st century – 60 years later – actually resembles very closely: Today large transmission networks cover every region of the earth's surface, even the innumerable islands; today we live in a "world nervous system so comprehensive in terms of time and space and intelligence-carrying capacity as eventually to enable one anywhere to keep in touch, by more or less all his senses and to the extent desired, with his environment and fellow man".[8] Just what is that reference to Whitman's words supposed to tell us?

1.1 Communication Network

Developments in communication technology since that time have been rich and varied. The telegraph, telephone and radio crystallized as applications from what had previously been a sub-area of physics and they eventually formed their own technical discipline within engineering. Moreover, new technological complexes such as the television, Internet and mobile radio have developed from the field of *electrical engineering* as it existed then, and large and small computer systems that are interconnected via machine communication components are now also included under the term communication

Fig. 1.2. E. A. Guillemin

---

[7] [167] Espenschied, Communications (1943), p. 402
[8] [167] Espenschied, Communications (1943), p. 402

networks. Today cars, apartment houses and entire building complexes can be equipped with this kind of technology, and networked computers have access to the worldwide Internet that includes satellite links; distances in space and time essentially play only a minor role for communication!

In the early '30s when Espenschied wrote his article, technology such as this was inconceivable. When people spoke of communication networks, they were referring only to telegraph, telephone and radio networks! A first book bearing the title *Communication Networks* was written by Ernst Adolf Guillemin[9] in 1931, and in January of 1932, seven lectures were delivered for a Lowell Institute course in Boston, Massachusetts. These lectures appeared later the same year as a book entitled *Modern Communication*.[10] The subject of the first lecture in this series and the book was "Social Aspects of Communication Development", and Arthur W. Page[11], who was the vice-president of the American Telephone and Telegraph Company (AT&T) at the time, also provided a survey of the entire history of communication. Until recently, he stated, if a man wanted to say something to his neighbor, he would first have to find him, and he would be able to reach no more listeners than the power of his voice allowed. Then scientists discovered much about sound and light waves and about electricity. It wouldn't be iconoclastic, he said, to claim that philosophers would have been very pleased if the third type of waves had been discovered earlier: electric or radio waves. Sound, light and radio waves were perfectly suited to general communication purposes, since they could spread out in all directions from their point of origin, but sound and light waves did not extend great distances over the earth's surface. Radio or electric waves on the other hand, could do so easily, despite the curvature of the earth![12]

> Electric waves predicted mathematically by Maxwell in 1865, experimentally produced by Hertz, and adapted to commercial uses by Marconi in 1895, provide the most direct use of electricity in communication. The wire telegraph and the wire telephone are additional steps to that fundamental discovery.[13]

The second lecture was delivered by Harold De Forest Arnold[14], who was then director of research at Bell Telephone Laboratories. Since it was still

---

[9] Ernst Adolf Guillemin (1898–1970), American electrical engineer
[10] [432] Page et al., Communication (1932)
[11] Arthur Wilson Page (1883–1960) was vice-president of "public relations" at the American Telephone and Telegraph Company (AT&T) from 1927 to 1946
[12] [433] Page, Aspects (1932), pp. 14f
[13] [433] Page, Aspects (1932), p. 16
[14] Harold De Forest Arnold (1883–1933), American physicist. Arnold's research on X-ray technology at Western Electric led to successful developments in telephony and radio technology. Arnold was significantly involved in the first transcontinental demonstration of radio telephony on September 29, 1915, from New York City to Arlington, Virginia, San Francisco and Honolulu

completely unknown at that time if outer space was filled with an *ether*, a material medium in which electromagnetic waves are propagated[15], Arnold, having previously spoken about waves in the air, said:

> Ether waves are obviously more perplexing subjects of research. Perhaps there is no ether; then of course there are no ether waves, there is no doubt about the usefulness of light and radio waves, and whether or not we can completely comprehend what they are, we must at any rate bend effort to understand all the rules which must be followed in using them, and all the difficulties with which their use is surrounded.[16]

Only once the nature of these waves had been fully researched would it be possible also to fathom what electricity actually is: "There is perhaps nothing simpler in the world than electricity, for we have come to think that it is the aggregation of little units which are always the same and quite unchangeable."[17] It was meanwhile also known, however, that electricity could behave like a wave, and the laboratories had proven this peculiar fact by studying electricity for the purposes of communication. It was not yet possible to say what this knowledge could mean for communication technology, yet there was no doubt that it was important and that its discovery was a further sign of the fact that there was ever more to find out about those elementary things which we cannot change but which we must use.[18]

Arnold then spoke about the role of mathematics in this technology, about its precision and the possibility of discovering analogous relationships in the various branches of the natural sciences. This paragraph shall be cited here in its entirety:

> But it is in the problems of shape and association of the materials which we have at our disposal that we find the most complicated and the most immediately important of all our tasks. The telephone transmitter, for instance, is composed of some thirty different parts. The test of the adequacy of this assemblage of parts is that it must take from the air the minute and highly complicated energy flow that is speech and must translate this into a flow of electricity which retains all the voice's complexity; and it must do this reliably and cheaply, under vicissitudes of location and climate, and throughout a long period of years. Its design is mathematically exact in an unusual sense of the words. Our fundamental studies of the voice have made it possible for

---

[15] Electromagnetic waves were first identified as "ether waves" by the Scottish mathematician and physicist Lord William Thomson Kelvin (1824–1907) when he translated Hertz's 1888 work into English

[16] [22] Arnold, Introduction (1932), p. 30

[17] [22] Arnold, Introduction (1932), p. 30ff

[18] [22] Arnold, Introduction (1932), p. 30ff

us to describe the air-borne form of words in terms of Fourier's analysis – a mathematical description of wave form in terms of frequency, amplitude, and phase – with very great accuracy and completeness. We have also developed methods for analyzing the electrical currents which flow out from the transmitter and can describe them in the same mathematical terms. And so we may think of the transmitter as transforming the mathematical equation which, in terms of mass, elasticity, and viscosity, represents the motion of the air, into the corresponding equation which in terms of inductance, capacity, and resistance represents the motion of electricity in the telephone wire. It is just because this whole problem has been considered as a mathematical one, and every detail of the transformation has been traced through all the parts of the instrument, that we have our present-day quality of telephone, phonograph, and sound-picture reproduction. The promise for the future rests in the existence of just such exact mathematical analysis to every problem in which there is a translation of energy between mechanical and electrical forms.[19]

The foundations of mankind's communication structures, structures which still exist today, were created during the 1930s. Much may have changed, as well, yet the technological basis forged back then is still valid today, and the trust placed in its mathematical exactness has also been placed in the computability of physical systems. These successes proved that the scientists and engineers were right!

When James Clerk Maxwell[20] succeeded in 1873 in demonstrating electromagnetic waves and mathematically proving their possible existence, it quickly became clear that a medium for transmitting information had been found that could rely upon current-carrying conductors, for Maxwell had also shown that the electromagnetic waves were temporal oscillations of electromagnetic fields which spread out in a waveform and that said waves could induce electric current in an electrical conductor, the strength of which oscillates synchronously to the wave field. Eleven years later, Heinrich Hertz[21] successfully sent and received corresponding carrier waves. In 1894, Guglielmo Marconi was able to transmit messages via the first wireless, nine meter long transmission path when he made a bell ring using an oscillator as a transmitter and a coherer as a receiver.[22] In 1896, Marconi demonstrated a wireless telegraphy system

---

[19] [22] Arnold, Introduction (1932), p. 33ff
[20] James Clerk Maxwell (1831–1879), British physicist. Maxwell was a professor at Aberdeen, 1856–1860, and at King's College in London, 1860–1865, after which he was a private scholar
[21] Heinrich Rudolf Hertz (1857–1894), German physicist. See also the biography [175] Fölsing, Hertz (1997)
[22] Guglielmo Marconi (1874–1937) founded the Wireless Telegraph Company (later the Marconi Company) in 1897. He used radio to bridge the channel between England and France in 1899 and the Atlantic between Cornwall and

over a distance of three kilometers for members of the army and navy, and in 1899, he succeeded in establishing the first wireless link across the English Channel.

Although Marconi did attend lectures by Augusto Righi[23] at the University of Bologna in which he demonstrated experiments with the "Hertzian waves", never completed a mathematically-based university education in the natural sciences or engineering; there had been no mathematization of communication technology to speak of at that point. Marconi had labored by trial and error and had chalked up a number of failed attempts.[24] Thanks to his great experimental skill, he was nevertheless blessed with great and lucrative success.

The mathematization process of communication technology really began with the discovery of filter circuits, or electrical filters, in the second decade of the 20th century. Because of problems that were observed in both conducted and wireless message transmission, scientists increasingly analyzed the *behavior* of the communication technology systems using the mathematical tools they had acquired during their studies, although the engineers' knowledge of them left quite a lot to be desired. Count von Arco[25], who had been Adolf Slaby's assistant[26], even characterized the initial experiments with wireless telegraphy as *electric alchemy*, and he stressed: "Thought experiments were excluded and only the purest and rawest empiricism prevailed".[27] Owing to the fact that basic mathematical-physical knowledge was often lacking or insufficient, those involved sought analytical methods that would lead them to their goal although they had not mastered the Maxwellian equations and could not employ corresponding calculations. The network and system theories represented a similar concept. Since electrical engineers were interested merely in the input and output behavior of the basic building blocks of communication engineering and less in a physical explanation of what was happening inside such components of electrical circuits, they were content with a corresponding

---

Newfoundland in 1901. He used microwaves in 1932 and radar in 1935. In 1909, he and Carl F. Braun (see footnote 27) were jointly awarded the Nobel Prize in Physics for their contributions to the development of wireless telegraphy

[23] Augusto Righi (1850–1920), Italian physicist, studied at the University of Bologna and later taught at the universities in Palermo (1880–1885) and Padua (1885–1889) and then another 32 years in Bologna

[24] [86] Blumtritt, Nachrichtentechnik (1997), p. 66

[25] Count Georg von Arco (1869–1940), German engineer, co-founder and first Technical Director of Telefunken AG (1903)

[26] Adolf Slaby (1849–1913), German electrical engineer. From 1886 he was the first full professor of theoretical mechanical engineering and electrical engineering at the Technical College of Charlottenburg, and was a co-founder of Telefunken AG (1903)

[27] [426] Noll, Nachrichtentechnik (2001)

black box theory with which they could analyze the real systems modeled in this way.

Marconi's rival on German soil, the physics professor Carl F. Braun[28] discovered the *closed oscillating circuit* in 1928, which, in contrast to the *open aerial circuit*, consisted of a capacitor and a coil and which slowly decayed in the high-frequency oscillations stimulated by electrical discharges without electromagnetic waves penetrating outwardly. However, if an oscillating circuit such as this is supplied with electromagnetic waves of its own *eigenfrequency*, then special calibrated oscillations are enhanced while others are inhibited. This effect could be used to adjust certain frequencies on the transmitter and receiver side; a short time later it proved to be extraordinarily important to the further development of communication engineering.

Karl Wilhelm Wagner[29] in Germany and George Ashley Campbell[30] in the United States discovered at the same time in 1915 that it was possible during the transmission of electromagnetic waves to use particular circuits to favor not only a particular frequency or an unlimited number of frequencies but even entire predetermined frequency ranges. In the German milieu, these were referred to as 'Siebschaltungen', while the term 'electric filter' was preferred among Americans. Wagner was able to show "how it is possible to achieve a filtering of predetermined frequency ranges by starting with a relatively simple quadrupole resonance circuit and connecting it repeatedly in series

---

[28] Carl Ferdinand Braun (1850–1918) became a teacher of mathematics and physics at the Thomasschule in Leipzig in 1874, an extraordinary professor of physics in Marburg in 1877 and an ordinary professor of physics in Strasbourg in 1880. Prof. Braun's company Telegraphie GmbH was founded in 1898 and merged with Siemens & Halske in 1901. In 1903, the Telefunken Company emerged as a subsidiary of Siemens and AEG. Braun traveled to the US on company business in 1914, but due to a German naval blockade and his own bout with cancer, he never returned home. Braun and Marconi were jointly awarded the Nobel Prize in Physics in 1909

[29] Karl Wilhelm Wagner (1883–1953) became professor and director of the High-Voltage Department of the Imperial Institute of Natural and Engineering Sciences in 1913 and president of the Imperial Office of Telegraphic Engineering in 1923. In 1927, he became professor of general vibration studies at the Technical University of Berlin

[30] George Ashley Campbell (1870–1954) was one of the pioneers in the development and application of mathematical methods in telegraphy and telephony over great distances. Following his graduation from the Massachusetts Institute of Technology (MIT) in 1891, he studied at Harvard, where he earned a Master's degree in 1893 and a Ph.D. in 1901. He then spent three years studying in Göttingen, Vienna and Paris before being employed as a research engineer at AT&T in 1897. From 1934 to 1935, he worked at Bell Telephone Laboratories, and in 1936 he received the IRE Medal of Honor "for his contributions to the theory of electrical networks". See also [142] Dietzold, Network (1948), [368] McKeen et al., American (1938), p. 217

1.1 Communication Network 15

with itself (known as a'Kettenschaltung', ladder network). If the number of links in the chain is allowed to increase indefinitely, the transceiver will no longer permit certain predeterminable frequency ranges to pass through at all".[31]

Campbell had proceeded from the groundbreaking 1887 investigations conducted by Oliver Heaviside[32], in which he developed electrically charged coils that maintain a signal while it is being propagated. Vannevar Bush[33], who published Campbell's collected works, wrote 30 years later:

> In Heaviside's day the time was not ripe for great communication systems, nor were men of his time receptive to what appeared to be radical departures from convention. Unless there had been prophets to see the light and to carry it over into the new day, the great work of Heaviside would have been lost. These men who carried the torch can be numbered on the fingers of one hand, and of them Campbell stands out as the predominant figure, not alone because his grasp was sound and persistent, but also because he was associated with those who by transportation and communication developments were beginning the great work of joining America into a single nation. Into this early advance, to meet a commercial exigency, came the loading of telephone lines. The vision was Heaviside's. The design formulas were Campbell's. Out of a mathematical argument come coils of the right size and in the proper places to enable wires to go underground and to carry speech to long distance.[34]

However, the greater the distances between sender and receiver grew, the less practicable the process became of testing the communication systems for functionality only after they had been installed. The trial and error method needed to be replaced by something better:

> Empiricism here [continued Bush] would have been utterly hopeles, and unjustifiably expensive. One cannot temporise with a long transmission line running across actual country by trying this and replacing that. Haphazard work of that nature could have but resulted in discouragement and failure. Procedure by thoroughly rigorous mathematical treatment to the condition where operation is understood is the necessary step if the result is to be practically accomplished. For

---

[31] [119] Cauer, Siebschaltungen (1930), p. 27f

[32] Oliver Heaviside (1850–1925), British physicist and autodidact. He worked for the Great Northern Telegraph Company in Newcastle, though he resigned in 1874 because of increasing deafness

[33] Vannevar Bush (1890–1974), American electrical engineer, went to the Massachusetts Institute of Technology (MIT) in 1919 and was a professor there from 1923 to 1932. There he also constructed the Differential Analyzer analog computer

[34] [109] Bush, Analyzer (1931), quoted in [169] Fagen, History (1975), p. 894

this, Campbell, and Campbell almost alone, deserves the credit. He gave us the loaded line.[35]

Campbell's studies led to the invention of the wave filter as circuits and a network that permit the passage of signals (waves) of a particular range of frequencies while those with frequencies outside this range can be attenuated and suppressed. Owing to its similarity to a chemical process in which large particles can be separated from small ones by passing the solution through a membrane with pores that are smaller than the large particles, this process was referred to as "electrical filtering".

Fig. 1.3. G. A. Campbell

Without going into too many of the details of electrical engineering, electrical filters are specially adapted electrical oscillating circuits, resonators. If we extend the analogy of mechanical oscillation, we can regard electrical filters as oscillating circuits in which the current passing through a capacitor corresponds to the restoring force of the mechanical resonator and the inductivity of a coil corresponds to the inert mass. Since the electric current through the coil in an electrical oscillating circuit constantly changes direction, the capacitor is continually charged or discharged. Depending on the choice of capacitors and coils, it is possible to create low-pass, high-pass and band-pass filters that accordingly allow the passage of waves with frequencies below a threshold frequency f_1 or above a threshold frequency f_2 or in a range between two threshold frequencies f_3 and f_4; *filter attenuation bands*, which interrupt all frequencies outside a particular frequency band between f_5 and f_6, are also possible. Electrical filters transmit alternating currents and alternating voltages at frequencies within a particular *frequency band*. Oscillations within this range are allowed to pass while oscillations outside this pass band are blocked (Fig. 1.5).[36]

---

[35] [109] Bush, Analyzer (1931), quoted in [169] Fagen, History (1975), p. 894f
[36] This is thus also referred to as the "blocking zone" of the filter

Between 1924 and 1931, research on electrical filters was continued primarily at Bell Telephone Laboratories and AT&T by Otto J. Zobel[37], Ronald M. Foster[38] and John R. Carson[39] as well as by A. C. Bartlett, Wilhelm Cauer[40], Hendrik W. Bode[41], W. Brandt, Hans Piloty[42], Otto Brune[43] and Sidney Darlington[44], who applied the theories of linear algebra, complex variables and combinatorial topology to network synthesis and analysis: "It was a time

---

[37] Otto Julius Zobel (born 1887) studied at Wisconsin, graduated from there and worked from 1916 to 1934 at American Telephone and Telegraph Company (AT&T), then at Bell Telephone Laboratories. He worked on electrical transmission techniques. [368] McKeen et al., American (1938), p. 1598

[38] Ronald Martin Foster (born 1896) studied at Harvard and was then with AT&T and Bell. Foster worked with symbolic logic, Fourier integrals and electrical network theory. [368] McKeen et al., American (1938), p. 474

[39] John Renshaw Carson (1886–1940), American electrical engineer, studied at Princeton University and at the Massachusetts Institute of Technology (MIT). He earned his M.S. at Princeton in 1912, where he was employed as an instructor of electrical engineering from 1912 to 1914. Later he worked at the American Telephone and Telegraph Company (AT&T) on experiments on telephony. Carson was responsible for the first telephone link between Pittsburgh and Baltimore, and he developed the mathematical methods for the calculations required to conduct waves by metallic tubes. In 1924, he was awarded the Liebman Memorial Prize by the Institute of Radio Engineers. From 1925 until his death, Carson worked as a mathematician and electrical engineer for Bell Telephone Laboratories. He received an honorary doctorate from the Brooklyn Polytechnic Institute in 1937 and the Elliot Cresson Medal from the Franklin Institute of Philadelphia

[40] Wilhelm Cauer (1900–1945), German mathematician and engineer. Cauer was an extraordinary professor at the Technical University of Berlin from 1939 and was killed by Soviet soldiers in 1945

[41] Hendrik Wade Bode (1905–1982), American electrical engineer, studied in such places as Urbana, Illinois, Tempe, Arizona and Ohio, where he finished his studies in 1926. After that he was employed at Bell Telephone Laboratories. While he was working there, he studied at Columbia University in New York, where he earned his Ph.D. in 1935

[42] Hans Piloty (1894–1969), German communication engineer, worked from 1925 to 1931 as a scientist at the Allgemeine Electricitätsgesellschaft (AEG) and in 1931 became a professor at the Technical University of Munich

[43] Otto Walter Heinrich Oscar Brune (born 1901 in South Africa) studied at MIT and was awarded a B.S. and M.S. in electrical engineering there in 1929. He remained there until 1935, then returned to South Africa. (J. McKeen Cattell and Jacques Cattell (eds.): *American Men of Science. A Biographical Directory*. New York: The Science Press 1938, p. 183)

[44] Sidney Darlington (1906–1997), American physicist and electrical engineer. Studied physics at Harvard until 1928 and at MIT in Boston as well as at Columbia University in New York, where he earned his Ph.D. in 1940. From 1929 he worked at Bell Telephone Laboratories

when the intuition of the electrical engineer needed support from the rigorous analysis of the professional mathematician".[45]

Fig. 1.4. W. Cauer

In Germany, Wilhelm Cauer was working on a monograph about filter circuits, which he finished in 1931 during a one–year research residence in the US with Vannevar Bush's working group at MIT, a trip he was able to undertake thanks to a Rockefeller scholarship. The book contained a complete theoretical description as well as listings of the various filter types. Additionally, it was distinguished by the fact that it publicized a new method of designing electrical wave filters that had a number of advantages over the principle espoused by Zobel, who had produced filters of the Campbell type. As Sidney Darlington remembered:

> At Bell Laboratories a number of us first learned about Cauer's canonical circuits and his Chebyshev approximations at a conference on Cauer's proposed sale of some of his patents. It was an important event in my professional life.[46]

However, because of Cauer's mathematically challenging procedure, which was highly unusual for electrical engineers in the United States back then, his work was largely ignored. Only a few network theoreticians made any attempt at all at the time to look into Cauer's treatment of electrical filters.[47] In the US, Ernst Adolf Guillemin, an ambitious young electrical engineer, praised:

> ... the form of Cauer's publication, which is quite mathematical and in general not in accord with the manner in which similar material is presented in this country.[48]

In the same year, the first of Guillemin's two volumes on *Communication Networks* was published. In it the author became one of the first in the

---

[45] [169] Fagen, History (1975), p. 899
[46] [136] Darlington, History (1984)
[47] See [118] Cauer, et al., Life (2000)
[48] [210] Guillemin, Design (1931)

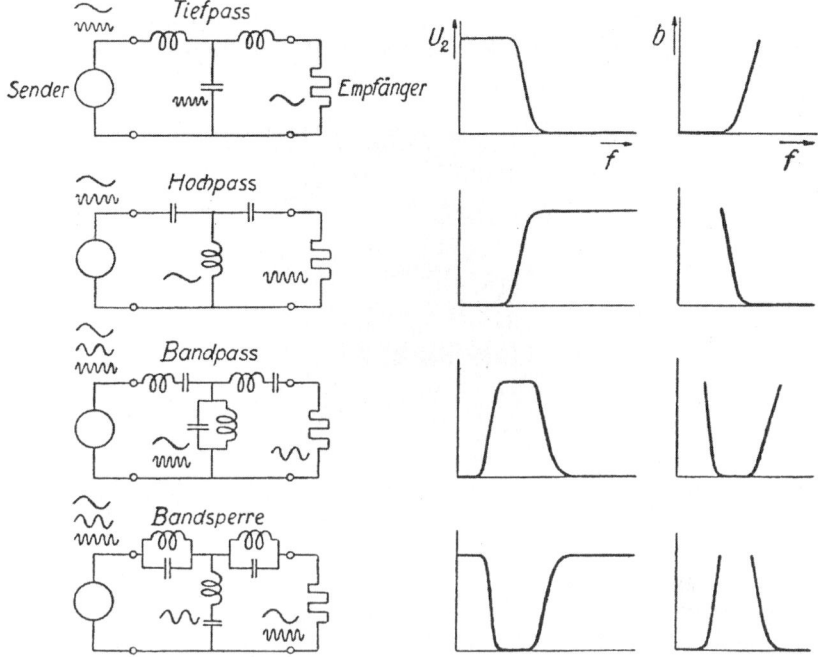

Fig. 1.5. Various filter types

United States to stress the necessity of mathematical abstractions for future research in electrical engineering:

> The existing texts on communication principles deal with the theory and practical significance of special types of networks commonly used in or telephone circuits. Progressive research, however, demands a more thorough grasp of the fundamental methods of attack on network problems in general.[49]

At the conclusion of the foreword to volume one of his book, Guillemin's first expressions of gratitude are to the German physicist Arnold Sommerfeld[50], "through whose teaching I gained my first real insight into

---

[49] [213] Guillemin, Networks I (1931), p. iii

[50] Arnold Johannes Wilhelm Sommerfeld (1868–1951), German mathematician and physicist, studied mathematics in Göttingen. In 1897, Sommerfeld became professor of mathematics at the Bergakademie Clausthal-Zellerfeld and, in the year 1900, full professor of technical mechanics at the Technical University of Aachen. He was given the chair of theoretical physics at the University of Munich in 1906. Beginning in 1910, Sommerfeld advocated the quantum concept and in 1919 presented a first summary of quantum theory in the book *Atombau und Spektrallinien*

the philosophy of oscillatory systems, and particularly in the use of complex methods as compared to the trigonometric representation."[51]

Fig. 1.6. Arnold Sommerfeld reading a letter in Berkeley in 1923

It can be assumed not only that Guillemin had actually read Cauer's work in the original German but also that a close teacher-pupil relationship can be observed here between a famous professor and his former student, a relationship that should be illuminated somewhat more closely.

1.2 Network Analyses

Early in the 20th century, quantum theory and quantum mechanics were discovered in Germany as the new physical theory of the atom, and for the most part they were also further developed in Germany. Alongside the Nobel Prize-winning physicists Max Planck[52], Albert Einstein[53](1921),

---

[51] [213] Guillemin, Networks I (1931), p. v

[52] Max Karl Ernst Ludwig Plank (1858–1947), German physicist, studied in Munich and Berlin and became a professor in Munich in 1880. He became an extraordinary professor in Kiel in 1885 and in Berlin in 1889, where in 1892 he was awarded the chair in theoretical physics. Beginning in 1894 he was a member of the Prussian Academy of Sciences. From 1905 to 1909 (and again in 1915–16), he was chairman of the German Physics Association. From 1930 to 1937, Planck was president of the Kaiser Wilhelm Society for the Advancement of Science. He is the discoverer of the natural constant h, known as Planck's quantum of action, and he received the Nobel Prize in Physics in 1919, retroactive to the year 1918

[53] Albert Einstein (1879–1955), German-Swiss-American physicist. In 1905, as a "technical expert, 3rd class" at the patent office in Bern, Einstein published three

Niels Bohr[54] (1922), Max Born[55] (1954), Louis, Duc de Broglie[56] (1929), Erwin Schrödinger[57] (1933), Paul Dirac[58] (1933) and the next generation

---

treatises on the Theory of Relativity, Brownian motion and quantum theory, which are the basis for his renown. In 1911, he became professor of theoretical physics in Prague and entered the Prussian Academy of Sciences in 1914 as a regular full-time member. Einstein established the General Theory of Relativity in 1915. In 1933, he resigned his position in the Academy and established a new domain at the Institute for Advanced Study in Princeton, N. J. in the United States

[54] Niels Henrik David Bohr (1885–1962), Danish physicist. Having earned his doctorate (1911), Bohr worked at the University of Cambridge and with the physicist Ernest Rutherford (1871–1937) in Manchester. In 1916, Bohr became professor of theoretical physics in Copenhagen and, in 1918, he established the correspondence principle as a link between classical physics and quantum theory. In 1926–27, he and Werner Heisenberg brought quantum theory to a preliminary conclusion (Copenhagen Interpretation)

[55] Max Born (1882–1970), German-British physicist. Born earned his doctorate (1907) and his professorial qualification (1909) in Göttingen, and in 1914, became professor of physics in Frankfurt am Main. He was a professor in Göttingen from 1931 to 1933, but was laid off in 1933 because of his Jewish heritage. He emigrated, via Cambridge and Bangalore, to Edinburgh, where he remained until 1953. In 1954, Born returned to Germany. Together with Werner Heisenberg and Ernst Pascual Jordan (1902–1980), Born formulated the closed mathematical theory of quantum mechanics in 1925

[56] Louis Victor Pierre Raymond, Duc de Broglie (1892–1987), French physicist. In 1923–24, he discovered the idea of wave-particle duality. He showed that each particle can have a wave associated with it, the wavelength of which is inversely proportional to the particle impulse. He also linked Max Planck's equation with Albert Einstein's energy-mass relation

[57] Erwin Schrödinger (1887–1960), Austrian physicist. In 1927, Schrödinger became Max Planck's successor as professor of theoretical physics at Berlin University. He emigrated to Oxford in 1933 and to Graz in 1936, where he was fired one year later, following Austria's *Anschluss* by Germany. In 1939, Schrödinger obtained a lifelong appointment at Dublin's Institute for Advanced Studies. During his years of emigration, he published works on the statistical interpretation of wave mechanics, which contradicted Heisenberg's matrix mechanics

[58] Paul Adrien Maurice Dirac (1902–1984), British physicist. In 1926, Dirac developed a version of quantum mechanics that contained Heisenberg's and Schrödinger's matrix and wave mechanics as special cases. He established the equation named after him in 1928 and predicted the existence of the positron as an antiparticle of the electron; this was detected in cosmic radiation in 1932 by the American physicist Carl David Anderson (1905–1991). Dirac also explained the quantum spin as a relativistic effect. From 1932 to 1969, Dirac was the Lucasian Professor of Mathematics at the University of Cambridge

including Werner Heisenberg[59] (1932) and Wolfgang Ernst Pauli[60] (1945), it is also important to mention Arnold Sommerfeld. The latter two had studied under him, though the Prize remained elusive to Sommerfeld himself. Talk of this new atomic theory spread quickly, and scientists in other countries were naturally very interested in following this scientifically revolutionary "quantum leap". "A great effort was made, particularly in America, not to miss out on the chance to be involved in the developments achieved in Europe. The prominent nuclear theorists from overseas were invited to deliver guest lectures, or else this or that European theorist was wooed with a promising long-term job offer at one's own university. Summer schools were particularly important in allowing the American students to make contact with the eminent international experts. In the US, this type of informal meeting emerged very early on."[61]

Considered the most important German physicist of the age after Albert Einstein, Arnold Sommerfeld was offered the Carl Schurz Memorial Professorship[62] at the University of Wisconsin in Madison shortly after the First

---

[59] Werner Karl Heisenberg (1901–1976), German physicist. Heisenberg studied under Arnold Sommerfeld and received his doctorate in 1923, became Max Born's assistant in 1924 and worked 1924–25 with Niels Bohr in Copenhagen. In 1927, he became a professor at the University of Leipzig and head of its Theoretical-Physical Institute. He publicized the "Uncertainty Relations" that are named after him in 1927. An appointment to a professorship in physics at Munich was withdrawn in 1936; Heisenberg was defamed by the National Socialists. From 1941 to 1945, he was director of the Kaiser Wilhelm Institute and professor in Berlin, and in 1945–46 was detained in England with other nuclear physicists. From 1946 to 1951, Heisenberg led the Max Planck Institute of Physics in Göttingen, from 1958 until 1970 he was a professor at the University of Munich and head of its Max Planck Institute of Physics

[60] Wolfgang Ernst Pauli (1900–1958), Austrian-Swiss physicist. Pauli became familiar with relativity and quantum theory due to his contact with Sommerfeld and, after earning his doctorate (1921–22), became Max Born's assistant in Göttingen, then went to Hamburg and Copenhagen and, in 1928, became a professor at the Federal Institute of Technology in Zurich

[61] In German: „Vor allem in Amerika gab es viele Bemühungen, den Anschluss an die europäischen Entwicklungen nicht zu verpassen. Man lud die prominenten Atomtheoretiker aus Übersee zu Gastvorträgen ein oder lockte den einen oder anderen europäischen Theoretiker mit einem vielversprechenden Stellenangebot auf Dauer an die eigene Universität. Besonders wichtig fü r die Begegnung amerikanischer Studenten mit den internationalen Koryphäen waren Sommerschulen, eine in den USA schon sehr früh aufgekommene Form informeller Zusammenkunft." [151] Eckert, Atomphysiker (1993), p. 109f

[62] This guest professorship was established in 1911 by Americans of German descent to honor the German-born freedom fighter and American politician Carl Schurz (1829–1906). Schurz, a liberal Republican, was able to flee after the German Revolution in March of 1848 and traveled to North America in 1853, where he established himself as an attorney. In an election campaign five years later,

World War, in the winter of 1922–23. Other US universities followed suit by inviting Sommerfeld, who regarded his role during these visits as a kind of cultural ambassador, a role he felt honor-bound to fulfill.[63]

During this first trip to America, Sommerfeld made the acquaintance of the ethnic German Guillemin family, whose two sons Ernst Adolf and Victor traveled to Munich two years later to conduct their doctoral studies with Sommerfeld. Sommerfeld's "votum informativum" for the elder Guillemin brother's dissertation provides a bit more information about this arrangement:

> I would like to add that I came to know the Guillemin family as especially good "German-Americans" during my first visit to America and that I have repeatedly received monetary contributions from the father. The brother Victor Guillemin is likewise at our university and will likewise be submitting his doctoral dissertation shortly.[64]

In 1924, Ernst Adolf Guillemin, the older of the two Guillemin sons, had the opportunity, thanks to an assistant exchange between MIT's electrical engineering department and the University of Munich's Institute of Theoretical Physics, to travel to Germany on a Saltonstall Traveling Fellowship.[65] Not quite two years later, on July 7, 1926, he graduated *cum laude* with a dissertation entitled *On the Theory of Frequency Multiplication by Iron Core Coupling*. In addition to Sommerfeld (physics), his oral examination was administered by Professors Wilhelm (Gulielmo) Wien[66]

---

he supported Abraham Lincoln (1809–1865), who would later become the 16th president of the United States. Schurz was a major general in the US Civil War and publisher of the *Westliche Post* and the *New York Evening Post*. He was elected senator from Missouri and eventually became Secretary of the Interior

[63] [151] Eckert, Atomphysiker (1993), p. 77f

[64] Ich möchte hinzufügen, dass ich die Familie Guillemin bei meiner Amerika-Reise als besonders gute „Deutsch-Amerikaner" kennen gelernt habe und dass ich von dem Vater wiederholt Geldzuwendungen erhalten habe. Der Bruder Victor Guillemin ist ebenfalls an unserer Universität immatrikuliert und wird ebenfalls alsbald eine Doctor=Arbeit vorlegen. [695] Arnold Sommerfeld, Votum Informativum for the dissertation by Ernst A. Guillemin, Archive of the Ludwig Maximillian University in Munich (OCI 52p)

[65] [208] Guillemin, Autobiography (1968)

[66] Wilhelm Carl Werner Otto Fritz Franz Wien (1864–1928), German physicist, studied in Göttingen and Berlin, worked from 1883 to 1885 with the German physicist and physiologist Hermann von Helmholtz (1821–1894) and graduated in 1886. In 1889, Wien became ordinary professor in Giessen and one year later succeeded the German physicist Wilhelm Konrad Röntgen (1845–1923) as professor of physics in Würzburg. Wien again became Röntgen's successor in 1919, this time as a professor in Munich

as department head (also physics), Otto Hönigschmid[67] (chemistry) and Constantin Caratheodory[68] (mathematics).

In his dissertation, Guillemin described his investigations in which the frequency of an alternating current machine could be stepped up by energizing an iron core, an "essential task in the practice of telegraphy" at that time, which tied in with an older work by Sommerfeld and with which Jonathan Zenneck[69]

Fig. 1.7. Arnold Sommerfeld (1st row, 3rd from right) with physicists at the California Institute of Technology in Pasadena. This photo was taken in the winter semester of 1922–23 during Sommerfeld's stay in the US as the Carl Schurz Guest Professor at the University of Wisconsin [692]

---

[67] Otto Hönigschmid (1878–1945), Czech-German chemist. Working at the Institute for Radium Research in Vienna in 1910, Hönigschmid succeeded in determining the atomic weight of radium

[68] Constantin Caratheodory (1873–1950), German mathematician. Caratheodory was of Greek descent, but was born in Berlin and grew up in Brussels, where his father served as the Turkish ambassador. Caratheodory attended the universities in Berlin and Göttingen, and was a professor of mathematics at the University of Munich from 1924 until his death

[69] Jonathan Zenneck (1871–1959), German physicist, assistant to C. F. Braun in his attempts at wireless telegraphy; in 1906, Zenneck became an ordinary professor in Braunschweig, 1911 in Danzig and then in Munich until 1936

had experimented.[70] In mid-February, the very same month Guillemin's dissertation was submitted, Sommerfeld wrote to Walter Rogowski[71], the publisher of the *Archiv für Elektrotechnik* to request that Guillemin's work might be "published in the archive"[72], a request[73] that was granted.[74]

Sommerfeld had evaluated the dissertation as "*cum laude*" but not without noting that his student's education had been in electrical engineering and not in physics, for the

> author proceeds such that he calculates the current and voltage profiles piece by piece for the various branches of the characteristic (idealized to a linear stroke) and that he provides for the necessary consistency at the edges of the intervals. This method is in accordance with the electrical engineer-descriptive thought process of the author and was devised completely independently by him. I will soon be publishing along with him another, more mathematically elegant method using Fourier analysis, which leads to the same results.[75]

---

[70] [695] Arnold Sommerfeld, Votum Informativum for the dissertation by Ernst A. Guillemin, Archive of the Ludwig Maximillian University in Munich (OCI 52p)

[71] Walter Rogowski (1881–1947), German electrical engineer. Rogowski became ordinary professor for general theoretical electrical engineering at the Technical University of Aachen in 1920. The "Rogowski Institute" that was founded there in 1929 eventually became the independent Faculty for Electrical Engineering at the Rheinisch-Westfällische Technical University in Aachen (RWTechnische Hochschule)

[72] [689] Letter from Arnold Sommerfeld to Walter Rogowski, February 18, 1926, Archive: Munich German Museum (Archive NL 89, 003); facsimile, Sommerfeld Project: $http: //www.lrz-muenchen.de/ \sim Sommerfeld/gif100/02177\_01.gif$

[73] As early as a letter dated October 20, 1912, Rogowski had expressed to Sommerfeld his desire that the archive might publish, "from time to time, summarizing papers about this or that aspect of modern physics"; Rogowski was particularly interested at that time in a contribution by some "young force" about alternating current resistance. [690] Letter from Walter Rogowski to Arnold Sommerfeld, October 20, 1912, Archive: Munich German Museum (Archive: NL 89, 012); abbreviated version, Sommerfeld Project: $http: //www.lrz-muenchen.de/ \sim Sommerfeld/KurzFass/01208.html$

[74] [217] Guillemin, Theorie (1926)

[75] „Verfasser geht so vor, dass er den Strom- und Spannungsverlauf stückweise berechnet für die verschiedenen Aeste der (zu einem geradlinigen Zuge idealisierten) Charakteristik und dass er an den Grenzen der Intervalle für die nötige Stetigkeit des Überganges sorgt. Diese Methode entspricht der elektrotechnisch-anschaulichen Denkweise des Verfassers und ist von ihm selbstständig durchgebildet worden. Eine andere mathematisch elegantere Methode mittels Fourier-Analyse, welche zu denselben Resultaten führt, werde ich mit ihm gemeinsam demnächst publizieren." [695] Arnold Sommerfeld, Votum Informativum for the dissertation by Ernst A. Guillemin, Archive of the Ludwig Maximillian University in Munich (OCI 52p)

Guillemin himself also wrote in his dissertation[76] that "the Fourier method can be carried out with complete mathematical precision even with an idealized conception of the problem (as we shall also be doing here)", and he likewise announced a work to be published jointly with Sommerfeld, which was to appear in the *Zeitschrift für Technische Physik*.[77]

As early as the previous January, in a letter to Professor Frederick J. Norton[78], who intended to employ Guillemin at his institute following his return from Munich, Sommerfeld praised Guillemin's physics training, yet he also stressed here that he saw Guillemin's greater gift elsewhere:

> I believe, however, that his real interests lie more in the direction of electrical engineering and that it is first and foremost in this area where he will accomplish original achievements. I would therefore like to recommend that he be transferred to the electrical engineering department, if possible.[79]

When Guillemin returned to the United States and MIT in 1926, he first worked with Vannevar Bush as an instructor for two years. Starting in 1928, as an assistant professor in MIT's electrical engineering department, he taught the subject areas of electrical networks and system stability in the communication division, which was led by Edward L. Bowles. This position marked the beginning of Guillemin's brilliant career as a network and system theorist and as an inspired university instructor, during the course of which he penned six textbooks on different aspects of network analysis and synthesis. Writing for the publication of an anthology, Rudolf Emil Kalman[80], one of Guillemin's former students, commemorated his recently deceased teacher:

> Specifically his assignment was the development of a graduate course in "Advanced Network Theory, including the Design of Electromagnetic Wave Filters and Related Networks". It was a fortunate coincidence in the history of technology. For all was just right: the time, the place, the discipline and the man, giving rise to a life-long romance. A corner-stone of Electrical Engineering found a dedicated and exceptionally gifted builder. Guillemin immediately recognized the challenge and within a short period proved more than equal to it. In 1931

---

[76] [217] Guillemin, Theorie (1926)

[77] I was not able to locate any such work published jointly by Guillemin and Sommerfeld, however

[78] Frederick John Norton (born 1896), American physicist

[79] „Allerdings glaube ich, dass seine eigentlichen Interessen mehr in der elektrotechnischen Richtung liegen und dass er auf diesem Gebiete in erster Linie zu originellen Leistungen kommen wird. Ich möchte daher empfehlen, ihn später, wenn es möglich ist, in das elektotechnische Department zu versetzen." [691] Letter from Arnold Sommerfeld to Frederick Norton, January 27, 1926, Archive Munich German Museum (Archive: NL 89, 003); facsimile, Sommerfeld Project: $http://www.lrz-muenchen.de/ \sim Sommerfeld/gif100/02190\_01.gif$

[80] Rudolf Emil Kalman (born 1930), Hungarian-American electrical engineer

the first volume of *Communication Networks* appeared, followed by the second volume in 1935. This was the first, and for more than a decade, the only modern treatment of Network Theory, still regarded as a "classic".[81]

Once Guillemin had become an associate professor in 1936, he turned his attention to researching and teaching the theory of electrical circuits. In 1940, he was appointed to the Microwave Committee of the National Defense Research Committee[82], a function in which he spent half of his time advising various groups in the MIT radiation laboratory. Here he was charged with solving all manner of problems associated with the design of electrical networks for various applications; networks for producing radar pulses were developed in one of these projects – and scientists even began to speak of the *Guillemin line!* Beginning in 1941, when Edward Bowles became an advisor to the US War Department, Guillemin was initially the head of the communication option in the electrical engineering department at MIT[83] and he led the entire institution from 1944 as a full professor.[84]

Fig. 1.8. Ernst A. Guillemin in conversation with MIT professor Lan J. Chu (right) and the vice-president and director of Burnell & Co., Lewis G. Burnell, at the opening of he new Guillemin Research Laboratory in Cambridge, Mass

Guillemin's book *Synthesis of Passive Networks* was published in 1957. He dedicated it to the electrical engineer Otto Brune, which he explained in

---

[81] [261] Kalman et al., Network (1971), p. viii
[82] This body was later renamed the Office of Scientific Research and Development
[83] *Proceedings of the IRE*, April 1962, Notes to the Contributors, p. 1429
[84] In 1960, Guillemin became the first ever recipient of MIT's Edwin Sibley Webster Professorship for Electrical Engineering. When he became a professor emeritus in 1963, he published his final book: *Theory of Linear Physical Systems*. The IRE and AIEE conferred upon him the Medal of Honor "for outstanding scientific and engineering achievements" in 1961 and the Medal in Electrical Engineering Education in 1962. Guillemin died on April 6, 1970. See [261] Kalman et al. Network (1971), p. viii, and *Who Is Who in America 1966–67*, p. 848

the foreword by citing Brune's dedication in laying the foundation of network analysis by means of mathematics:

> It is not out of place to call attention to some of the early work in network synthesis following the initial contributions of O. J. Zobel, G. A. Campbell and R. M. Foster. It was through the interest and encouragement of Vannevar Bush and Norbert Wiener that the necessary climate for a deeper study of network analysis and synthesis was established and fostered among our group, and in my opinion the one primarily responsible for establishing a very broad and mathematically rigorous basis for realization theory generally was Otto Brune, to whose clear and penetrating insight we owe much more than is generally credited to him by those who know of his work only through his very meager publications. Everyone who studies network synthesis knows about Brune's contributions to the RLC driving-point impedance problem, but only those who have listened to my lectures know that it was his creative thinking that also laid the rigorous mathematical foundations for all realization theory. It is for this reason that I enthusiastically dedicate this volume to him.[85]

In the first volume of his *Communication Networks*, Guillemin had already mentioned *communication engineer* as a new career for electrical engineers. The job profile that could be expected for this career was also written about by Norbert Wiener about ten years later in the book *Extrapolation, Interpolation and Smoothing of Stationary Time Series*, although it was not published until 1949.[86] Wiener had written the book during the war, but its content was subject to secrecy, and so the newly developed methods and techniques for the design of communication systems could only be made available to the public in the reprint that appeared in 1949. Communication and control engineering had received an enormous boost thanks to concentrated efforts during the war. Wiener spoke in his foreword of an *impetus* that had elevated this technology to a level that could not have been predicted, and he indicated that the time for great changes had now arrived:

> Many perhaps do not realize that the present age is ready for a significant turn in the development toward far greater heights than we have ever anticipated. [87]

Wiener considered his *Cybernetics: or Control and Communication in the Animal and the Machine*[88], published the year before, to be a philosophical basis for this reorganization and standardization of communication and control

---

[85] [215] Guillemin, Synthesis (1957), p. xi
[86] [560] Wiener, Extrapolation (1949)
[87] [560] Wiener, Extrapolation (1949), p. v
[88] [558] Wiener, Cybernetics (1948)

technology and cybernetics, and now he could also deliver the mathematical-technical – and primarily statistical – platform he had been developing for so long: a basis for *communication engineering*, for an *information theory*, as the emerging scientific-technical discipline would come to be known in subsequent years[89]:

> This is the study of messages and their transmission, whether these messages be sequences of dots and dashes, as in the Morse code or the tele-typewriter, or sound-wave patterns, as in the telephone or phonograph, or patterns representing visual images, as in telephoto service and television. In all communication engineering – if we do not count such rude expedients as the pigeon post as communication engineering – the message to be transmitted is represented as some sort of array of measurable quantities distributed in time. In other words, by coding or the use of the voice or scanning, the message to be transmitted is developed into a time series. This time series is then subjected to transmission by an apparatus which carries it through a succession of stages, at each of which the time series appears by transformation as a new time series. These operations, although carried out by electrical or mechanical or other such means, are in no way essentially different from the operations computationally carried out by the time-series statistician with slide rule and computing machine.[90]

Wiener thought cybernetics and communication engineering were inseparably linked to one another, as is already evident in the subtitle of the cybernetics book and as some of his other publications show, e. g. the keynote article *A New Concept of Communication Engineering*, which had appeared in the journal *Electronics* back in January of 1949. In print, the text was supplemented by two "boxes" in which the journal's editors provided a brief rundown of the content as well as other information and quotes from Wiener's *Cybernetics*, which had just been published a few months earlier. In this text, too, Wiener did not differentiate between philosophical foundation and technique. In order to achieve his goal of re-establishing and standardizing all of the subareas of communication engineering, he had to contend with the difference arising in the German language between *Starkstromtechnik* ("heavy current engineering") and *Schwachstromtechnik* ("weak current engineering"), which illustrated the differentiation between power engineering and communication engineering. "Electrical Engineering is divided into two main branches, termed Power Engineering and Communication Engineering."[91] Power engineering (*Starkstromtechnik*) was used for the transport of energy, while the transfer

---

[89] See [25] Aspray, Information (1985).
[90] [560] Wiener, Extrapolation (1949), p. 2
[91] [557] Wiener, Communication (1949). Also in [568] Wiener, Works (1985), pp. 197–199

of information fell within the scope of communication engineering (*Schwachstromtechnik*), termed thusly because the need to use great amounts of energy for the purposes of communication engineering was now a thing of the past:

> Powerful low frequencies by transmitters, which characterized the early days of transatlantic telegraphy, are being supplanted by transmitters of relatively high frequency, moderate power and astounding range of reliable reception.[92]

The fundamental unit of communication engineering as a whole was covered by this tradition, he said, for "in that moment in which circuits of large power are used to transmit a pattern or to control the time behavior of a machine, power engineering differs from communication engineering only in the energy levels involved and in the particular apparatus used suitable for such energy levels, but is not in fact a separate branch of engineering from communications."[93] However, the standardized basis for what would become communication engineering required more than just creating an artificial division between power and communication engineering. Thus in *Extrapolation, Interpolation and Smoothing of Stationary Time Series*, Wiener pursued the goal of uniting communication engineering with the field of statistical time series in theory and practice, and the relevant methods for this purpose included filtering in addition to extrapolation and prediction.[94]

Predicting a time series (or a message) could certainly not simply consist of its constant continuation. This would instead be a matter of statistical prediction, estimating the continuation of the time series (or communication), its most probable future pattern while minimizing random error. The details of Wiener's methods for mastering the prediction of time series do not need to be discussed at this point; a number of other hints emanating from his pen should suffice[95]:

> To predict the future of a curve is to carry out a certain operation on its past. The true prediction operator cannot be realized by any constructible apparatus; but there are certain operators which bear it a certain resemblance, and are in fact realizable by apparatus which we can build.[96]

The solution Wiener developed along with his colleague Julian Bigelow lay in self-correcting target tracking. It led them to the principle of feedback, which acts not only on the gun mount but also on the person operating the device. Working with Mexican physiologist Arturo Rosenblueth, whom Wiener had known very well since the early 1940s and whose interdisciplinary discussion

---

[92] [557] Wiener, Communication (1949), p. 74
[93] [560] Wiener, Extrapolation (1949), p. 3
[94] See also Chap. 3 for Wiener's Theory of Prediction
[95] See Chap. 3 for more details, however
[96] [558] Wiener, Cybernetics (1948), p. 12

groups[97] he regularly attended, Wiener and Bigelow developed the original ideas of cybernetics, which will be treated in greater detail in Chap. 3. In his book about this research program, Wiener also characterized the problem of the filter design, and he came right to the point with respect to its use in communication engineering:

> We often find that a message has been distorted by outside disturbances, which we call "noise". We then examine the problem of reestablishing the original message – the message with a certain phase lead or a certain delay – and we do so with the aid of an operator that is applied to the distorted message. The optimal design of said operator and of the device that realizes it depends on the statistical nature of the message and the noise - regardless of whether the errors are considered individually side by side or whether they are combined into a collective error. In designing the filters, we have thus replaced empirical and fairly random methods with processes that are completely founded in science.[98]

In *Extrapolation, Interpolation and Smoothing of Stationary Time Series*, Wiener described the problem in greater detail:

> We have a message which is a time series, and a noise which is also a time series. If we seek that which we know concerning the message, which is not bound to a specific origin in time, we shall see that such information will generally be of a statistical nature; and this will likewise be true of our information of the same sort concerning the noise alone, or the noise and the message jointly. While this statistical information will in fact never be complete, as our information does not run indefinitely far back into the past, it is a legitimate simplification of the facts to assume that the available information runs back much further into the past than we are called upon to predict the future. The usual electrical wave filter attempts to reproduce a message "in its purity", when the input is the sum of a message and a noise.[99]

Wiener had been working with the filter problem on a very mathematical-abstract level, yet he knew very well that there was quite a lot to do below this level in order to implement his standardization program. He thus referred to the work of appropriate experts, naming his MIT colleague Ernst Guillemin by name: "The problem of realization takes one into the theory of equivalent

---

[97] At that time, Rosenblueth hosted a monthly "dinner party" in Vanderbilt Hall at Harvard University. "These meetings were deliberate attempts to further communication between the various sciences. They were attended by scientists from many different specialisms and on each occasion one of them would introduce a discussion on his own topic". See [526] Stewart, Origins (1959), second section

[98] [558] Wiener, Cybernetics (1948), p. 30

[99] [560] Wiener, Extrapolation (1949), p. 10

networks as developed by Guillemin and others." As we shall see, the research conducted by Wiener and Guillemin in the late 1940s would be consolidated by members of a new generation of communication engineers, which included in particular the electrical engineer Lotfi Zadeh.[100]

In contrast to power engineering[101], which Wiener had no problem classifying in the new technological discipline of *communication engineering* because the definition of power was completely clear to him, doing the same for communication engineering proved to be much more problematic at first. The information technology content of communication engineering was far from simple to characterize, since very few books on the subject of communication had dealt with the concept of information at that time, and the average communication engineer had had to get by up to this point without a clearly defined body of information. This problem, too, had been solved by the scientific efforts put forth during the Second World War, on one hand by Wiener using the tool of his statistical theory of time series, and on the other hand by Claude E. Shannon, with his *Mathematical Theory of Communication*.[102] The next chapter will return to both developments. It should only be mentioned here that Claude E. Shannon's theory was also based on statistical assumptions and that it is therefore entirely possible to interpret the following sentence from *Cybernetics* in a spirit of commonality; nevertheless, in using the plural form of the personal pronoun – we – Wiener was referring not to Shannon but to his colleague Bigelow:

> In so doing, we have made communication engineering into a statistical science, a branch of statistical mechanics.[103]

1.3 Ernst A. Guillemin, Lotfi A. Zadeh, Dr. Jekyll and Mr. Hyde

Having graduated with a Bachelor of Science in electrical engineering from the University of Tehran[104] in 1942 and after working for a year as a technical contractor with the United States army forces in Iran, the student Lotfi Aliasker Zadeh came to the USA in 1944.[105] He had left Iran in order to pursue his

---

[100] See the later sections in this chapter
[101] Power engineering is sometimes also called "drive engineering"
[102] [507] Shannon, Communication (1948)
[103] [507] Shannon, Communication (1948), p. 30
[104] Quoted from information listed under "The Author" in: *Columbia Engineering Quarterly*, January 1950, p. 13
[105] Lotfi A. Zadeh was born Lotfi Aliaskerzadeh on February 4, 1921 in Baku, the capital of Azerbaijan, which was part of the Soviet Union at the time. At the age of ten, he moved to Iran. His father was a business man and newspaper correspondent; his mother was a doctor

scientific work unhindered.[106] He spent three months in Cairo waiting for a neutral ship that would take him to the United States and eventually wound up on a Portuguese freighter to Philadelphia. Zadeh worked at International Electrical Laboratories in New York City for a short time, but then applied in 1944 to the Massachusetts Institute of Technology (MIT) in Cambridge, Massachusetts, and was accepted to continue his electrical engineering studies. For the thesis he completed with Robert M. Fano[107], he was awarded the degree of Master of Science in 1946.[108]

Zadeh had originally planned to then earn his doctorate at MIT, but in 1945 his parents also moved to America – albeit to New York – and so he decided to apply for a teaching contract and the opportunity to write a dissertation at Columbia University, although Ernst Guillemin urged him to remain at MIT. His application letter to Professor Walter Curry[109], chair of the Columbia electrical engineering department, in which he reported on his graduation from MIT, was met with success. Guillemin wrote at least one letter to Curry:

> I regard Mr. Zadeh as one of the most brilliant students that it has recently been my pleasure to know. You will be fortunate indeed if he decides to join your staff. I say this with feeling for I had hoped to be able to get Mr. Zadeh to take part in research activities here at MIT and shall very definitely regret losing him.[110]

Zadeh did move to New York, but he remained very close to his teacher at MIT, both personally and professionally. They were in agreement on many points regarding technical issues of communication engineering, and were vociferous in their calls for a greater mathematical foundation for electrical engineering in research and teaching. When different specialties began to emerge within electrical engineering in the 1950s, various "professional groups" were formed within the tradition-rich IRE, which then also published their own "transactions". The contributions to the Symposium on Information Theory, which had been held in September of 1950, were thus published in February 1953 in the first edition of the *IRE Transactions on Information Theory*.[111]

---

[106] "I could have stayed in Iran and become rich, but I felt that I could not do real scientific work there ... Research in Iran was nonexistent." Lotfi Zadeh quoted in the *IEEE Spectrum* (June 1995)

[107] Robert Mario Fano (born 1917), Italian-American electrical engineer. Fano studied engineering sciences in Turin, emigrated to the US in 1939 and from 1963 to 1968 directed the MAC Project at MIT (MIT Project on Mathematics and Computation). Fano was awarded the Claude E. Shannon Award in 1976

[108] [705] R.S. interview with L.A. Zadeh (1999)

[109] Walter Andrew Curry (born 1894) was an associate professor from 1942 to 1947, then a professor and fellow in the Department of Electrical Engineering at Columbia University in New York

[110] Quoted from [574] Zadeh F., Life (1998)

[111] In this first edition of the *IRE Transactions of Information Theory*, Dennis Gabor wrote an article entitled *Communication Theory, Past Present, Future*, that had

The first volume of the *IRE Transactions on Circuit Theory* had already been published in December 1952, and its editorial board included Guillemin and Zadeh. When a special issue on "Fourier integral papers"[112] was to be published in September of 1955, Guillemin, who had been an early and vocal proponent of incorporating the Fourier integral theory into the electrical engineering curriculum, took on the task of writing an introduction to this field[113], while Zadeh was credited as co-author on a paper about the generalization of the Fournier integral.[114,115]

There were also, however, some fundamental differences regarding the theory and practice of the scientific discipline advocated by Guillemin and Zadeh, which would continue to inform later developments, all the way up to Zadeh's founding of Fuzzy Set Theory. These different views can be illustrated very nicely using a literary allusion, and credit for this example must go to Ernst Adolf Guillemin.

When Robert Louis Stevenson published his novel *The Strange Case of Dr. Jekyll and Mr. Hyde* in 1886, he could surely have had no idea that 50 years later it would serve as a pattern of argument for a matter having to do with electrical engineering, yet Guillemin cited "Dr. Jekyll and Mr. Hyde" in the foreword of his 1952 textbook *Introductory Circuit Theory* in order to demonstrate the "dual character of network theory".[116] A brief rundown of the content, plus a few quotes from the novel, will be useful before Guillemin's analogical argument follows.

---

apparently not been delivered as a lecture, while *A History of the Theory of Information*, the article by British electrical engineer Colin Cherry (1914–1979), also appears as the first paper in the proceedings. Dennis Gabor (1900–1979), Hungarian-British electrical engineer, studied physics at the Technical University of Budapest beginning in 1920, and from 1921 to 1924 at the Technical University of Berlin-Charlottenburg, where he graduated with a Master's in physics. Gabor emigrated to England during the National Socialist period. In 1971, he received the Nobel Prize in Physics in recognition of his invention and development of the holographic method

[112] *IRE Transactions On Circuit Theory*. Published by the Professional Group on Circuit Theory under the aegis of the *Institute of Radio Engineers*, Vol. CT-2, September 1955, No. 3

[113] [211] Guillemin, Integral (1955)

[114] [401] Miller et al., Generalization (1955)

[115] Other texts about the Fourier integral in this issue include the foreword by J. G. Brainerd, p. 226, as well as: C. H. Page, Applications of the Fourier Integral in Physical Science, pp. 231–237; W. R. Bennett, Applications of the Fourier Transforms in Circuit Theory and Circuit Problems, pp. 237–243; W. K. Linvill, R. E. Scott, E. A. Guillemin, Evaluation of Fourier Transforms, pp. 243–250; T. Murakami, Murlan S. Corrington, Applications of the Fourier Integral in the Analysis of Color Television Systems, pp. 250–255

[116] [212] Guillemin, Introductory (1953), p. xx

1.3 Ernst A. Guillemin, Lotfi A. Zadeh, Dr. Jekyll and Mr. Hyde

Dr. Jekyll, an esteemed physician, is convinced "that man is not truly one, but truly two".[117] He experiments with chemicals of many kinds and is even willing to try them out on himself in order to separate his competing natures: "I had long since prepared my tincture; I purchased at once, from a firm of wholesale chemists, a large quantity of a particular salt which I knew, from my experiments, to be the last ingredient required; and late one accursed night, I compounded the elements, watched them boil and smoke together in the glass, and when the ebullition had subsided, with a strong glow of courage, drank off the potion."[118] This is the birth of his alter ego Edward Hyde, who combines all of the bad personality traits. The medicine required to transform him back into Dr. Jekyll eventually begins to run short: "My provision of the salt, which had never been renewed since the date of the first experiment, began to run low. I sent out for a fresh supply and mixed the draught; the ebullition followed, and the first change of colour, not the second; I drank it and it was without efficiency."[119] Mr. Hyde begins to dominate Dr. Jekyll, and the drama ends in suicide. In a final letter, Dr. Jekyll reports on all of these events, and also expresses his supposition "that my first supply was impure, and that it was that unknown impurity which lent efficacy to the draught."[120] Dr. Jekyll can state neither the measure nor count of the amount of the salt required, for he does not know and never has, but has been calculating and toiling with inexact quantities.

Guillemin assumed that the readers of his textbook were familiar with this story. It served as an analogy for an important distinction:

> One final point. In the teaching of this subject I regard it as important to remind the student frequently that network theory has a dual character (no connection with the principle of duality); it is a Dr. Jekyll-Mr. Hyde sort of thing; it is two-faced, if you please. There are two aspects to this subject: the physical and the theoretical. The physical aspects are represented by Mr. Hyde – a smooth character who isn't what he seems to be and can't be trusted. The mathematical aspects are represented by Dr. Jekyll – a dependable, extremely precise individual who always responds according to established custom. Dr. Jekyll is the network theory that we work with on paper, involving only pure elements and only the ones specifically included. Mr. Hyde is the network theory we meet in the laboratory or in the field. He is always hiding parasitic elements under his jacket and pulling them out to spoil our fun at the wrong time. We can learn all about Dr. Jekyll's orderly habits in a reasonable period, but Mr. Hyde will continue to fool and confound us until the end of time. In order to be able to tackle him at all, we must first become well

---

[117] [525] Stevenson, Jekyll (2004), p. 59
[118] [525] Stevenson, Jekyll (2004), p. 60
[119] [525] Stevenson, Jekyll (2004), p. 73
[120] [525] Stevenson, Jekyll (2004), p. 73

acquainted with Dr. Jekyll and his orderly ways. This book is almost wholly concerned with the latter. I am content to leave Mr. Hyde to the boys in the laboratory.[121]

Even as an MIT student attending Guillemin's course on network theory, Lotfi Zadeh had not wanted to adhere to his professor's opinion on this matter. There were already irreconcilable differences between the two at that time when it came to what network theory should be and what it should accomplish. The network theory taught by Guillemin was a theory of perfect objects that differed considerably from reality. Zadeh recalled a telling discussion from this period:

> Everything was idealized: resistors, inductors, they were all perfect elements. So even at that point I had some discussions with him [Guillemin]. I said this is unrealistic. The real world is not like that. I mean resistors are not pure resistors, capacitors are not pure capacitors, and so forth, and I said, I told him that I think that at some point in the future circuits will be designed and analyzed using computers even when I was a student at MIT.[122]

Guillemin was just as unable at that time to comprehend the view of his student as Zadeh was of his mentor. They saw different worlds: "... because his world, it was not a world of mathematics, it was a world of real things, but it was an idealized world".[123] As Zadeh sums it up today:

> He was happy with that world ... He constructed the world by himself. It was a perfect world. Everything was perfect in that world. He was happy and so he never considered noise, he never considered nonlinearities, he never considered imprecision; he never considered those things, so it was an idealized world. He was happy.[124]

In 1946, Zadeh obtained a position at Columbia University in New York as an instructor responsible for teaching the theories of circuits and electromagnetism. He found a supervisor for his dissertation in Associate Professor John Ralph Ragazzini[125], who was about 10 years older, and the two were soon good friends. In 1949, he received his Ph.D. from Columbia University for his work entitled *Frequency Analysis of Variable Networks*.[126] Zadeh

---

[121] [212] Guillemin, Introductory (1953), p. xx
[122] [707] R. S. interview with L. A. Zadeh (2001)
[123] [708] R. S. interview with L. A. Zadeh (2001)
[124] [708] R. S. interview with L. A. Zadeh (2001)
[125] John Ralph Ragazzini (born 1912) studied at City College in New York. In 1941, he earned a Ph.D. from Columbia University and was given a position at the Columbia University School of Engineering. Ragazzini eventually became a professor there
[126] [632] Zadeh, Networks (1950). The original manuscript R 143 was received by the Institute of Radio Engineers (IRE) on April 11, 1949 and was published in the *Proceedings of the IRE* in 1950

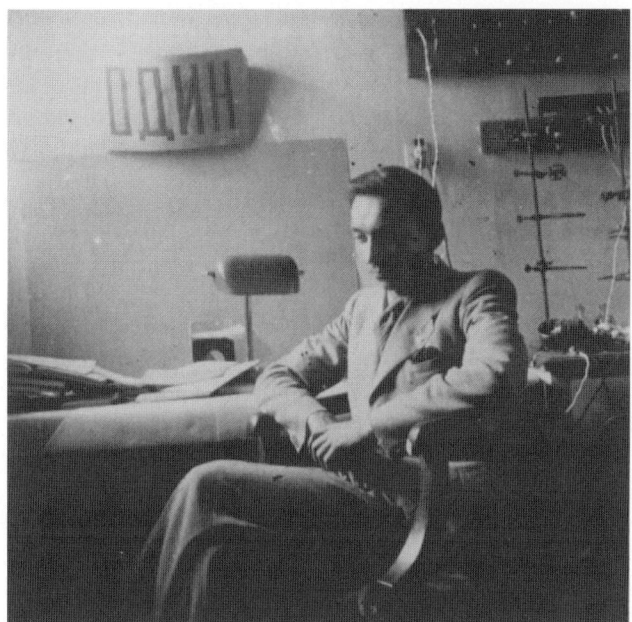

Fig. 1.9. Lotfi A. Zadeh at his desk in Tehran in 1937. A number of experimental structures can be seen beside him: theory and reality in immediate proximity

presented the results of his doctoral dissertation on September 1st of the same year at the IRE West Coast Convention in San Francisco, California. In it, he had developed a method for analyzing linear variable networks, a formulation that utilized the mathematical operator theory in order to formulate and deal with system functions of linearly variable systems. This was a generalization of the standard method with which time-invariant networks were treated. This method and the resulting calculus were also the subject or starting point of other works that Zadeh published in the following two years.[127] Finally, he also wrote a condensed version for mathematicians and physicists[128], oriented to the mathematical operator theory, which was followed by similar works in 1951.[129]

After finishing his dissertation, Zadeh turned his attention to the problems associated with the electrical filter. He first authored an internal report about his

---

[127] [576] Zadeh et al., Extension (1950), [669] Zadeh, Band-Pass (1950), [602] Zadeh, Correlation (1950), [596] Zadeh, Circuit (1950), [620] Zadeh, Impulsive (1950), [630] Zadeh, Networks (1951), [657] Zadeh Stability (1951), [621] Zadeh, Initial (1951), [607] Zadeh, Excitation (1951), [609] Zadeh, Filtering (1951)

[128] [636] Zadeh, Operators (1951)

[129] [634] Zadeh, Operational (1951), [609] Zadeh, Filtering (1951), [616] Zadeh, Heaviside (1951), [611] Zadeh, Filtration (1951)

study[130], though the by-line of the journal article *An Extension of Wiener's Theory of Prediction*[131], which was published shortly thereafter, included Ragazzini's name in addition to his own – "just courtesy!"[132] In this work, which expanded Wiener's prediction theory, Zadeh indicated that the foundations for his work could be found both with Wiener and with Andrei N. Kolmogorov.[133] Both of them, Wiener and Kolmogorov, had proceeded from this problem: If knowledge about the past and present of the physical system are given, how can its future behavior be predicted?

Fig. 1.10. J. Ragazzini

Zadeh generalized Wiener's theory in two ways: 1) The signal component of a given time series was separated into two parts, of which the first is a non-random function in time that can be represented as a polynomial, while the other part functions as a stationary and statistical random function represented by a given correlation function. In Wiener's theory, by contrast, a non-random portion of the signal occurred only when it consisted of a known function in time. 2) The response behavior of the predicting system or the weighting function used to make the prediction should disappear outside of a finite time interval. In Wiener's theory, on the other hand, this time interval was assumed to be infinitely long.
Zadeh showed that determining the weighting function leads to the solution to a modified Wiener-Hopf equation, for which he could provide an explicit solution. This publication was a milestone in the theory of network synthesis.

---

[130] The author of this internal report was Zadeh alone. [707] R. S. interview with L. A. Zadeh (2001)

[131] [576] Zadeh, Extension (1950)

[132] See e. g. [707] R. S. interview with L. A. Zadeh (2001) among others. In several conversations with the author, Zadeh stressed that he wrote all of his texts alone; the co-authorships had always been a matter of courtesy – with one exception. This exception [123] Chang, Mapping (1972) will be discussed in Chap. 6

[133] [294] Kolmogorov, Interpolation (1941), [560] Wiener, Extrapolation (1949)

Working with Ragazzini, who was advising on a number of third-party research projects that were being conducted by scientists in his department and at other institutions, Zadeh also published the paper *Analysis of Sampled-Data Systems*.[134] One paper cited in it had been produced by R. J. Schwarz and Kenneth S. Miller at New York State University in New York City.[135] The mathematician Miller[136] worked at Columbia University's Electron Research Library (ERS) on one of the research projects being conducted from outside the university; Zadeh also published a few papers with him on the subject of the problem of filters, which shall be discussed in the next section.[137]

1.4 Ideal and Optimum Filters

The differences in opinion between Zadeh and Guillemin on the matter of electrotechnical systems had become conspicuous and can be illustrated quite well using the example of electrical filters. Guillemin saw an ideal world, while Zadeh viewed the real world before him, with real capacitors, resistors and coils – real electrical circuits, which could only ever be highly imprecise approximations of the ideal entities in Guillemin's world of electrotechnical theory. Yet Guillemin had known very well that this world of non-ideal things existed; he had simply preferred to leave it to the "boys in the laboratory" and not deal with it himself. Zadeh did not want to follow him in this respect. His interest lay not only in these real things that didn't exactly obey the theory. He was also especially interested in the relationships of real objects to their theoretical counterparts. This interest is expressed in the work he produced during the 1950s. He made extensive use of modern mathematical methods for his general theory of linear signal transmission systems[138]; here he employed Fourier analysis as well as Hilbert space and operator calculus, which had been developed almost 20 years earlier for use, for example, in the quantum mechanics of John von Neumann. In the tradition of Norbert Wiener, Zadeh applied these mathematical methods to problems in communication engineering in order to represent general principles in the transformation of signals and in the characterization of input-output relationships in nonlinear systems.[139]

---

[134] The paper was presented at the AIEE (American Institute of Electrical Engineers) Summer General Meeting, which took place in Minneapolis June 23–27, 1952, [575] Zadeh et al., Analysis (1952)
[135] [402] Miller, Analysis (1950)
[136] Kenneth Sielke Miller (born 1922) studied mathematics at Columbia University in New York and earned his Ph.D. in mathematics from there
[137] [578] Zadeh et al., Ideal (1952), [577] Zadeh et al., Filters (1952), [580] Zadeh et al., Multiplexing (1952), [585] Zadeh et al., Solution (1956)
[138] [613] Zadeh, General (1952)
[139] [630] Zadeh, Networks (1950), [620] Zadeh, Impulsive (1950), [592] Zadeh, Application (1950), [613] Zadeh, General (1952)

Nevertheless, his theory also paid tribute to the distinctive features of the technical components of real communication systems:

> Many of the mathematical techniques used in the theory are commonly employed in quantum mechanics, though in a different form and for different purposes. The dissimilarities are due largely to the special character of signal transmission systems and the nature of the problems associated with the transmission and reception of signals.[140]

Here Zadeh considered a signal transmission system to be the system N shown in the following figure, without any further restrictions, into which an input signal $u(t)$ enters and which emits an output signal $v(t)$ (Fig. 1.11).
N is linear if it satisfies the following additivity properties:

$$u(t) = c_1 u_1(t) + c_2 u_2(t), \quad \text{und} \quad v(t) = c_1 v_1(t) + c_2 v_2(t),$$

where c_1 and c_2 are any constants and $v_1(t)$ and $v_2(t)$ are the system responses to $u_1(t)$ and $u_2(t)$.

In his theory, Zadeh represented the signals as vectors in an infinite-dimensional signal space Σ.[141] Vectors of such a space Σ can be projected onto a subspace M (and along its complement M'), and Zadeh interpreted such a projection in his mathematical theory "in physical terms as the filtering of the class of signals representing Σ with an *ideal filter* N which passes without distortion (or delay) all signals belonging to M and rejects all those belonging to M'". [142]

A short time later, he continued the comparative analysis between the projections in a function space on the one hand and filtering with ideal filters on the other hand[143] when he borrowed some of the properties of ideal filters from algebraic projection combination and additionally employed the "functional symbolism" seen the following diagram (Fig. 1.12).

Thus if two ideal filters N_1 and N_2 are given, then $N = N_1 + N_2$ is a filter that combines the two original filters in parallel, whereas $N = N_2 N_1$ represents the serial combination (series connection) of the two filters, while the third combination $N = N_1 | N_2$ signifies their "separation".

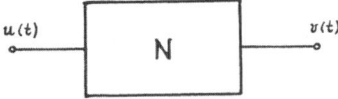

Fig. 1.11. Schematic representation of a signal transmission system

---

[140] [613] Zadeh, General (1952), p. 293f
[141] In particular, Zadeh included the Hilbert space L_2 of all quadratically integratable functions
[142] [613] Zadeh, General (1952), p. 305. Emphases in the original
[143] [578] Zadeh et al., Ideal (1952)

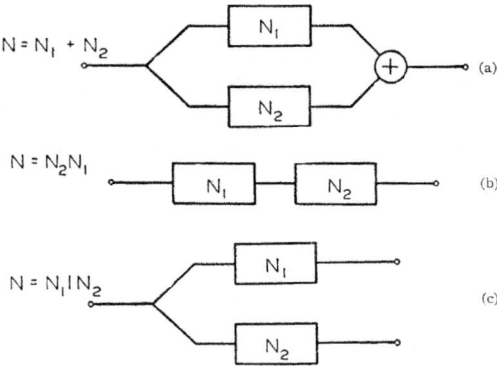

Fig. 1.12. Filter combinations (Zadeh's functional symbolism)

In this paper, Zadeh further defined the filter $1 - N$, which is complementary to N, as well as other algebraic properties which do not need to be discussed here. Instead, his final paragraph on applications for the theory should be mentioned. It begins as follows:

> The analogy between ideal filters and projections, and the important role played by the latter in the theory of linear transformations, leads one to believe that ideal filters should eventually occupy an important place in the theory of linear signal transmission systems. In so far as the practical applications of ideal filters are concerned, a preliminary study indicates that such filters might be advantageously used in several applications, particularly in connection with the simultaneous transmission (multiplexing) and filtration of signals.[144]

Zadeh presented a somewhat more favorable view in his thoughts on the two problems in question at the meeting of the Section of Mathematics and Engineering of the New York Academy of Sciences on February 15, 1952:

> It appears that communication theory offers a very fertile field for function space methods, perhaps even more fertile than that of quantum mechanics.[145]

Proceeding from Shannon's information theory, Zadeh pleaded here, too, for the use of mathematical methods and theories of function spaces, particularly the Hilbert space representation of signals, and he also referred to a number of his own printed works.[146] In this lecture, he first discussed the following problem: $X = \{x(t)\}$ is a set of signals from which any given signal $x(t)$ is selected, transmitted over a "noisy" channel Γ and received at the other end

---

[144] [578] Zadeh, et al., Ideal (1952), p. 227
[145] [645] Zadeh, Problems (1952), p. 201
[146] [611] Zadeh, Filtration (1952), [613] Zadeh, General (1952)

as $y(t)$. As a result of noise and distortions – effects represented by Γ – $y(t)$ is thus different from the input signal $x(t)$. It is nevertheless possible under certain circumstances to re-obtain the signal $x(t)$ (or a time-delayed copy of it) from $y(t)$, and Zadeh introduced the appropriate methods in his lectures. If he briefly characterized the relationship between $x(t)$ and $y(t)$ as $y = \Gamma x$, then the process by which $x(t)$ was re-obtained could be written inversely as follows: $x = \Gamma^{-1} y$.[147]

After embedding the signal space Σ in a function space whose points consist of ordered pairs $(x(t), y(t))$ (short form: (x, y)), Zadeh assigned each such pair to a distance function $d(x, y)$, which can be seen as a measure for the inequality of the two signals and had the usual metric properties; Zadeh also provided the usual examples of this type of metrics in function spaces. In order to re-obtain a transmitted signal x_k from the received signal y, the receiver must calculate the "distance" from y to all of the possible transmitted signals $x_1, x_2, ..., x_n$ in terms of the metric $d(x, y)$ and then select that signal that is closest to y (Fig. 1.13): $d(x_k, y) < d(x_i, y), i \neq k$, for every k and i. Zadeh referred to the use of this principle in many communication systems, though not necessarily in the quantitative form shown above! He noted that it is very difficult or even impossible in many practical situations to define a quantitative measure for a distance function in order to calculate the inequality of two signals; in such cases, a *neighborhood concept* from the theory of topological spaces could be used. Spaces such as these, Zadeh surmised, could be very interesting with respect to applications in communication engineering.

The second point Zadeh discussed in his lecture was the multiplex transmission of two or more signals. The system had two channels for this purpose, he said, and the corresponding signal sets were $X = \{x(t)\}$ and $Y = \{y(t)\}$. On the receiver side of the communication system, a signal $u(t) = x(t) + y(t)$ would then be received; the task for the receiver was therefore to extract the two signals $x(t)$ and $y(t)$ from $u(t)$. Thus the problem was that of finding two

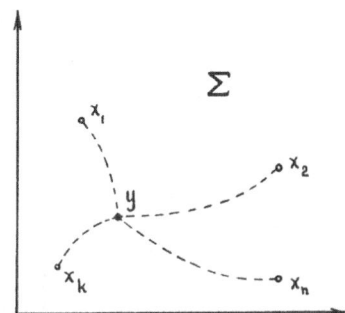

Fig. 1.13. Recovery of the input signal by comparing the distances between the obtained signals and all possible transmitted signals

---

[147] Γ^{-1} is the inverse function of Γ on the set $\{y(t)\}$, if it exists

filters N_1 and N_2 such that the following were true for all x in X and all y in Y:

$$N_1(x+y) = x \quad \text{und} \quad N_2(x+y) = y.$$

These two equations correspond to the separation process $N = N_1|N_2$, which is shown in Fig. 1.11 as the third filter combination.

Since the signals $x(t)$ and $y(t)$ are defined on an interval $(0,T)$ and in this respect can be seen as vectors of an n-dimensional signal space Γ, there are corresponding subspaces M_x and M_y in Γ, each of which can be characterized by k and l relations of the following form:

$$\begin{aligned} f_i(x_1, x_2, ..., x_n) = 0, & \quad i = 1, 2, ..., k, \\ g_j(y_1, y_2, ..., y_n) = 0, & \quad j = 1, 2, ..., l, \end{aligned}$$

wherein $x_1, x_2, ..., x_n$ and $y_1, y_2, ..., y_n$ are the coordinates of x and y and wherein f_i and g_j, respectively, are special functions of these coordinates. Further n relations among the coordinates of x, y and u are obtained from the sum $u(t) = x(t) + y(t)$:

$$x_\nu + y_\nu = u_\nu, \quad \nu = 1, ..., n.$$

All told, there are thus $k + l + n$ equations with $2n$ unknowns $x_1, x_2, \ldots, x_n$ and $y_1, y_2, \ldots, y_n$. If $k + l = n$ is true, then the equations can generally be solved and the solution to the equation system can be indicated as follows:

$$x_\nu = H_\nu(u_1, u_2, ..., u_n) \quad \nu = 1, 2, ... n$$

or briefly: $x = H(u)$.

The functions H_ν and H thus provide the desired characterization of each ideal filter.

Zadeh concluded his presentation with the observation that, in the case of the linearity of the coordinate equations, the operation of the ideal filter corresponds exactly to a projection of the signal space Γ on the corresponding subspace M_x along M_y. He illustrated ideal filtration in the linear and in the nonlinear instance for a two-dimensional example in the following Fig. 1.14:

When considering similar diagrams characterizing electrical filters, in which the current is applied against the frequency, it is readily apparent that there aren't any ideal filters in reality at all. One always observes *tolerance ranges*, there is always a surrounding area of cut-off frequencies that form the transition between the passband and the stopband; these are not sharp boundaries; in reality they are always fuzzy (see Fig. 1.5)!

"From the mathematical point of view, the theory of filtering is essentially a study of certain types of mappings of function and sequence spaces." Zadeh

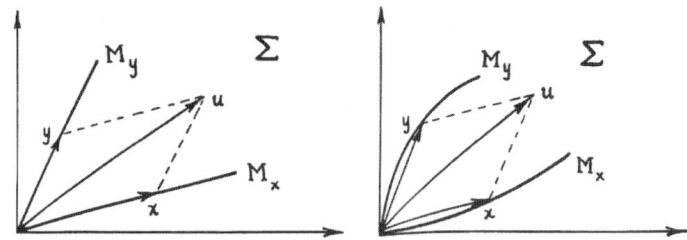

Fig. 1.14. Geometric representation of linear (left) and nonlinear (right) filtering

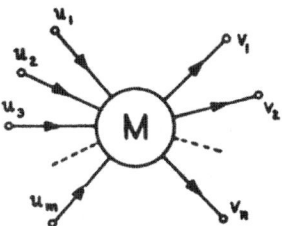

Fig. 1.15. Representation of a multipole ($m + n$-pole)

wrote this in one of his last texts on the filter theory[148], but the mathematically/analytically sophisticated results of his publications on this subject do not need to be discussed any further here.[149] The obvious discrepancy between mathematical theory and electrotechnical communication systems eventually led him to completely new ideas on how to expand the theory in order to apply it. By forming new concepts and using the subsequent results, Zadeh tied directly into Wiener's statistical basis of filter theory but then immediately surpassed it. If filters that were realized as electrical circuits did not operate according to the mathematical theory, then one must be content with less – *optimum* filters:

> A distinction is made between ideal and optimum filters, the former being defined as filters which achieve a perfect separation of signal and noise.[150]

If ideal filtration is not possible, though, which is often the case when the signal is mixed with noise, then one must accept that the filtration can only be incomplete. In such cases, a filter that delivers the best possible approximation of the desired signal – and "a particular meaning" of "best approximation" is used here – is called an optimum filter. Zadeh stressed that an optimum filter depends upon a reasonable criterion:

---

[148] [666] Zadeh, Theory (1953), p. 35
[149] See: [613] Zadeh, General (1952), [611] Zadeh, Filtration (1952), [645] Zadeh, Problems (1952); [581] Zadeh et al., Optimum (1952), [640] Zadeh, Optimum (1953), [633] Zadeh, Nonlinear (1953), [666] Zadeh, Theory (1953)
[150] [666] Zadeh, Theory (1953), p. 35

Because of its complex nature, the question of what constitutes a "reasonable" criterion admits of much argument. Since $x(t)$ and $y(t)$ are, in general, random functions, it is natural to formulate such a criterion in statistical terms.[151]

However, there were usually not enough statistical data to determine an optimum filter by means of statistical calculations. For this reason, Zadeh now proposed an alternative and thus a practicable approach to evaluating the performance of a filter F that operated not by statistical methods, but rather once again used a distance in the function space:

... we shall say that a filter F is optimum relative to a distance function $d(v, x)$ if
$$\langle d[F(x * y), x] \rangle_{A_v} = Q = \text{a minimum for all } t,$$
where the average is taken over the ensemble $x(t), y(t)$.[152]

Without going any deeper into mathematical details, it should be stated very clearly here that Zadeh had turned away from statistical methods during the course of his work on optimum filters in the mid-50s, and he recognized that a more promising approach was that of finding an optimum filter relative to a distance to be minimized in the function space of the signals. It will later become apparent that this renunciation of statistics was a great deal more complete than this.

1.5 Generalizations

Filters are specific communication systems, and Zadeh generalized his thoughts about electrical filtration into a general theory of signal transmission systems:

Ideal filters have many uses, of which only a few have thus far been exploited in practice. In particular, it can readily be shown that any signal transmission system (fixed or variable) may be approximated (to within a constant time delay) by a weighted parallel-series combination of ideal filters.[153]

As he had done with ideal filters and projection operators in the function space, he now devised for the theory of general signal transmission systems an analogy between them and more general operators, paying special attention to the so-called Heaviside operators. He then also employed algebraic considerations, which led to a product relationship of the operators that he interpreted physically as a series connection of communication systems.

---

[151] [666] Zadeh, Theory (1953), p. 47
[152] In this case, the symbol ∗ signifies the "mixing" of the two signals $x(t)$ and $y(t)$
[153] [613] Zadeh, General (1952), p. 306

Finally, he was able to establish correlations to "Volterra's composition rule" and to the mathematical operation of convolution.

A year later, he condensed his ideas even more generally in *A Contribution to the Theory of Nonlinear Systems*[154], in which he expanded his theory to include the nonlinear. He studied communication systems having only one input pole and one output pole, which were thus called "two-poles" (or "2-poles"), and Zadeh then defined parallel and series combinations for nonlinear two-poles. He employed a comparative functional analysis and eventually generalized Wiener's theory of prediction and filtration to a particular class of linear two-poles. At the same time, his paper on *Nonlinear Multipoles* was being published in the *Proceedings of the National Academy of Sciences*.[155] Here Zadeh used the still relatively new graph theory, proposed by Mason[156] in 1951, and the mathematical network theory according to Robert Duncan Luce[157] to bridge the gap from electrical networks, which had long been described as multipoles, to a very much more abstract theory of nonlinear multipoles. A multipole, or more precisely an $m + n$-pole, was understood here to be an entity M with m input poles $u_i (i = 1, ..., m)$ and n output poles $v_j (j = 1, ..., n)$ (see Fig. 1.15).

An $m+n$-pole is characterized by a set of n input-output relationships of the form

$$v_j = F_j(u_1, u_2, ..., u_m), \qquad j = 1, 2, ..., n$$

where F represents an operator acting on the ordered m-tuple

$$(u_1, u_2, ..., u_m).^{158}$$

Zadeh defined equivalent multipoles M_1 and $M_2 (M_1 = M_2)$ as those with identical input-output relationships ($F_j^{(1)} = F_j^{(2)}$). The sum of two multipoles he defined by the sum of their input-output relationships, and the product of a $k + m$-pole M_2 and an $m + n$-pole $M_1 (M = M_2 M_1)$ that Zadeh introduced corresponds to their composition. He concluded his algebraic investigations with the definitions of the left-inverse and right-inverse multipole as well as of the unit multipole and the inverse multipole of a given multipole, ultimately the definition of the linearity of multipoles and with that their delimitation over nonlinear multipoles.

---

[154] [633] Zadeh, Nonlinear (1953)
[155] [629] Zadeh, Multipoles (1953)
[156] Samuel Jefferson Mason (1921–1974), American electrical engineer, professor at MIT from 1954 until his death. For more on Mason's graph theory, see: [354] Mason et al., Circuits (1970), [355] Mason, Graphs (1951), [356] Mason, Theory (1953)
[157] [330] Luce, Theorems (1952). Robert Duncan Luce (born 1925), American mathematician
[158] [629] Zadeh, Multipoles (1953), p. 274f

Zadeh then demonstrated that each linear system can be represented as interconnected linear two-poles and adders, and after a number of functional analyses like those he had also employed in the previously published articles, he stated that nonlinear multipoles can be approximated by given linear multipoles. When Zadeh again performed a detailed treatment of his analysis of multipoles in 1957, he emphasized the usefulness of this method, which had become popular within the field of network engineering, although some electrical engineers among "the boys in the laboratory" dealt with it only reluctantly:

> In many respects, the basic concepts and theorems of the multipole approach are merely natural generalizations of their counterparts in the conventional two-pole method of circuit analysis. However, the greater generality and abstractness of the multipole method give it a considerably broader range of applicability than that possessed by the two-pole method. On the debit side, the same characteristics make it somewhat more difficult to understand on first exposure and lessen its appeal to those who like to deal with circuits on the physical level.[159]

Yet as before, a discrepancy still existed between theory and practice that made it difficult to reconcile the multipole method or similar mathematical techniques with the experiments. In this regard, Zadeh placed his hope in the future developments that had been poised conspicuously on the horizon for several years:

> ... in general, nonlinear systems are not susceptible of a strictly analytical treatment and therefore do not provide the system theorist with a fruitful field for purely theoretical investigation. In recent years, however, the advent of large scale digital computer and other mechanized means of computerization has profoundly influenced the basic philosophy of system design and analysis. Thus, in virtue of the availability of machine computers, it has become sufficient to carry the analytical treatment only to a point where the problem is reduced to mathematical operations which can be handled by such computers. The development has made it practicable to study nonlinear systems which cannot be completely analyzed by purely analytical means.[160]

The digital computer did not come any closer to solving the fundamental problem separating mathematical theory from theoretical practice, but Zadeh would not come to this realization until about ten years later. The enthusiasm for the digital computer that resonated in these latter words traces back to the period of his move from MIT to Columbia.

---

[159] [628] Zadeh, Multipole (1957), p. 97
[160] [640] Zadeh, Optimum (1953), p. 396

1.6 Let's go digital!

As a member of the department of engineering in New York, Zadeh soon had the opportunity to attend guest lectures by Norbert Wiener and Claude E. Shannon about their recent research in communication engineering. As early as 1946, Shannon had spoken here about his mathematical information theory[161], which he had yet to publish; it was published in two parts in June and October of 1948.[162]

> It was a very interesting time because that was the time when Shannon described his information theory and Wiener cybernetics and things like this. It was a very exciting period of time.[163]

Zadeh had met Norbert Wiener when the latter was an MIT professor and had spoken with him a number of times during his studies. Developments in cybernetics played a part in the subject he chose for his dissertation, but he wasn't yet familiar with Claude E. Shannon at that point. So Zadeh was hearing about Shannon's "information theory" for the first time, as the *Mathematical Theory of Communication*. He was impressed, he was fascinated, "because my training was all in frequency analysis".[164] Shannon's theory opened up a new world to the electrical engineer who, until this point, had been socialized by the theory of continuous analog systems: "Something that nobody talked about in that time!"[165] For Zadeh it was

> ... the first time also about the sort of the world of the digital than the analog. The world of the digital! So I became very much interested in that. In the digital, so even though with the force of that I talked was all continuous: circuit systems and so forth. Nevertheless that was my interest because on that time there were no courses on digital. No other courses! And so I became interested in the digital thing. [...] So I said, today what's important is in all frequency, Fourier analysis, this, that, but in the future that might be important. So I was convinced early on and I always said: digital – that's it! That's what the future is. There were many discussions at that time in the forties whether we should go this or go that, but I always said: Let's go digital! [166]

However, since the new digital view of information and communication engineering was, in Zadeh's opinion, not yet a suitable subject for a dissertation, his scientific career continued for a while longer in the "analog

---

[161] [707] R. S. interview with L. A. Zadeh (2001)
[162] [507] Shannon, Communication (1948)
[163] [707] R. S. interview with L. A. Zadeh (2001), [706] R. S. interview with L. A. Zadeh (2000)
[164] [705] R. S. interview with L. A. Zadeh (1999)
[165] [705] R. S. interview with L. A. Zadeh (1999)
[166] [706] R. S. interview with L. A. Zadeh (2000)

world". When he became an assistant professor at Columbia University in 1950, he was searching for new research topics. Both information theory and digital technology interested him more and more in addition to time-variable networks. Even while he was still working with the theory of optimum filters and pondering the general principles of signal transformation to characterize input-output relationships, he turned his attention to digital systems:

> My training was in numeral continuous analog systems, analog systems! But I always felt that the future belongs to digital systems and I was very much influenced by Shannon's talk that he gave in New York in 1946 in which he described his information theory.[167]

He began to deliver lectures on the automata theory, and in 1949 he organized and moderated a discussion meeting about digital computers, *thinking machines*, at Columbia University in New York in which Claude E. Shannon, Edmund Berkeley[168] and Francis J. Murray[169] took part. It was probably the first public debate on this subject ever![170] Zadeh understood that he was treading new ground with every step:

> So what I want to say is: I was very much involved but at the same time I was conscious that there were two camps, one was the digital camp and that was the "Shannon camp", and the other one was the continuous camp and what was the "Wiener camp". And although I did not do some work on digital systems at that time, nevertheless I felt that the digital, absolute digital, came and that was the feeling that was done at MIT also. So MIT chose to go with digital, with Shannon and so forth. Even though Wiener was a member of the cycle of MIT, nevertheless at that time his influence was not as great as Shannon's.[171]

Zadeh's enthusiasm for digital systems and digital computer was evident in a seminar on system and information theory that he offered at Columbia

---

[167] [705] R. S. interview with L. A. Zadeh (1999)

[168] Edmund Callis Berkeley (1923–1988), American mathematician, studied mathematics and logic at Harvard University until 1930. He later worked for various life insurance companies. He joined the U.S. Navy in 1942 and worked at Dahlgren Laboratory. In 1947, he was a co-founder of the Association for Computing Machinery. His book *Giant Brains* was the first about electronic computers

[169] Francis Joseph Murray (1911–1996), American mathematician, became professor of mathematics at Columbia University in New York in 1949 and Director of Special Research in Numerical Analysis at Duke University in Durham, North Carolina in 1957. From 1974 to 1980, Murray was the director of undergraduate studies at Duke

[170] [707] R. S. interview with L. A. Zadeh (2001). This discussion meeting is mentioned in: [699] Zadeh, Metrics (2000)

[171] [707] R. S. interview with L. A. Zadeh (2001)

THINKING MACHINES
A New Field in
Electrical Engineering
DR. LOFTI A. ZADEH
ELECTRICAL ENGINEERING DEPT.

Fig. 1.16. Illustration accompanying Zadeh's article *Thinking Machines. A New Field in Electrical Engineering*. The author's first name was misspelled here, an error that was often repeated over the decades that followed

starting in 1951. Most notable, however, is an article he penned for the January 1950 edition of the student publication *Columbia University Quarterly*.

This was a paper that did not resemble his other published works at all. He entitled it *Thinking Machines. A New Field in Electrical Engineering*[172], and this article inspired not only the accompanying illustration (Fig. 1.16) but also the title page of the journal (Fig. 1.17). This text was intended to describe the state of affairs for the guild of electrical engineers at the start of the second half of the 20th century. What Zadeh was trying to do was clarify the possibilities open to electrical engineers in those early days of the "digital age" of electronic machines, called "electronic brains", against a background of headlines from popular publications:

> "Psychologists Report Memory is Electrical", "Electric Brain Able to Translate Foreign Languages is Being Built", "Electronic Brain Does Research", "Scientists Confer on electronic Brain" – these are some of the headlines that were carried in newspapers throughout the nation during the past year. What is behind the headlines? How will "electronic brains" or "Thinking Machines" affect our way of living? What is the role played by electrical engineers in the design of these devices? These are some of the questions that we shall try to answer in this article.[173]

With these words, Zadeh, the young assistant professor of electrical engineering, made himself heard right at the halfway point of the 20th century. The influence of Shannon[174] and Wiener is impossible to miss. The future – he was convinced – belonged to digital systems and new computers: "That was digital!"[175] At the beginning of the text, he cited Wieners *Cybernetics*, which had been published two years before and in which the author considered animate and inanimate systems from the viewpoint of control, cybernetic and communication theory.

---

[172] [667] Zadeh, Thinking (1950), pp. 12f, 30f
[173] [667] Zadeh, Thinking (1950), p. 12. See also: [352] Martin, Myth (1993)
[174] Shannon had been a research assistant in MIT's department of electrical engineering since 1936, where he came into contact with Vannevar Bush's Differential Analyzer
[175] [705] R. S. interview with L. A. Zadeh (1999)

Fig. 1.17. Title page of the *Columbia Engineering Quarterly*, January 1950

On the one hand, this dealt with the mathematization of control and cybernetic processes in the living system of man himself but, on the other hand, with the corresponding processes in the devices created by man, particularly the apparatus that, in a certain sense, approximated the higher forms of human activity. It was the principles and organization of such machines, which in this respect behave like human brains, that he wanted to address in his

Relay Circuit Element	Symbolic Logic Interpretation
Circuit A	Statement A
Closed circuit	A is false
Open circuit	A is true
Series connection of A and B	A and/or B (A∨B)
Parallel connection of A and B	A and B (A·B)

Fig. 1.18. Zadeh's table showing the similarity between statements and circuits

article. Thus while Zadeh was paying generous tribute to Wiener's merits, he did not mention Shannon in the article at all, although he made reference to the content of Shannon's first and seminal work, namely the fact that the operations of Boolean algebra could be carried out by means of electrical circuits. He even illustrated this analogy with a table resembling the one Shannon had published over 10 years earlier in his Master's thesis, *A Symbolic Analysis of Relay and Switching Circuits* (Fig. 1.18).

Claude E. Shannon had come to the MIT department of engineering from Michigan in 1936 with a Master's degree in mathematics. At MIT he worked with Vannevar Bush on the Differential Analyzer project.[176] The relay circuits of this then-unique computer, an analog computer for solving differential equations up to the 7th order, had fascinated him. In Michigan he had studied symbolic logic and Boolean algebra, and he was able to apply this knowledge to provide a theoretical description of the two states of electromechanical relay circuits (on/off; power flowing/power not flowing).[177] He found Boolean algebra to be a calculus beautifully suited to the analysis and synthesis of circuit systems. His eventual doctoral advisor Vannevar Bush encouraged him to search for similarities between the theory of electric networks and this logic, and Shannon showed in his Master's thesis in 1937[178], which was published in the *Transactions of the American Institute of Electrical Engineering* the following year, that all Boolean statements can be represented by electrical circuits (Fig. 1.19).[179]

> We shall limit our treatment to circuits containing only relay contacts and switches, and therefore at any given time the circuit between any two terminals must be either open (infinite impedance) or closed (zero

---

[176] The Differential Analyzer was an analog computer that was designed and built at MIT in the 1920s under the direction of Vannevar Bush. See: [109] Bush, Analyzer (1931), [431] Owens, Bush (1986), pp. 63–95

[177] An extensive biography of Shannon can be found in the American edition of his collected papers: [514] Sloane et al., Shannon (1993), pp. xi–xvii. German translations of selected writings by Shannon were published in [280] Kittler et al., Shannon (2000), [514] Sloane et al., Shannon (1993)

[178] [508] Shannon, Relay (1938), pp. 713–723

[179] [508] Shannon, Relay (1938), p. 713

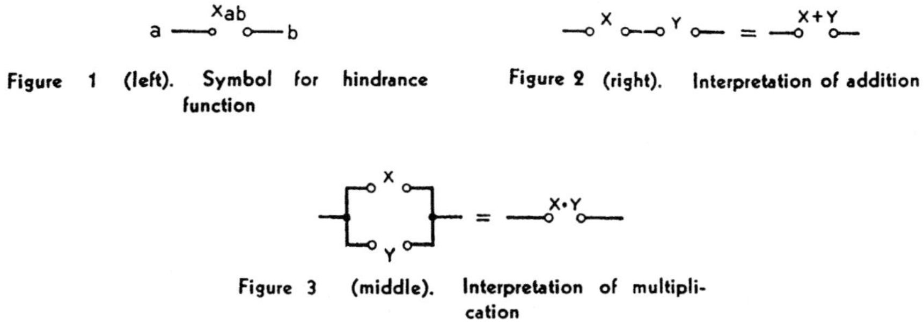

1.6 Let's go digital! 53

Figure 1 (left). Symbol for hindrance function

Figure 2 (right). Interpretation of addition

Figure 3 (middle). Interpretation of multiplication

Fig. 1.19. Shannon's representation of Boolean statements using electrical circuits

impedance). Let us associate a symbol X_{ab}, or more simply X, with the terminals a and b. This variable, a function of time, will be called the hindrance of the two-terminal circuit $a - b$. The symbol 0 (zero) will be used to represent the hindrance of a closed circuit, and the symbol 1 (unity) to represent the hindrance of an open circuit. Thus when the circuit $a - b$ is open $X_{ab} = 1$ and when closed $X_{ab} = 0$. The hindrances X_{ab} and X_{cd} will be said to be equal if whenever the circuit $a - b$ is open, the circuit $c - d$ is open, and whenever $a - b$ is closed, $c - d$ is closed. Now let the symbol + (plus) be defined to mean the series connection of the two-terminal circuits whose hindrances are added together. Thus $X_{ab} + X_{cd}$ is the hindrance of the circuit $a - d$ when b and c are connected together. Similarly the product of two hindrances $X_{ab} \ldots X_{cd}$ or more briefly $X_{ab}X_{cd}$ will be defined to mean the hindrance of the circuit formed by connecting the circuits $a - b$ and $c - d$ in parallel.[180]

With his popular scientific article, Zadeh wanted first to clarify how a thinking machine differed from other machines. To do so, he used an old and very simple example:

However, an idea of the principles involved in a thinking machine can be obtained from the description of a Tit-Tat-Toe playing device which was recently demonstrated by Robert Haufe at Caltech before a meeting of the American Institute of Electrical Engineers.[181]

In addition to chess, Tit-Tat-Toe, also known as Tic-Tac-Toe, was one of the games that scientists wanted to teach machines to play early on.[182] The game

---

[180] See also [193] Gardner, Denken (1985)
[181] [667] Zadeh, Thinking (1950), p. 12
[182] Even Charles Babbage (1791–1871), the English mathematician, Cambridge professor and first designer of programmable machines, was reported to have wanted to make a machine that could play Tic-Tac-Toe and chess in order to be able to finance the construction of his Analytical Engine. See: [143] Dorf, Computers (1981), p. 396

Symbol	Interpretation in Relay Circuits	Interpretation in the Calculus of Propositions
X	The circuit X	The proposition X
0	The circuit is closed	The proposition is false
1	The circuit is open	The proposition is true
$X + Y$	The series connection of circuits X and Y	The proposition which is true if either X or Y is true
$X\,Y$	The parallel connection of circuits X and Y	The proposition which is true if both X and Y are true
X'	The circuit which is open when X is closed and closed when X is open	The contradictory of proposition X
$=$	The circuits open and close simultaneously	Each proposition implies the other

Fig. 1.20. Shannon's table showing the similarity between statements and circuits

is played by two players on a board of three by three squares. The players take turns filling a square with their respective symbols (usually X and 0). The game is won by the player who manages to place three of his symbols in a row (horizontal, vertical or diagonal).[183]

Haufe's machine functioned with relay circuits. It saved information about which of the individual fields were filled with the players' symbols, it could make sensible moves and could indicate the result at the end of the game. When it was the machine's turn, it classified all nine fields according to whether or not filling them was strategically desirable. These classes were then searched for empty fields. An empty field with the highest strategic value was then filled.[184] Robert Haufe's Tic-Tac-Toe machine (Fig. 1.21), which Zadeh displayed in his article, was naturally much simpler than other machines that were referred to as "thinking machines". Yet the MIT Differential Analyzer, for example, was not a thinking machine because it could not make decisions unless they were exceedingly trivial. However, Zadeh considered the ability to make decisions to be a characteristic feature of thinking machines:

> Despite its simplicity, Haufe's machine is typical in that is possesses a means for arriving at a logical decision based on evaluation of a number of alternatives. More generally, it can be said that a thinking machine is a device which arrives at a certain decision or answer through a process of evaluation and selection.[185]

Since the newest digital computers of the day, such as BINAC[186] and UNIVAC[187], possessed the ability to arrive at certain non-trivial decisions,

---

[183] For more on Tic-Tac-Toe, see: [240] Hillis, Computerlogik (1999), pp. 16–23
[184] [232] Haufe, Tit-Tat-Toe (1949), p. 885
[185] [667] Zadeh, Thinking (1950), p. 13
[186] BINAC (Binary Automatic Computer) was an early electronic computer built by electrical engineers John Presper Eckert (1919–1995) and John Mauchly (1907–1980) at the Moore School in Philadelphia in 1949 for the Northrop Aircraft Company
[187] UNIVAC (Universal Automatic Computer) was the first commercial computer made in the USA. This computer was also designed by John Presper Eckert and John William Mauchly

Fig. 1.21. The two units comprising Robert Haufe's Tit-Tat-Toe machine

Zadeh also pronounced them "thinking machines", even though their "ability to think" was very limited! He then presented the following diagram of thinking machines to demonstrate the "thinking" process of such machines (Fig. 1.22): Incoming input data are sorted and processed in the *processor*. Some of this processed data is then sent to *storage* to be saved for later use (this storage can be in the form of punch cards, tapes or cathode ray tubes), "and it has the same function as memory in a human brain".[188]

Another portion of the processed data as well as some of the saved data are called up into the unit known as the *computer*, where necessary calculations are performed. The computer is not the essential component of the thinking machine, however, unless the calculation either is the end result itself or will be needed at the end in order to make the decision. More important is the *decision maker*, for it is here that decisions are reached. All of the relevant information coming from the computer and from storage is evaluated and weighted according to the commands and criteria present within the machine. The final answer or decision is formed on this basis as *output*. Dashed lines lead from the decision maker to all three elements of the machine: These are the so-called *feedback connections*. This feedback allows the three elements to operate as a function of the data obtained from the *decision maker* as needed.

1.6.1 A View from 1950 into the Future of 1965

Zadeh illustrated his argumentation by peering forward into the year 1965, which was then 15 years in the future. Three years earlier, in this version of

---

[188] [667] Zadeh, Thinking (1950), p. 13

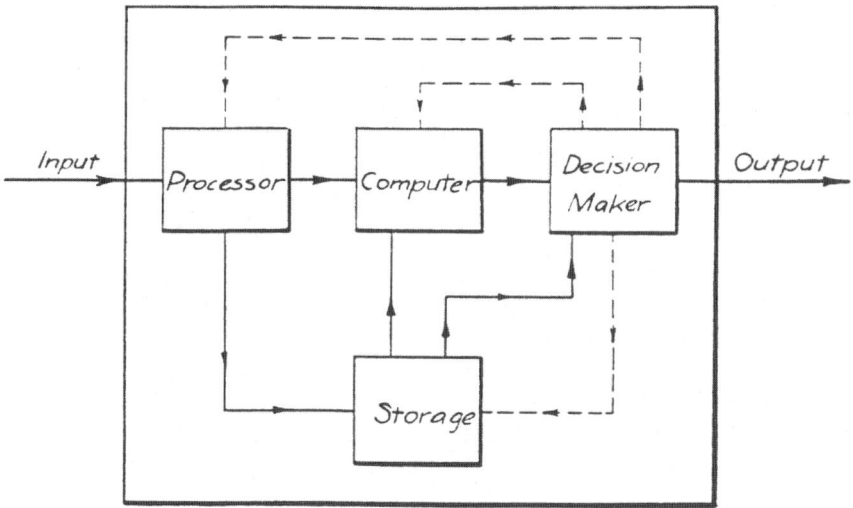

Fig. 1.22. Zadeh's chart for the basic elements of a "Thinking Machine"

the future, the administration at Columbia University had decided, for reasons of economy and efficiency, to close the admissions office and install in its place a thinking machine called the "Electronic Admissions Director". The construction and design of this machine had been entrusted to the electrical engineering department, which completed the installation in 1964. Since then, the "director" has been functioning perfectly and enjoying the unqualified support of the administration, departments and students. This thinking machine functions as follows:

1. Human secretaries convert the information from the list of applicants into series of numbers $a_1, a_2, a_3, ..., a_n$; each number represents a characteristic, e. g. a_1 could stand for the applicant's IQ, a_2 for personal character, and so on.
2. The lists coded thusly are provided to the *processor*, which processes them and then relays some of the data to the *computer* and another part of the data to *storage*. On the basis of applicant data as well as university data, the *computer* calculates the probabilities of various events, such as the probability that a student will fail after the first five years. This information and the saved data are sent to the *decision maker* to come to final decision on whether to accept the applicant. The decision is then made based on directives, such as these two:
    - accept if the probability of earning the Bachelor's degree is greater than 60%;
    - reject if the probability that the applicant will not pass the first year of college is greater than 20%.

Zadeh didn't consider the machine sketched out here to be as fanciful as student readers (and surely others, as well) may have thought: Machines such as this could be commonplace in 10 or 20 years and it is already absolutely certain that thinking machines will play an important role in armed conflicts that may arise in the future.[189] Now, in the year 1950, though, there was still much to be done so that these or similar scenarios of the future could become reality.

> Thinking Machines are essentially electrical devices. But unlike most other electrical devices, they are the brain children of mathematicians and not of electrical engineers. Even at the present time most of the advanced work on Thinking Machines is being done by mathematicians. This situation will last until electrical engineers become more proficient in those fields of mathematics which form the theoretical basis for the design of Thinking Machines. The most important of these fields is that of symbolic logic.[190]

The fundamental principles of thinking machines, Zadeh stressed, were developed by mathematicians, but the physical realization, the construction of the thinking machines, was the task of electrical engineers, who design and build the memory chips, processors, "computors", decision makers, etc. Until now, if electrical engineers had come into contact at all with such heretofore far-removed subjects as Boolean algebra, polyvalent logic, etc., it was through friendly relations with mathematicians. Now, however, the time didn't seem so far off when it would be just as important for post-graduate electrical engineers to take classes in mathematical logic as classes on complex variables: "Time marches on."[191]

1.6.2 System Theory

Zadeh wrote a second article for the New York student publication *Columbia Engineering Quarterly* in 1954. This text, too, was directed programmatically to a new development. It had to do with the *system theory*, which concerned the general principles of characterizing input-output relationships. Zadeh wanted this text to be an easily understandable introduction to this new theory, but also to hint at the specific characteristics of nonlinear systems. The fact that he himself had been doing research in the fields of linear and nonlinear systems[192] identified him as an expert, and the fact that he had initiated a seminar on system theory earlier in the year, "which is believed to

---

[189] [667] Zadeh, Thinking (1950), p. 30
[190] [667] Zadeh, Thinking (1950), p. 31
[191] [667] Zadeh, Thinking (1950), p. 31
[192] [630] Zadeh, Networks (1950), [620] Zadeh, Impulsive (1950), [592] Zadeh, Application (1950), [596] Zadeh, Circuit (1950), [657] Zadeh, Stability (1951), [613] Zadeh, General (1952)

be the first of its kind offered by any university"[193], instilled in the reader the great timeliness of the subject. Zadeh commented on this fact with assuasive words at the beginning:

> If you have never heard of system theory, you need not feel like an ignoramus. It is not one of the well-established branches of sciences. In fact, it has not yet been officially recognized as a scientific discipline. It does not appear on programs of meetings of scientific societies nor in indices to scientific publications. It does not have well-defined boundaries, nor does it have settled objectives.[194]

He calls upon *Webster's Dictionary* for a definition of *system concept*: a system is "*an aggregation or assemblage of objects united by some form of interaction or interdependence*".[195] Examples he cites of such systems are particles that are mutually attractive, a group of people who form a community, a complex of interwoven branches of industry, electrical networks and digital computers.

The above image of a "black box" appears proudly between the title and the text and is perhaps intended to suggest the clarity and simplicity of the system theory (Fig. 1.23).

More abstract than a black box is the concept of the *multipole*, which is a device with an unspecified structure that operates on a set of inputs, causes, actions, etc. and results in a set of outputs, effects, results, etc. A multipole is most often represented as a circle or a box, with each pole corresponding to an input or output, depending on where the arrows are pointing.

A multipole with m *inputs* and n *outputs* is referred to as an $m + n$-pole; if $m = 0$, then it is also called a source. The relationship between the inputs and the outputs characterizes each multipole. Multipoles whose input variables $u_1, u_2, u_3, \ldots, u_m$ and output variables $v_1, v_2, v_3, \ldots, v_n$ can be represented as time-dependent functions are interesting because, in this case, it

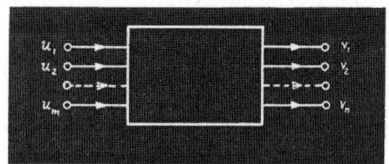

Fig. 1.23. Illustration of a system as a black box for Zadeh's article *System Theory*

---

[193] [661] Zadeh, System (1954), p. 34
[194] [661] Zadeh, System (1954), p. 16
[195] [661] Zadeh, System (1954), p. 16

is possible to mathematically examine not only the statistical behavior but also the dynamic behavior. The input-output relationship of the multipole can then be symbolically expressed thusly:

$$v_1 = f_1(u_1, u_2, ..., u_m)$$
$$v_2 = f_2(u_1, u_2, ..., u_m)$$
$$\vdots$$
$$v_n = f_n(u_1, u_2, ..., u_m)$$

or more simply:

$$v = f(u).$$

To illustrate, Zadeh looked at an induction coil A to which a time-variable voltage $v(t)$ is applied. The system can then be considered a *two-pole*, with the voltage forming the input $v(t)$ while the output $i(t)$ is the current flowing through the induction coil at time t. The input-output relationship $i = f(v)$ corresponds to the following transformation v in i (where L is the inductivity of A):

$$i(t) = \frac{1}{L} \int_{-\infty}^{t} v(t)dt.$$

System theory is nevertheless a scientific discipline "to the study of systems per se, regardless of their physical structure". Abstract systems, the elements of which did not need any particular physical identity, formed the basis of system theory, and they owe their great scientific power to this abstractness, for the abstraction process made it possible to leap from the specific to the general, from a mere set of data to broadly applicable theories.

> Unfortunately, this abstractness is also an impediment to a wide acceptance of the system-theoretic approach in engineering, since the present day engineer is, in general, inadequately trained to think in abstract terms. Nevertheless, there is little doubt that it is only a matter of time before system theory attains the acceptance it merits.[196]

A *system* can be represented as an *aggregation* of interrelated *objects*, and each *object* can be represented as a multipole. The fact that two or more objects are related is equivalent to the fact that the corresponding multipoles are linked, as is shown in the following Fig. 1.24. Zadeh put it succinctly: "A system is a collection of interconnected multipoles."[197]

While the illustration above shows the block diagram of a system, we can obtain the same information about the system when we identify the input-output relationship for each system component as well as the links between

---

[196] [661] Zadeh, System (1954), p. 16
[197] [661] Zadeh, System (1954), p. 17

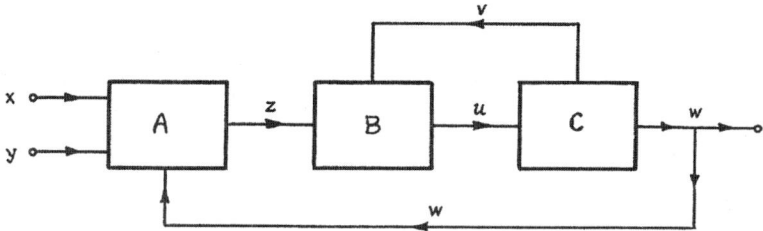

Fig. 1.24. Block diagram with input-output relationships

these multipoles. The input-output relationships for multipoles A, B, C in the above figure are (with undefined functions f, g, h, k):

$$\begin{array}{ll} A & z = f(x, y, w); \\ B & u = g(z, v); \\ C & v = h(u); \\ & w = k(u). \end{array}$$

Systems can also be described with the aid of linear graphs. A graph is a set of nodes that are linked by oriented lines. Each node represents a system variable, while the lines stand for their relationships to each other. In the following figure (Fig. 1.25), the relationship in the form $z = f(x, y)$ starts, for example, at x and y and ends at w.

The input variables of the system are the nodes at which a line does not end; in other words, the output variable for x and y is every node w from which no other line leads away.

Since this graphic method of describing system structures becomes less effective as the number of system variables increases, when this number reaches about 10 it is better to use other options, such as matrices in which the rows and columns can be identified with the system variables. Those entries not equaling 0 in a matrix specify the input-output relationship among the corresponding variables. In the Fig. 1.26, the entries in the first column indicate $u = g(z, v)$ and those in the last column signify the relationship $z = f(x, y, w)$.

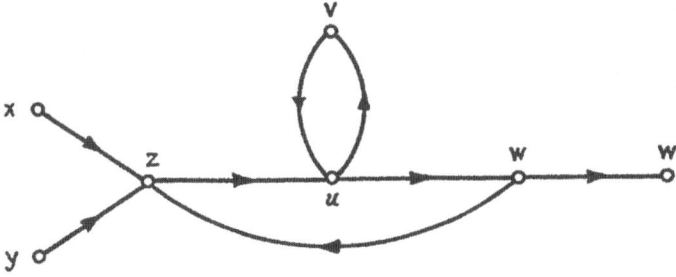

Fig. 1.25. Oriented linear graph

	u	v	w	x	y	z
u	0	1	1	0	0	0
v	1	0	0	0	0	0
w	0	0	0	0	0	1
x	0	0	0	0	0	1
y	0	0	0	0	0	1
z	1	0	0	0	0	0

Fig. 1.26. Matrix representation of a system

In the early '50s, at least those methods that employed the directed graph and the matrices had not yet become widely used tools for representing, analyzing and synthesizing electrical networks. Zadeh made reference to the writings on graph theory by Samuel J. Mason[198], who was working with electronic feedback systems at MIT, and to John von Neumann's work on automata theory.[199]

In the final paragraph of his article, Zadeh noted that there is a functional characterization for every multipole. For example, if output $y(t)$ of a two-pole M is linked to input $x(t)$ as follows:

$$y(t) = \frac{dx}{dt} + x^2$$

then the functional characterization can be seen in the following equation:

$$\frac{d^2y}{dt^2} + 2\frac{dy}{dt} + y = \frac{dx}{dt} + 2x.$$

However, functional characterizations of multipoles like this can also be figured indirectly using the response behavior of the systems to individual inputs; a corresponding procedure is wholly sufficient for linear systems:

> By reason of this additive property of linear systems, their functional characterization is a relatively simple matter. Indeed, all that is necessary for a complete characterization is the knowledge of how the system responds to a family of "elementary" time-functions (such as sine waves) into which any given input can be resolved.[200]

Yet as has already been seen with regard to nonlinear filters, searching for functional characterizations of nonlinear systems presents great difficulties; Zadeh referred here to his article about nonlinear multipoles[201], a paper which

---

[198] [356] Mason, Theory (1953). Zadeh had already cited Mason's work [355] Mason, Graphs (1951) in [629] Zadeh, Multipoles (1953). For more on Mason, see footnote 155
[199] [418] Neumann, Logic (1952)
[200] [661] Zadeh, System (1954) p. 19
[201] [629] Zadeh, Multipoles (1953)

had been written for students, but did not discuss the conclusions he had come to in it. It was perhaps in deference to these readers that he attempted to treat the problem in a descriptive style. For this purpose, he chose the example of a toggle switch that can be switched "up" ($= a_1$) or "down" ($= a_2$), which were the possible inputs of the system. Such an input will result either in the familiar sound of a switch or not; ($= b_2$). The possible switch states are then $S_1 =$ "on" and $S_2 =$ "off" (when the switch is "down").

- If the switch is in state S_1 and input a_1 is entered, then the output is b_2, whereas the output would be b_1 for input a_2.
- The situation is similar in the other instance: If the switch is in state S_2, then input a_1 will bring out output b_1, while input a_2 would result in output b_2.

Thus one input can trigger two outputs: a) The appropriate output signal appears or b) the system state changes from (e. g.) S_i to $S_j (i \neq j), i, j \in \{1, 2\}$.

In order to characterize the system, it is necessary to specify not only the dependence between the output symbol on the one hand and the input symbol on the other hand, but also the way in which the system state at a point in time depends on the previous system state. If there are only a few possible system states and the system is based on a small alphabet, then oriented graphs are well-suited to represent the transitions from one state to another (see Fig. 1.27). In the figure, the nodes correspond to the states and the arrows represent the state transitions, the symbols in parentheses are the input signals while the symbols outside the parentheses stand for the corresponding outputs. Therefore, an arrow from S_i to S_j labeled with the symbols $b_n(a_m)$ has this meaning:

> When the system is in state S_i the application of input symbol a_m yields the output symbol b_n and shifts the system into state S_j.[202]

Figure 1.27 is another illustration of the graph of this switch:

At the end of the text, Zadeh listed a number of interesting problems that arise when such systems, which he also called "finite-state transducers", are characterized by graphs:

> For instance, how should one combine the graphs of two or more transducers connected in tandem? How can one find the graph of an inverse to a given transducer, i. e., a transducer which acting on the output of the given transducer would yield its input? How can one determine by inspection of the graph whether or not a transducer is stable? And so on.[203]

---

[202] [661] Zadeh, System (1954), p. 34
[203] [661] Zadeh, System (1954), p. 34

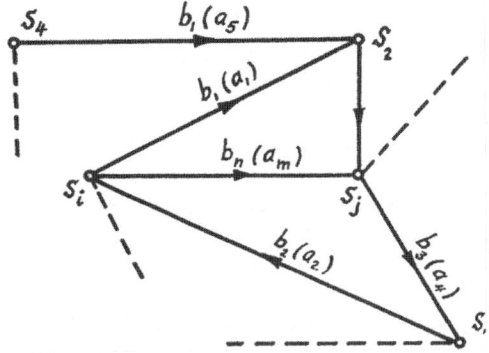

Fig. 1.27. Graphic representation of the toggle switch system

1.6.3 From System Theory to the Theory of Fuzzy Systems

When an anniversary edition of the *Proceedings of the IRE* appeared in May 1962 to mark the 50th year of the Institute of Radio Engineers (IRE), Zadeh submitted a landmark article: *From Circuit Theory to System Theory*.[204] This paper amounted to a substantial expansion of his above-mentioned text from 1954[205], for he was now describing in great detail the problems and fields of application of system theory and how they related to network, control and communication theory as well as operations research. Yet these considerations were also far-reaching in a completely different sense, since reference was made to the differences between mathematical and non-mathematical system treatment and between animate and inanimate systems.

How these reflections came about will be discussed in Chap. 4. What should be mentioned here is Zadeh's participation in the Second Systems Symposium at Case Western Institute of Technology in April 1963, for which his contributions were published under the title *Views on General Systems Theory*. The title suggests an interdisciplinary event, and half of those presenting were selected representatives of inter- and trans-disciplinary *cybernetics* or of *General Systems Theory*.[206] These scientific areas will be covered more extensively in Chap. 3.

---

[204] [662] Zadeh, System (1962)
[205] [661] Zadeh, System (1954)
[206] In addition to Zadeh, the presenters at the symposium included: Mihajlo D. Mesarovic, Kenneth E. Boulding, Russell L. Ackoff, the American mathematicians Abraham Charnes (1917–1992) and William Wager Cooper (born 1914), the American philosophers and logicians Hilary Putnam (born 1926) and John R. Myhill (1923–1987), Ralph W. Gerard, the American electrical engineer (and former doctoral student of Ernst A. Guillemin) William Kirby Linvill (born 1919), Robert Kalaba, who was then the director of the Department of Electrical Engineering at the Polytechnical Institute of Brooklyn, where this symposium took place, Rudolf F. Drenick (born 1914), William Ross Ashby, Anatol Rapoport and West C. Churchman. [399] Mesarovic, Views (1964). Information on the

In his 1962 article *From Circuit Theory to System Theory*, Zadeh mentioned this kinship explicitly and specifically cited Ludwig von Bertalanffy, the founder of the *General Systems Theory*. To close the chapter, this paragraph will be quoted in its entirety because it is here *and expressly in this context* that we find Zadeh's earliest postulation of "fuzzy mathematics":

> Among the scientists dealing with animate systems, it was a biologist – Ludwig von Bertalanffy – who long ago perceived the essential unity of systems concepts and techniques in various fields of science and who in writings and lectures sought to attain recognition for "general systems theory" as a distinct scientific discipline. It is pertinent to note, however, that the work of Bertalanffy and his school, being motivated primarily by problems arising in the study of biological systems, is much more empirical and qualitative in spirit than the work of those system theorists who received their training in the exact sciences. In fact, there is a fairly wide gap between what might be regarded as "animate" system theorists and "inanimate" system theorists at the present time, and it is not at all certain that this gap will be narrowed, much less closed, in the near future. There are some who feel that this gap reflects the fundamental inadequacy of the conventional mathematics – the mathematics of precisely-defined points, functions, sets, probability measures, etc. – for coping with the analysis of biological systems, and that to deal effectively with such systems, which are generally orders of magnitude more complex than man-made systems, we need a radically different kind of mathematics, the mathematics of *fuzzy* or cloudy quantities which are not describable in terms of probability distributions. Indeed, the need for such mathematics is becoming increasingly apparent even in the realm of inanimate systems, for in most practical cases the *a priori* data as well as the criteria by which the performance of a man-made system is judged are far from being precisely specified or having accurately-known probability distributions.[207]

---

scientists mentioned by name here can be found in the appropriate footnotes in Chap. 3

[207] [662] Zadeh, System (1962), p. 857. Emphasis added

2

Logical Tolerance, *Ensembles Flous* and *Probabilistic Metrics*

Where of one cannot speak, thereof one must be silent.[1]

2.1 The Vienna Circle

Vienna at the turn of the century – the so-called *Vienna Modern* (Wiener Modern(e)) – was not a uniform movement in art, music, literature; on the contrary, a pluralism of lifestyles and intellectual currents had developed. "Every key figure in the *Vienna Modern* – according to Edward Timms[2] – gathered around him a group of gifted followers and discussion partners."[3]

Whereas the literary and musical coteries maintained by the society ladies had been the meeting places for composers, virtuosi, actors, painters and poets[4] who were products of the 19th century[5], at the turn of the 20th century the novel ways of getting together were decidedly "modern"; these were private or partially public circles in which younger, interested men (participants were

---

[1] „Wovon man nicht sprechen kann, darüber muss man schweigen." [570] Wittgenstein, Tractatus (1921), p. 7

[2] [535] Timms, Kreise (1993). Edward Timms (born 1937), British Germanist

[3] Quoted from: [325] Lorenz, Moderne (1998), p. 21

[4] Those frequenting these circles included the Czech-Austrian historian of philosophy Theodor Gomperz (1832–1912), the Austrian physician, author and commentator Arthur Schnitzler (1862–1931) and the Czech-Austrian physicist and philosopher Ernst Mach (1838–1916) (see footnote 25), the Austrian-American composer Arnold Schönberg (1874–1951), the Austrian-British painter, illustrator and writer Oskar Kokoschka (1886–1980) and the Austrian dramatist and lyricist Hugo von Hoffmannsthal (1874–1929)

[5] Such as the coterie led by Josephine von Wertheimstein (1820–1894), Theodor Gomperz's sister, which she maintained in her villa in the Döbling district of Vienna, or that of political activist Sophie Baroness Todesco (1825–1895) or the writer and journalist Berta Zuckerkandl (1864–1945), whose salon played a role in the founding of the Vienna Secession

less often female) came together to engage in intellectual exchange. They headed for the coffeehouses!

The coffeehouses featured a flair for the random, the unpredicted, the irregular. Here people sat together with no seating arrangement and one was just as likely to wind up next to a potential patron as an intellectual opponent or comrade-in-arms. Customers could also read all of Vienna's daily papers free of charge or consult an encyclopedia.[6] There was a circle of young architects surrounding Otto Wagner[7], the Zionist movement coalesced around Theodor Herzl[8], the Wiener Werkstätte ("Vienna Workshop") gathered around Josef Hoffmann[9] and the psychoanalysis community (Wednesday Society) around Sigmund Freud.[10] Hermann Bahr[11], Arthur Schnitzler, Karl Kraus[12] and Adolf Loos[13] should likewise be mentioned as the centers of other discussion groups. Some members of these groups, such as the publisher Hugo Heller[14], who belonged to the "Freud circle", held salon evenings with poetry readings, musical performances and scientific lectures. A few personalities were associated with several circles, and it was through such men that fruitful communication arose between the circles and the avant-garde.

It is possible – and it was Carl E. Schorske[15] who formulated this theory – to interpret the *Vienna Modern* as the reaction by a frustrated generation who had been robbed of their ability to act in the political and social arena.[16]

---

[6] This service is still offered today by Café Griensteidl, which reopened at the Michaelerplatz in Vienna in 1990!

[7] Otto Wagner (1841–1914), Austrian architect and art historian

[8] Theodor Herzl (1860–1904), Hungarian-Austrian jurist and author. Herzl was concerned with anti-Semitism, which he initially saw as a social problem. As a result of the anti-Semitic rioting he witnessed in Paris during the Dreyfus Affair, the trial of Jewish officer Alfred Dreyfus (1859–1935), Herzl developed the idea of an organized emigration of the Jews. Herzl became the initiator of political Zionism with his book *Der Judenstaat* ("The Jewish State")

[9] Josef Hoffmann (1870–1956), Czech-Austrian architect and visual artist who worked with everyday objects. Cofounder of the Vienna Secession in 1897

[10] Sigmund Freud (1856–1939), Austrian neurologist and psychopathologist, founder of psychoanalysis. Freud was appointed lecturer in neuropathology in Vienna in 1885 and in 1902 became a professor of neuropathology at Vienna University. Freud emigrated to Great Britain in 1938 and died in London a year later

[11] Hermann Bahr (1863–1934), Austrian poet and literature critic, theorist of impressionism

[12] Karl Kraus (1874–1936), Czech-Austrian writer and critic

[13] Adolf Loos (1870–1933), Czech-Austrian architect and critic. Loos was strongly influenced by the theories of Otto Wagner and he later came out against art nouveau and the Wiener Werkstätte

[14] Hugo Heller (1870–1923), Austrian bookseller and publisher

[15] Carl E. Schorske (born 1915), American historian and cultural scientist

[16] [495] Schorske, Wien (1982)

This modern movement was also going on in places other than Vienna. It didn't even begin here but in Paris, where the *avant-garde* in art and literature was much more energetic. Yet unlike in Paris, the *Vienna Modern* was much more critical, even more self-critical. Allan Janik[17] characterized it by this very quality: the *Vienna Modern* was "critical modernism".[18] The *avant-garde* in Vienna reflected the *avant-garde* in Paris and Vienna. Jacques Le Rider[19] writes that, in this respect, we can say that there was an anticipation of postmodern subjects in the *Vienna Modern*.

In this period following the collapse of the Dual Monarchy, Friedrich Stadler sees a "phase of the late-bourgeois and highly capitalist society which – until the takeover by the fascists – had been molded on one hand by the societal democratization processes that followed a failed revolution and on the other hand by the tendencies of a "conservative monarchy".[20]

Vienna's population had tripled between 1880 and 1910 and had now reached about two million. This was the result of both the incorporation of suburban neighborhoods into the city and the streams of immigrants from Bohemia, Moravia, Slovakia and the Jewish settlements of Galicia. Alongside the new Jewish demographic, the class of industrial laborers grew dramatically and soon amounted to 52% of the population. They lived for the most part on the outskirts of the city, which were then incorporated in 1890. Established Viennese citizens were forced to "squeeze in" together with them – something they saw as a potential threat. According to Allan Janik[21], the heterogeneous structure of the populace, their cultural milieus, linguistic communities and ethnic groupings, all in a relatively small space was substantially responsible for the birth of Vienna's creative milieu. It was necessary for people to learn to deal with contrasts, and it also meant that the political structure of the city changed quickly and for a long time to come, which contributed to the feeling of insecurity felt by the bourgeoisie. Their reaction was shaped by ignorance and subterfuge: Hermann Broch[22] termed this culture of using entertainment and decoration to ignore impending crises a "gay apocalypse". In the 1880s, mass parties were founded as a countermovement to the parties

---

[17] Allan Janik (born 1941), Austrian-American philosopher and historian
[18] [256] Janik, Milieus (1990), p. 156
[19] [460] Rider, Illusion (1990). Jacques le Rider (born 1954), French Germanist, has been a professor of German studies at the Université Paris VIII Vincennes-Saint-Denis since 1990. Rider is a specialist in Vienna at the turn of the 20th century
[20] [520] Stadler, Kreis (1997), p. 211f
[21] [256] Janik, Vienna (1990)
[22] Hermann Broch (1886–1951), Austrian-American author and universal scholar, ran his family's factory in Vienna for 20 years, but became a freelance writer in 1927. He emigrated to the US as a Jew in 1937, where he lived primarily in New York. His most well-known works are *Die Schlafwandler (The Sleepwalkers)* and *Der Tod des Vergil (The Death of Virgil)*

tailored to the bourgeois liberals. In addition to the Social Democrats, who united at the Hainfeld party congress of 1888–89, this period saw the organization of the German nationalist *Alldeutschen* – forerunner of the National Socialists – and the likewise anti-Semitic Christian Socialists, who eventually undermined the predominance of the Liberals. Using a nickname for the empire, philosopher Manfred Geier[23] called this atmosphere "Kakanien's Confusion"[24], and it was also described by Robert Musil[25] in *The Man Without Qualities*:

> According to its constitution it was liberal, but the system of government was clerical. The government was clerical, but people's lives were liberal. By law all citizens were equal, but, of course, not everyone was a citizen.[26]

Little changed in this regard after the First World War, when the Social Democrats built up the city as "Red Vienna" and the Christian Socialists governed in the provinces:

> Cultural currents like humanism, pacifism, scientism, life and social reform – which were represented even during the monarchy by Ernst Mach[27], Josef Popper-Lynkeus[28], Albert Einstein, Ludwig

---

[23] Manfred Geier (born 1943), German philosopher

[24] [194] Geier, Kreis (1992)

[25] Robert Musil (1880–1942), German engineer and writer. Musil worked as a theater critic, essayist and freelance writer primarily in Vienna from 1918 to 1931, from 1931 to 1933 in Berlin before returning to Vienna after the National Socialists came to power. From there he fled to Zurich in 1938 and then to Geneva. His most well-known work is *Der Mann ohne Eigenschaften (The Man Without Qualities)*

[26] „Es war nach der Verfassung liberal, aber es wurde klerikal regiert. Es wurde klerikal regiert, aber man lebte freisinnig. Vor dem Gesetz waren alle Bürger gleich, aber nicht alle waren eben Bürger."[410] Musil, Mann (1970), p. 33f

[27] Ernst Mach became professor of mathematics in Graz in 1864 and then also professor of physics at the same university two years later. In 1867, he took on a professorship for experimental physics in Prague and became a professor of inductive philosophy at the University of Vienna in 1895. In 1897 Mach resigned from his professorship due to ill health. Mach was appointed to the Austrian parliament in 1901

[28] Josef Popper(-Lynkeus) (1838–1921), Czech-Austrian writer and engineer who was active in the peace movement. Popper – Lynkeus was a pseudonym – studied mechanical engineering at the German polytechnic institute in Prague and later became a railroad official in Vienna

Boltzmann[29], Berta von Suttner[30], Alfred H. Fried[31], Rudolf Goldscheid [32], Friedrich Jodl[33], Sigmund Freud and others – continued to exert their influence after 1918, possibly in an even more pointed and uncompromising way than in the aestheticizing, impressionistic epoch of "Kakanien"[34].

At the universities, bourgeois liberals collided with fascists, nationalists and clericalists; the Social Democrats were in the minority here.

"Circles" were forming in the scientific community, as well. Economists and jurists came together in an intellectual circle (*Geistkreis*) around philosopher and economic scientist Friedrich August von Hayek.[35] There was the *Mises Circle* of Ludwig von Mises[36], the circles presided over by philosophers

---

[29] Ludwig Boltzmann (1844–1906), Austrian physicist, became professor of theoretical physics in Graz in 1869, later in Munich, Vienna and Leipzig. In 1902, he was granted the professorship in Vienna previously held by Austrian mathematician and physicist Josef Stefan (1835–1893), whose assistant he had been early on in his career

[30] Berta von Suttner (1843–1914), Czech-Austrian aristocrat, became private secretary to Swedish chemist Alfred Nobel (1833–1896) in Paris. Her first contact with the peace movement came in 1887, and two years later she published the novel *Die Waffen Nieder! (Lay Down Your Arms!)*, which made her famous. In 1881 she founded the Austrian, and in 1882 the German Peace Society. Berta von Suttner was awarded the Nobel Peace Prize in 1905

[31] Alfred Hermann Fried (1864–1921), Austrian writer and Esperanto textbook author. He received the Nobel Peace Prize in 1911

[32] Rudolf Goldscheid (1870–1931), Austrian writer, philosopher and economist. He founded the "Sociological Society" in Vienna in 1907 and became president of the Peace Society in Vienna in 1923

[33] Friedrich Jodl (1849–1914), German philosopher and psychologist, studied in MunichL, where he habilitated in 1870 and became an ordinary professor in Prague shortly thereafter. In 1896, he was granted a professorship in Vienna, a position he held until his death

[34] „Kulturelle Strömungen wie Humanismus, Pazifismus, Szientismus, Lebens- und Sozialreform – bereits in der Zeit der Monarchie durch Ernst Mach, Josef Popper-Lynkeus, Albert Einstein, Ludwig Boltzmann, Berta von Suttner, Alfred H. Fried, Rudolf Goldscheid, Friedrich Jodl, Sigmund Freud und andere vertreten – wirkten auch nach 1918 weiter, vielleicht sogar in einer pointierteren und kompromissloseren Weise als in der ästhetisierenden, impressionistischen Epoche «Kakaniens»." [520] Stadler, Kreis (1997), p. 212

[35] Friedrich August von Hayek (1899–1992), Austrian-British economic and social philosopher. He went to the London School of Economics in 1931, obtained British citizenship in 1938 and moved to the University of Chicago in 1950. In 1974, he was awarded the Nobel Prize in Economics

[36] Ludwig Edler von Mises (1881–1973), Austrian economist. Von Mises taught at the University of Vienna for no pay from 1913 to 1934. From 1909 to 1934, he also advised the Austrian Chamber of Commerce and the Austrian government. He fled to Geneva in 1934 and then to the USA in 1940. He taught at New York University from 1945 to 1969

Robert Reininger[37] and Heinrich Gomperz[38], as well as the circle surrounding the philosopher and psychologist Karl Bühler[39], who had been summoned to Vienna at the same time as Moritz Schlick.[40],[41]

Schlick's circle went down in the history of philosophy and science as the "first Vienna circle"[42], in which a handful of Vienna's young scientists met regularly between 1907 and 1912 to discuss issues of philosophy and scien-

Fig. 2.1. O. Neurath

---

[37] Robert Reininger (1869–1955), Austrian philosopher, natural scientist and jurist. In 1922, Reininger became an ordinary professor of the history of philosophy at the University of Vienna, where he taught until 1940

[38] Heinrich Gomperz (1873–1942), Austrian philosopher, philologist and jurist. Gomperz earned his doctorate in 1869 under Ernst Mach, habilitated in 1900 in Bern and was an ordinary professor of philosophy in Vienna from 1924 to 1934. In 1934, Gomperz was forced to retire; he emigrated to the United States in 1935, where he was a guest professor at the University of Southern California

[39] Karl Bühler (1879–1963), German doctor and psychologist, studied at Freiburg in Breisgau and Würzburg and elsewhere and earned his doctorate in Strasbourg. In 1907, he qualified as a professor in Würzburg. After stints as a professor in Bonn, Munich and Dresden, he was an ordinary professor of psychology at the University of Vienna from 1922 to 1938. After emigrating, he was a professor of psychiatry at the University of California in Los Angeles from 1945 to 1955

[40] Friedrich Albert Moritz Schlick (1882–1936), earned his doctorate in mathematical physics under Max Planck in Berlin in 1904 with a paper entitled *Über die Reflexion des Lichtes in einer inhomogenen Schicht (On the Reflection of Light in an Inhomogeneous Layer)*. Afterward he studied in Göttingen, Heidelberg and Berlin, wrote the book *Lebensweisheit (Worldly Wisdom)* and then went to Zurich for two years to study psychology; in 1911 he was promoted to private lecturer in philosophy in Rostock with the paper *Das Wesen der Wahrheit nach der modernen Logik (The Essence of Truth according to Modern Logic)*. He became a professor in Rostock in 1917, an ordinary professor in Kiel in 1921 and in 1922, at the behest of Hans Hahn, he became the successor to Boltzmann and Mach in the professorship for natural philosophy (philosophy of inductive sciences) in Vienna

[41] Bühler would later become Karl Raimund Popper's doctoral adviser

[42] [225] Haller, Neopositivismus (1993), in particular pp. 45–60

tific theory. Its members included Hans Hahn[43], who was already a professor, Philipp Frank[44], who had just received his doctorate, and the economics Ph.D. Otto Neurath.[45] Also to be found alongside these regular attendees were the engineering Ph.D. Richard von Mises[46] and Hahn's sister Olga, who, despite being totally blind, managed to complete her dissertation in mathematics with Neurath's help and in 1912 became his second wife.

The group met on Thursday evenings in a Vienna coffeehouse and their discussions usually carried on past midnight. The main subjects were:

- the search for ways to avoid the traditional ambiguity and incomprehensibility in philosophy and
- the search for the greatest possible rapprochement between philosophy and science, by which they meant natural, social and psychological sciences.

Many years later, Philipp Frank wrote:

> Our group didn't have any particular preference for a certain political or religious credo back then. We did, however, tend toward empiricism on one hand and long and clear chains of logical conclusions on the other.[47]

Otto Neurath recalled:

> We were primarily dealing with problems of physics having to do with cognitive theory and methodology, such as Poincaré's conventionalism, Duhem's concept of the aim and structure of physical theories.[48]

---

[43] Hans Hahn (1879–1934), Austrian mathematician, taught from 1905 in Vienna and Innsbruck as well as in Chernivtsi (1909) and Bonn (1916). From 1921 to 1934, Hahn was a professor at Vienna University

[44] Philipp Frank (1884–1966), Austrian-American physicist

[45] Otto Neurath (1882–1945), Austrian sociologist, philosopher and educational policy expert

[46] Richard von Mises (1883–1953), mathematician and engineer, born in Lwow (Lemberg), was a professor in Istanbul from 1933 to 1939 and became a professor at Harvard University in the US in 1944

[47] „Unsere Gruppe hatte damals keine besondere Vorliebe für ein bestimmtes politisches oder religiöses Credo. Wir hatten jedoch eine Neigung zum Empirismus einerseits und zu langen und klaren Ketten von logischen Schlussfolgerungen andererseits." [182] Frank, Hintergrund (1949), p. 245

[48] „Es ging vor allem um erkenntnistheoretische und methodologische Probleme der Physik, zum Beispiel Poincarés Konventionalismus, Duhems Auffassung von Ziel und Struktur der physikalischen Theorien." See [424] Neurath, Schriften (1981), p. 695

72   2 Logical Tolerance, *Ensembles Flous* and *Probabilistic Metrics*

In addition to the French conventionalists Henri Poincaré[49], Pierre Duhem[50], Édouard Le Roy[51], Abel Rey[52] and Gaston Milhaud[53], this Vienna circle had also been most strongly influenced by Ernst Mach. From this "Ur-circle" arose in the autumn of 1924 the so-called "Schlick circle" as a private Thursday evening discussion group. They met at the rear of the mathematics institute at Boltzmanngasse 5 in Vienna. One version of the genesis of the Vienna Circle credits as its founder the newly ordained professor Moritz Schlick, who had thus seized upon an idea by his students Friedrich Waismann[54] and Herbert Feigl.[55] Another version ascribes the founding to Hans Hahn, who 1) played a large part in bringing Schlick to Vienna as a philosophy professor, 2) as a mathematics professor managed to convince by far the most scientists who were interested in logic, cognitive and scientific theory to attend and 3) provided the meeting place. Finally, Otto Neurath offers a third version: When Schlick, along with Rudolf Carnap, happened upon the Vienna "Ur-Circle" during the academic year 1923–24, there was an initiative to found a "lasting cooperation of the like-minded".[56]

Fig. 2.2. M. Schlick

---

[49] Jules Henri Poincaré (1854–1912), French mathematician and physicist, was a professor in Caen from 1879 and at the Sorbonne in Paris from 1881 to 1912. Poincaré was a member of the Académie des Sciences from 1887, president in 1906, and became a member of the Académie Française in 1909
[50] Pierre Duhem (1861–1916), French physicist, science historian and philosopher
[51] Édouard Le Roy (1870–1954), French mathematician
[52] Abel Rey (1873–1940), French philosophy historian
[53] Gaston Milhaud (1858–1918), French philosopher and professor in Paris
[54] Friedrich Waismann (1896–1959), Austrian mathematician and philosopher. After his emigration (1937), he became a professor for the philosophy of mathematics (later philosophy of science) at Oxford
[55] Herbert Feigl (1902–1988), Austrian philosopher, later professor of philosophy in the USA (University of Minnesota)
[56] [225] Haller, Neopositivismus (1993), p. 68

In addition to the aforementioned, further initiators of the Schlick circle were the legal philosopher Felix Kaufmann[57] and the mathematician Kurt Reidemeister[58], though he left Vienna soon thereafter, plus Friedrich Waismann, who was Schlick's scientific assistant and librarian and dean of the community college, Rudolf Carnap[59], who came to Vienna in 1925, the mathematician Gustav Bergmann (from 1927), Richard von Mises, though he was no longer working in Vienna, Edgar Zilsel, Kurt Gödel (from 1926) and Karl Menger, whose life and work will be addressed separately.

Fig. 2.3. H. Hahn

These discussion evenings were continued uninterrupted from 1924 to 1936. Even when Schlick was in America. Always on Thursday evening. Often late into the night. Usually, some subject would be introduced by means of a report and a discussion would then ensue. The leader was Schlick, who once presented a report on causality himself.[60]

---

[57] Felix Kaufmann (1895–1949), Austrian-American jurist and philosopher, earned the Dr. jur. in 1920 and the Dr. phil. in 1926. Kaufmann taught legal philosophy from 1922 to 1938 as a private lecturer in Vienna. He also represented Vienna in the Anglo-Iranian Oil Company. Due to his Jewish heritage, he emigrated to the US in 1938, where he held a professorship for legal philosophy at the graduate faculty of the New York School for Social Research until his death

[58] Kurt Werner Friedrich Reidemeister (1893–1971), German mathematician, became professor of geometry at the University of Vienna on Hans Hahn's recommendation in 1923. He became a professor in Königsberg in 1927 but was fired in 1933. In 1934 he was called to Marburg and from 1935 was a professor in Göttingen

[59] Rudolf Carnap (1891–1970), German philosopher, professor in Vienna (1926–1931), Prague (1931–1935), Chicago (1936–1952), at the Institute for Advanced Study in Princeton (1952–1954) and at the University of California in Los Angeles (UCLA) (1954–1970)

[60] „Die Diskussionsabende wurden ununterbrochen fortgesetzt, von 1924 bis 1936. Auch als Schlick in Amerika war. Immer am Donnerstagabend. Oft bis spät in die

It was mostly publications by Carnap, Schlick, Hahn, Menger and Gödel concerning the foundations of mathematics that were discussed during this "non-public phase" of the Vienna Circle; to these were added books and articles by Bertrand Russell[61], Gottlob Frege[62] and Ludwig Wittgenstein.[63] Wittgenstein's *Tractatus logico-philosophicus*[64] was discussed very intensively, as was Carnap's *The Logical Structure of the World*[65] after 1930. Finally, the circle covered the subjects of the physicalism described by Carnap and Neurath and the writings on quantum theory by Planck, Bohr and Heisenberg.

The "public" debut of the Vienna Circle was staged! The publication of the manifesto *Wissenschaftliche Weltauffassung – Der Wiener Kreis (Scientific World View – The Vienna Circle)* and the founding of the Ernst Mach Society on November 23, 1928 in the ballroom of Vienna's Old Town Hall provided the opportunity. The society's name even had an explanatory addition: Society to Spread Awareness of the Exact Sciences. The informal Vienna Circle integrated itself into this society and Moritz Schlick became its chairman. In early 1929, Schlick received a lucrative offer from the University of Bonn, and thus both the Vienna Circle and the Ernst Mach Society faced a bitter loss. The board asked Schlick to stay in Vienna and he consented, before then traveling to California to accept a guest professorship at Stanford University.

Nacht. Meistens war es so, dass eine Sache durch ein Referat eingeleitet wurde und sich dann eine Diskussion entspann. Der Leiter war Schlick, der selbst auch einmal referiert hat über die Kausalität." See the remembrances by Viktor Kraft from 1973 and the interview with him (WKA Haarlem) in [520] Stadler, Kreis (1997), p. 239

[61] Earl Bertrand Arthur William Russell (1872–1970), British philosopher, logician, essayist and social critic, co-wrote the three-part *Principia Mathematica* (1910, 1912, 1913) with the British mathematician, logician and philosopher Alfred North Whitehead (1861–1947). He received the Nobel Prize for Literature in 1950. Russell was a fellow at Trinity College in Cambridge from 1895 to 1901 and a lecturer from 1910 to 1916. He once again became a fellow there in 1944

[62] Friedrich Ludwig Gottlob Frege (1848–1925), German mathematician, logician and philosopher. Frege became an extraordinary professor in Jena in 1879 and was an extraordinary honorary professor there from 1890 to 1917

[63] Ludwig Wittgenstein (1889–1951), Austrian philosopher, first studied engineering from 1906 to 1908 in Berlin and in Manchester in 1912. Bertrand Russell's writings on the foundations of mathematics inspired Wittgenstein to study mathematics and logic at Cambridge. In 1921, he published the *Tractatus logico-philosophicus*. From 1920 to 1926, he worked as a village school teacher in Austria, after which he lived in Vienna, where he encountered the Vienna Circle. He returned to Cambridge in 1936 and published *Philosophical Investigations*. Wittgenstein was made a professor of philosophy in Cambridge a year later

[64] [570] Wittgenstein, Tractatus (1921)

[65] [112] Carnap, Aufbau (1928)

Fig. 2.4. K. Gödel

At that time, the Ernst Mach Society was preparing to co-host a conference on the epistemology of the exact sciences with the Berlin Society for Empirical Philosophy, which took place in Prague on September 15–17, 1929. Hahn, Neurath and Carnap also authored a programmatic piece as a statement of the principles of the Vienna Circle, which was completed on August 15 and put out as Volume 1 in the series of publications by the Ernst Mach Society. It was to be presented to Schlick upon his return from America as a token of thanks for staying.[66]

Another notable moment in the Vienna Circle's public debut came in 1930 when Carnap and Hans Reichenbach took over publication of the journal *Erkenntnis* for the Ernst Mach Society and the Berlin Society for Empirical Philosophy.[67] The Sword of Damocles hung over the journal at the time, but it flourished under its new publishers, and it soon had over 400 subscribers.[68] However, due to the changing political climate in Germany and Austria, this new public face of *Logical Empiricism*, as the philosophy of the Vienna Circle and Berlin Society for Empirical Philosophy became known, was not blessed with long-lasting success, and the lives of the members themselves came under threat.

It was not least of all the "racially" motivated emigration starting in 1933 and 1934 that began the process of dissolution, after the departure of Feigl in 1931 and the unexpected death of Hahn in 1934. In 1935, Carnap left Prague, where he had lived since 1931. On 22 June 1936, Moritz Schlick (who had been reputed to be a Jew but actually stemmed from an old Austrian noble

---

[66] Schlick was not very happy with this manifesto, since, he said, he could "not approve of either the advertisement-like style or the somewhat dogmatic-sounding formulations of the brochure". Quoted from [408] Mulder, Weltauffassung (1968), p. 390. Menger and Wittgenstein were also not thrilled with this publication

[67] This magazine emerged from the *Annalen der Philosophie*, which had been founded by the German philosopher Hans Vaihinger (1852–1933) in 1919 and was published by the Felix Meiner Verlag in Leipzig until 1937, after which it was published by van Stockum & Zoon in The Hague until 1940

[68] See [235] Hegselmann et al., Erkenntnis (1991)

family) was shot to death by his former doctoral student Johann Nelböck on Vienna University's so-called "Philosopher's Steps". Karl Raimund Popper[69] and Karl Menger fled from Vienna to the United States in 1937. In 1938, Viktor Kraft was forced into retirement and Friedrich Waismann emigrated (with Karl Popper's help) to Oxford, where Wittgenstein was already living.

Fig. 2.5. Illustration depicting the murder of Moritz Schlick in Vienna's *Illustrierten Kronenzeitung* newspaper on June 22, 1936

---

[69] Sir Karl Raimund Popper (1902–1994), Austrian-born English philosopher and scientific theorist, earned his doctorate under Karl Bühler in 1928 with the dissertation *Zur Methodenfrage der Denkpsychologie (On the Problem of Method in the Psychology of Thinking)*, became a secondary school teacher in Vienna in 1930 and came into contact with the Vienna Circle. In 1934, he published the book *Logik der Forschung (Logic of Scientific Discovery)*. He and his wife emigrated to New Zealand in 1937, where he became a lecturer at Christchurch University. Here Popper wrote *Das Elend des Historizismus (The Misery of Historicism)* and *Die offene Gesellschaft und ihre Feinde (The Open Society and its Enemies)*. In 1946, Popper was made extraordinary professor at the London School of Economics and in 1949 professor of logic and scientific methodology at the University of London

Oscar Morgenstern[70] and Abraham Wald[71] went to the US. And in 1940, Kurt Gödel succeeded in escaping on the Trans-Siberian Railroad to Manchuria, and from there by ship to San Francisco.[72]

2.2 Karl Menger: "From the Criterion of Meaning to the Principle of Tolerance"[73]

Karl Menger[74] had studied at Vienna University from 1920 to 1924 and eventually earned a doctorate in mathematics. His original field of study had been physics (theoretical physics with Hans Thirring[75]), though mathematics (with Hans Hahn) and philosophy (with Moritz Schlick) followed soon thereafter. Menger had completed his first semester of physics in March of 1921 when Hans Hahn became a lecturer in the mathematics department in Vienna and offered a spring semester seminar called *New Developments in the Concept of the Curve*.

Menger had already written some important papers about the concept of dimension between 1921 and 1923.[76] It was during this time that he was

---

[70] Oscar Morgenstern (1902–1977), German economist, became a professor at the University of Vienna in 1935. He was removed from the university when the National Socialists occupied Austria in 1938. Morgenstern became a professor at Princeton University, where he remained until his retirement in 1970

[71] Abraham Wald (1902–1950), Hungarian mathematician and statistician, studied in Vienna under Menger and, following his emigration, taught at Columbia University in New York. Wald was killed in a plane crash

[72] Gödel arrived at Princeton in March 1940, where he remained for the second half of his life

[73] See [520] Stadler, Kreis (1997), p. 32

[74] Karl Menger (1902–1985) was the son of economist Carl (von) Menger (1840–1921) and his wife Mina, née Andermann. When Karl was born, his father was already relatively old and famous as the creator of the theory of marginal utility. He had studied in Prague, Vienna and Krakow, had worked in the prime minister's press office in Vienna for a number of years beginning in 1867 and was Crown Prince Rudolf's tutor from 1876. In 1873, he became a professor of political economics in Vienna. He received particular honors when, as a member of the Currency Commission in 1892, he played a part in the introduction of the gold standard in Austria-Hungary. Two years before the birth of his son Karl, he was made a lifetime member of the House of Lords

[75] Hans Thirring (1888–1976), Austrian physicist, was professor of theoretical physics at Vienna University from 1921 to 1938. Thirring was forced to retire in 1938, but became a professor again in 1946. From 1957 to 1963, he was a member of the *Bundesrat* (upper house of parliament) and was associated with the peace movement. The "Thirring Plan" demanded the unilateral disarmament of Austria. Hans Thirring was the father of Viennese physicist Walter Thirring (born 1927)

[76] Menger's father gave these papers to the editors of the monthly *Monatshefte für Mathematik und Physik* (MhMPh), which was published in Leipzig and Berlin, for

forced to miss three semesters due to a pulmonary disease, which required over a year in a sanitarium to cure. When he returned in April of 1923, Menger wrote a summary of his findings, the first part of which he submitted to Hahn as a dissertation entitled *Über die Dimensionalität von Punktmengen (On the Dimensionality of Point Sets)*.[77] This publication formed the basis for his subsequent scientific work in topology; it also contributed considerably to the worldwide renown of Vienna's school of topology in the 1920s.

Fig. 2.6. K. Menger

By far the most renowned topologist during this period was Luitzen Egbertus Brouwer[78] of the Netherlands, who had done important work on dimension theory as early as 1913, but had not finalized a theory. When Menger had started working in this area in the early '20s, he really had only one serious competitor: Pavel Urysohn[79] in the Soviet Union, although he drowned in

---

later publication; they did not appear until 1929: Karl Menger, *Zur Dimensions- und Kurventheorie (On the Dimension and Curve Theory)*. Unpublished papers from the years 1921–1923, published as [386] Menger, Punktmengen (1929)

[77] Menger earned the degree of Dr. phil. on November 13, 1924. The summary of his findings had already been published in two parts in 1923 and 1924: [394] Menger, Theorie (1923)

[78] Luitzen Egbertus Brouwer (1881–1966), Dutch mathematician and philosopher, founder of intuitionism in the philosophy of mathematics

[79] Pavel Samuilovich Urysohn (1898–1924), Russian mathematician

2.2 Karl Menger: "From the Criterion of Meaning to the Principle of Tolerance" 79

an accident without having published his findings. Menger's definition of the theory of dimension, which is considered to be basically equivalent to those of Brouwer and Urysohn, came about – independently of Urysohn's definition – without Menger having any knowledge of Urysohn's work.[80]

In 1925, having also earned the *Philosophicum* qualification under Moritz Schlick in addition to his Ph.D. with Hahn, Karl Menger was awarded a Rockefeller scholarship, which he used to go work with Brouwer at Amsterdam University. There he was appointed Brouwer's assistant on September 15; he was very soon joined from Vienna by his Polish student Witold Hurewicz[81], whose first papers on dimension theory had made a "brilliant" impression on Menger.

Menger qualified as a professor in Amsterdam in 1926 with his paper *Grundzüge einer Theorie der Kurven (Main Features of a Theory of Curves)*[82], which had appeared in 1925 in the *Proceedings of the Section of Sciences* published by the *Koninklijke Nederlandse Akademie van Wetenschapen* and then in the *Mathematische Annalen* in Berlin.[83] Amid an initially friendly but by autumn 1926 rapidly deteriorating relationship with Brouwer[84], whose intuitionism fascinated him but whose unorthodox ideas seemed to him to require elucidation, Menger earned great success in the field of geometry. He proved that every curve can be embedded topologically into the three-dimensional Euclidean space[85], he discovered a theorem of curves that could also be applied in electrical engineering[86] and he completed the first two of a series of studies of "general metrics" in which he introduced and studied convexity and geodesy in metric spaces and characterized subsets of Euclidean spaces within them.[87,88]

The disagreements Brouwer and Menger[89] had about priorities, as well as a letter from Hans Hahn encouraging him to take over the position

---

[80] See also [414] NDB 17 (1994), pp. 74–75
[81] Witold Hurewicz (1904–1956) was born in Poland and studied in Vienna under Hans Hahn and Karl Menger. He went with Menger to Amsterdam in 1927–28, and later succeeded Menger as Brouwer's assistant (1928–1936)
[82] [138] DBE (1998/1999), p. 62
[83] [382] Menger, Grundzüge (1925)
[84] The worsening of relations with Brouwer, according to Menger himself, was the result of Brouwer's claim to be the creator of dimension theory
[85] [393] Menger, Struktur (1926)
[86] [385] Menger, Kurventheorie (1926)
[87] [375] Menger, Dimensionstheorie (1928)
[88] See [381] Menger, Grundlagenfragen (1928), [384] Menger, Intuitionismus (1930). Both papers would influence Kurt Gödel
[89] Hans-Joachim Dahms, on the other hand, views the reason for Menger's departure from Amsterdam in the fact that "Brouwer's machinations in administrating his remuneration and his increasingly antisocialist and reactionary tirades had become intolerable". See [134] Dahms, Philosophie (1985), p. 340

of geometry professor in Vienna left vacant by Kurt Reidemeister's departure for Königsberg[90], led Menger to return to his hometown as a university professor of geometry[91] in the autumn of 1927 after two and a half years in Amsterdam and to go back to the Vienna Circle, for Schlick and Carnap were very interested in his work on dimension theory. He was thus engaged by Hahn and Schlick to take part in Schlick's colloquium.[92] Menger recalled that Hahn had addressed the subject of the *Tractatus logico-philosophicus* by Ludwig Wittgenstein.[93]

> He asked me if I had heard of the *Tractatus*. I answered that had started reading the book a while back but couldn't get past the first few pages. "That was also my experience at first," said Hahn, "and I didn't think I could take the book seriously at all. Only after I heard an excellent report about it in Reidemeister's circle three years ago and then carefully read it myself did it occur to me that it probably represents the most important contribution to philosophy since the publication of Russell's seminal writings. There were arguments about the book in the circle, though, and we had so many differences of opinion about details that Carnap suggested a year ago that, in order to clear up the confusion, we should dedicate as many consecutive

---

[90] Hans Hahn made a great effort to replace Reidemeister with Menger. See [134] Dahms, Philosophie (1985), p. 340. From 1928, he was an extraordinary professor of geometry. In the winter semester of 1927–28, he began at the University of Vienna with lectures about his dimension theory. The compilation of his findings was published a year later: [375] Menger, Dimensionstheorie (1928)

[91] Menger was professor of geometry in Vienna from 1927 to 1936. For a biography of Karl Menger, see $http://www-history.mcs.st-and.ac.uk/\sim history/ Mathematicians/Menger.html$ (accessed on July 13, 1998, 4:37 PM)

[92] While Manfred Geier ascribes this encouragement to Hahn alone ([194] Geier, Kreis (1992), p. 47), Menger himself remembered that Schlick also prompted him to take part. ([376] Menger, Einleitung (1988), p. 12). There are also inconsistencies regarding Menger's participation in Schlick's colloquium. While Geier claims, "His participation in the discussions is lively and extremely spirited", Menger himself wrote "Kurt Gödel and I, who tried to keep out of most of the arguments going on in the circle ..."([376] Menger, Einleitung (1988), p. 13)

[93] In an interview with Rudolf Haller, Karl Menger disputed the claim that the mathematician Hans Hahn had already given a seminar on the Tractatus in 1922 which had "deeply impressed" Professors Moritz Schlick (philosophy) and Kurt Reidemeister (mathematics), who had just come to Vienna in 1922. See [225] Haller, Neopositivismus (1993), p. 229, footnote 18. There is apparently no indication in the archives of conversations recorded by Friedrich Waismann that Hahn discussed the *Tractatus* before the year 1924, as mentioned by B. F. McGuinnes in his foreword as publisher ([546] Waismann, Wittgenstein (1984), p. 13)

2.2 Karl Menger: "From the Criterion of Meaning to the Principle of Tolerance"

sessions of the circle as necessary to reading the work in sections; and we did in fact devote all of the past academic year (1926–27) to this task." Hahn continued: "The *Tractatus* made the role of logic clear to me.[94]

Kurt Reidemeister had already delivered a lecture on the *Tractatus* in the Schlick circle on Christmas Day 1924; this was the colloquium that Moritz Schlick organized during the winter semesters of 1923–24 and 1924–25 within the framework of his philosophical seminar.[95] Rudolf Carnap reported in his autobiography that Wittgenstein's book had been "talked through sentence by sentence" at times. "Protracted deliberations were often necessary in order to figure out what was meant. Sometimes we could not find a clear-cut explanation. But we still understood a great deal of the book and had lively discussions about it."[96]

"Two *Tractatus* phases"[97] can be discerned from the remembrances of individual Vienna Circle members: The first phase lasted from 1923 to 1925, a second went from 1929 to 1931 and between them was a reading phase, which began after Carnap's arrival in Vienna. Several members had their first personal meetings with Wittgenstein during this intermediate period. Schlick had tried in vain since 1924 to contact the secretive philosopher, who at the time was an elementary school teacher in the villages of Puchberg and Otterthal.

---

[94] „Er fragte mich, ob ich vom *Tractatus* gehört hätte. Ich antwortete, ich hätte vor einiger Zeit angefangen, das Buch zu lesen, ohne jedoch über die ersten Seiten hinauszukommen. »Das war auch meine erste Erfahrung«, sagte Hahn, »und ich hatte nicht den Eindruck, das Buch sei überhaupt ernst zu nehmen. Erst als ich vor drei Jahren im Kreis Reidemeisters ein ausgezeichnetes Referat darüber hörte und dann selbst sorgfältig das ganze Werk las, ging mir auf, dass es wahrscheinlich den wichtigsten Beitrag zur Philosophie darstellt, seit der Veröffentlichung von Russells grundlegenden Schriften. Im Kreis kam es jedoch zu Auseinandersetzungen über das Buch, und es gab so viele Meinungsverschiedenheiten über Einzelheiten, dass Carnap vor einem Jahr den Vorschlag machte, wir sollten, um die Verwirrung aufzuklären, der abschnittsweisen Lektüre des Werks so viele aufeinanderfolgenden Sitzungen des Kreises widmen wie nötig; und in der Tat haben wir das gesamte vergangene akademische Jahr (1926/27) dieser Aufgabe gewidmet.«»Mir«, fuhr Hahn fort, »hat der Tractatus die Rolle der Logik klar gemacht."" [376] Menger, Einleitung (1988)

[95] These colloquia in the Schlick circle are considered a "first early circle", which later led to the regular meeting of the Vienna Circle at the Boltzmanngasse location. According to Reidemeister, those present during these first discussions were Kaufmann, Waismann, Feigl and Carnap in addition to Schlick and himself. Reidemeister also writes: "The Vienna Circle came together right after I moved to Königsberg." („Der Wiener Kreis konstituierte sich direkt erst nach meiner Übersiedlung nach Königsberg".) Reidemeister answered the call to go to Königsberg in the spring of 1925. (Questionnaire by Kurt Reidemeister (materials from Henk Mulder), Vienna Circle Archive Haarlem (WKA), Rijksarchief Noord Haarlem (Netherlands).) Quote from [520] Stadler, Kreis (1997), p. 267

[96] [117] Carnap, Weg (1963), p. 39

[97] [520] Stadler, Kreis (1997), p. 268

Fig. 2.7. L. Wittgenstein

His hope was revived only after he went with Charlotte and Karl Bühler to pay a visit to the home of Wittgenstein's sister Margaret Stonborough in 1927. This visit was followed by invitations to Wittgenstein's house on Prinz-Eugen-Strasse, which was also frequented by Friedrich Waismann, Herbert Feigl and his eventual wife Maria Kasper.

The discussions about Wittgenstein's work were quite controversial within the Vienna Circle. Schlick and Waismann were the "Wittgenstein faction", while Neurath, Carnap, Feigl, Menger and others took a more dissenting position. Particularly in 1926–27, the debate about elementary statements and tautology was very intense, and when Menger returned from Amsterdam in the fall of 1927, he found these discussions very disagreeable.

The Vienna Circle was supposed to carry on the analytical tradition of Ernst Mach.

The conceptual analysis of theories – it was believed – would bring with it advancements in the understanding of nature. The important questions were:

– What is the status of our knowledge?
– What is the status of statements about this knowledge?
– How are science and the world connected by logic?

The Vienna Circle was thus concerned with the three-way relationship of *world – language – science*. Its goal was to found a scientific philosophy with a new perspective and in particular to destroy metaphysics. To this end, metaphysical sentences first had to be demarcated from scientific sentences. Schlick had also introduced the criterion of verifiability, which Waismann described as follows: "If one cannot indicate any ways in which a sentence is true, then the sentence has no meaning whatsoever, for the meaning of a sentence is the method by which it is verified." In his *Tractatus*, Wittgenstein had espoused a logistical philosophy of language according to which the world can be clearly represented in the form of sentences, and Rudolf Haller sees three "basic concepts that the Vienna Circle adopted from Wittgenstein"[98]:

---

[98] [225] Haller, Neopositivismus (1993), p. 93

2.2 Karl Menger: "From the Criterion of Meaning to the Principle of Tolerance"

1. The essence of logical statements lies in their form (structure) and only therein. They are always true (or always false) statements. "The statements of logic are tautologies,"[99] they say nothing.[100] This means: They bear absolutely no relation to a representation of the real world and thus cannot be empirically confirmed or denied. "Outside logic, all is accident."[101] So the tautological statements have no empirical content; in Wittgenstein's terminology they are "sinnlos" (meaningless).
2. Wittgenstein conceived the empirical statements before a backdrop of the view that everything describable in reality could also be completely different (Hume). "No part of our experience is a priori ... There is no order of things a priori."[102] A statement is called empirical when it is possible to say what makes or would make it true, that is, when those conditions under which is can be true at all can be stated.
3. Thirdly, Haller mentions Wittgenstein's view of nature and the role of philosophy: Analysis and criticism of language must become the essential task of philosophy. "The object of philosophy is the logical clarification of thoughts."[103] "Philosophy limits the disputable sphere of natural science."[104] This criterion of demarcation can already be observed in the *Tractatus* – over ten years before Popper's *Logic of Scientific Discovery*!

The Vienna Circle was engaged not only in the formalization of mathematical or logical issues but also in the formalization of an empirical construct of the world and our knowledge of it. Menger concluded from this that the Vienna Circle had affiliated itself with Russell's idea of logicism from 1910, the fundamental theses of which stated that logic was based on mathematics and that all mathematical concepts and processes of reasoning trace back to those of logic.

The fact that logic reduces analysis to arithmetic is one of the results of the 19th century; Russell had merely added a definition of arithmetic terms and a proof of arithmetic axioms using logic. This last step was discussed both by Poincaré, and by the intuitionists, and it was no longer unequivocal due to the discovery of multi-valued logic, although the members of the Vienna Circle would not become aware of that until later.

---

[99] [570] Wittgenstein, Tractatus (1921), 6.1
[100] [570] Wittgenstein, Tractatus (1921), 6.11
[101] [570] Wittgenstein, Tractatus (1921), 6.3
[102] [570] Wittgenstein, Tractatus (1921), 5.634
[103] [570] Wittgenstein, Tractatus (1921), 4.112
[104] [570] Wittgenstein, Tractatus (1921), 4.113

Fig. 2.8. R. Carnap

Another question was related to language. One spoke of the language. What was this definiteness supposed to justify? Menger put this question to Carnap, Schlick and other Vienna Circle members, but received no satisfactory answer. Schlick, for example, did not take the question seriously. Carnap, on the other hand, was soon going to dismiss more than just this one question!

The intuitionism founded by Brouwer was a school of thought within the philosophy of mathematics that was counter to logicism. The controversy between these two movements became evident in set theory and in the concept of the actual infinite after it was introduced by Georg Cantor. Leopold Kronecker[105] had refuted this theory with the words: "Even the general concept of the infinite series (...) can in my estimation be accepted only with reservations, since in this specific case, on the basis of the arithmetic *law of formation* for members (...), certain conditions are shown to be satisfied that allow the series to be used as finite expressions."[106]

Brouwer had defended his two theses often since 1907:

1. The solvability of every problem Hilbert had formulated as an axiom in 1900 is equivalent to the logical law of the excluded middle. Yet since there is no adequate basis for Hilbert's axiom and since logic is based on mathematics (and not the other way around!), the law of the excluded middle is also an inadmissible proof.
2. It is inadmissible to combine things that possess one particular property into one set. Set theory is not justified in this way. Rather, it must be based on a *constructive* definition of sets.

Naturally, Brouwer had his own definition of sets:

> A set is a law on the basis of which, if a random complex of digits in the series 1, 2, 3, 4, 5, ... is selected again and again, each of these selections

---

[105] Leopold Kronecker (1823–1891), German mathematician, became an ordinary professor in Berlin in 1883. From 1861, he was a member of the Berliner Akademie and from 1868 a member of the Paris Academy. Kronecker is considered a pioneer of intuitionism

[106] Leopold Kronecker in: [41] Becker, Grundlagen (1964), p. 327

either generates a particular sign or nothing, or else it brings about the inhibition of the process and the definitive annihilation of its result, wherein a complex of digits can be indicated for each $n > 1$ after each uninhibited series of $n-1$ selections, and said complex, when selected as the nth complex of digits, does not cause the inhibition of the process. Each series of signs generated in this way from an unlimited selection series (which thus generally cannot be represented in finished form) is called an element of the set. The mode of formation common to the elements of set M is likewise briefly designated as M.[107]

2.2.1 Menger's Conventionalism

Menger had been influenced by Brouwer's intuitionism during his time in Amsterdam.[108] "Menger's attitude toward intuitionism became more critical later on, however."[109] A certain conventionalism seemed to him to be the right position in this situation and so he was skeptical toward "ultimate foundations", for which the Vienna Circle was apparently striving. When Hahn asked him in 1927 to report on Brouwer's intuitionism for the Schlick circle, he complied with the request the same year. He had also prepared a second portion of the lecture, however, in which he wanted to talk about his own mathematical-logical-philosophical beliefs; they included his criticism of intuitionism and referred to a plurality of logics and languages.[110]

> I presented the intuistic-fomalistic dictionary of set theory that I had devised as well as what just at that time had begun to take a firmer

---

[107] "Eine Menge ist ein Gesetz, auf Grund dessen, wenn immer wieder ein willkürlicher Ziffernkomplex der Folge 1, 2, 3, 4, 5... gewählt wird, jede dieser Wahlen entweder ein bestimmtes Zeichen oder nichts erzeugt oder aber die Hemmung des Prozesses und die definitive Vernichtung seines Resultates herbeiführt, wobei für jedes $n > 1$ nach jeder ungehemmten Folge von $n-1$ Wahlen wenigstens ein Ziffernkomplex angegeben werden kann, der, wenn er als n-ter Ziffernkomplex gewählt wird, nicht die Hemmung des Prozesses herbeiführt. Jede in dieser Weise von einer unbegrenzten Wahlfolge erzeugte Zeichenfolge (welche also im Allgemeinen nicht fertig darstellbar ist) heißt ein Element der Menge. Die gemeinsame Entstehungsart der Elemente der Menge M wird ebenfalls kurz als die Menge M bezeichnet." Luitzen Egbertus Brouwer, *Intuitionistische Mengenlehre*. Quoted from: [41] Becker, Grundlagen (1964), p. 329f

[108] Three years later, Brouwer traveled to Vienna at the invitation of Austrian physicist and Vienna University professor Felix Ehrenhaft (1879–1952) to deliver two lectures on the philosophy of mathematics. These were the first two of an entire lecture series by foreign guest scientists, which were financed by industrialists and delivered for a public audience in one of the physics institutes at the university. Menger suggested that Schlick and Waismann invite Wittgenstein to these lectures. Wittgenstein did actually attend at least one of the Brouwer lectures. According to Feigl, these lectures by Brouwer inspired Wittgenstein's return to philosophy after many years of abstinence

[109] [117] Carnap, Weg (1963), p. 76

[110] [387] Menger, Memories (1982), p. 88

shape in my mind – the epistemological consequences of my critique of intuitionism: the plurality of logics and languages entailing some kind of logical conventionalism.[111]

These ideas were very poorly received in the Vienna Circle.[112] Menger was contradicting not only Brouwer's intuitionism but also the views of the "Wittgenstein faction"[113] when he "showed – and some members of the circle were inclined to agree with him – that there was a degree of arbitrariness in the selection of the dividing line between admissible and inadmissible concepts and forms of deduction".[114] Menger, who may not have intended to play this role at all, became the "young savage" here and provoked "the old folks" with his opinion that there wasn't only one way to establish postulates of constructivity but instead that it was possible to have constructivity postulates that differed gradually:

1. An appropriate deductive mathematics could be developed for each of these different versions of constructivity, particularly for systems that are more restrictive than intuitionistic mathematics.
2. Insisting on a particular idea of constructivity and then characterizing the related developments as meaningful and any rejection of those findings as meaningless does not have the slightest amount of cognitive content. These views should be expelled from logic and mathematics and relegated to the proponent's biography.
3. With regard to mathematics and logic, the only important question is how it is possible to proceed from a particular set of statements to other statements according to certain rules, while the justifications of statements or transformation rules are nothing but empty words when it comes to intuition.[115]

Although Schlick assured him after the presentation that he now finally understood Brouwer's "several page long definition of a set" and although Hahn and Carnap concurred with this initial reaction, none of them offered a single word in response to his epistemological treatment in the second part.[116] He had directly contradicted their views, which can be summed up in these four points:

1. Consistent reference to *the* logic and *the* language;
2. the increasingly definitive use of the term tautology in the (absolute) sense (that statements are consequences of pure logical axioms) with reference to *the* logic;

---

[111] [387] Menger, Memories (1982), p. 88
[112] [387] Menger, Memories (1982), p. 88f
[113] See [387] Menger, Memories (1982), p. 88
[114] [117] Carnap, Weg (1963), p. 76
[115] See [380] Menger, Geometry (1970), p. 11
[116] [387] Menger, Memories (1982), p. 88f

2.2 Karl Menger: "From the Criterion of Meaning to the Principle of Tolerance" 87

3. the assertion (proposed by Schlick, Waismann, Hahn, Carnap and, Menger believed, Feigl) that all mathematical statements are tautologies;
4. a use of the term *meaningless* that Menger found to be random and imprecise, since it was not supported by any rules to indicate what should be considered *meaningful*.[117]

Menger spoke to the Vienna Circle again during the semester of 1928–29 about the basic concepts in set theory[118], and it was then that, spurred on by the conflicts between logicians and intuitionists, he developed what Carnap would later term a "principle of tolerance", which allowed different logics to coexist as long as they could be successfully instituted within their respective fields of application. Menger again spoke here of a plurality of different "logics". His colleagues' reactions were devastating: Carnap and Hahn were fairly skeptical at the time; Schlick, Kaufmann and Waismann rejected his ideas out of hand; Kraft and Neurath likely remained reserved and ambivalent.[119]

> Schlick soon began to shake his head, and I realized that I was receiving that mild punishment he meted out to loudmouths. And he actually tried to exchange furtive glances with the others. Only Waismann returned Schlick's look with a smile, at which point both of them launched into an emphatic attack on my arguments. [...] Carnap, who had always spoken about "the" logic and "the" language, was apparently absorbed in some deep meditation and said nothing. Neurath did not seem particularly interested. Victor Kraft listened but remained silent. I received no support from anyone, with the exception of a young man whom I had known up to that point only as one of the students in my course on dimension theory – and his name was Kurt Gödel.[120]

---

[117] See [387] Menger, Memories (1982), p. 88
[118] The content of this lecture corresponded to the first of four parts in a series of *Observations and Fundamental Questions*, which he wrote for the annual report of the German Mathematical Society in 1928, [381] Menger, Grundlagenfragen (1928). The paper *An Intuitionistic-Formalistic Dictionary of Set Theory*, reprinted as Chap. 5 of his *Selected Papers*, is a translation of this first "observation" undertaken by Menger himself
[119] [520] Stadler, Kreis (1997)
[120] „Schlick begann bald, seinen Kopf zu schütteln und ich realisierte, dass mich jene milde Strafe traf, die er Schwätzern zumaß. Und tatsächlich versuchte er auch bald flüchtige Blicke mit den anderen zu wechseln. Nur Waismann beantwortete Schlicks Blick mit einem Lächeln und dann begannen beide meine Argumente nachdrücklich anzugreifen. [...] Carnap, der ständig über «die» Logik und «die» Sprache gesprochen hatte, schien durch eine tiefe Nachdenklichkeit gefesselt zu sein und sagte nichts. Neurath war offensichtlich nicht besonders interessiert. Victor Kraft hörte zu, schwieg aber. Ich erhielt von niemandem Unterstützung, mit Ausnahme eines jungen Mannes, den ich bis zu diesem Zeitpunkt nur als einen meiner Studenten in meinem Kurs über Dimensionstheorie kannte – und dessen Name war Kurt Gödel." [387] Menger, Memories (1982), p. 89f

In another passage, Menger wrote about Gödel, who was about the same age and who had been a very reserved person all his life[121], and about the situation at the time:

> He indicated his interest merely by lightly shaking his head – to express agreement or skepticism ...; at Schlick's request, I reported during two sessions on the content of my first three "Observations on the Fundamental Questions". As I did, I noticed that Gödel was the only one in the circle who listened to myindexsrlogical tolerance principle remarks [from which the idea that Carnap called the principle of tolerance emerged] and nodded spiritedly in agreement.[122]

From that point on, Menger and Gödel formed a silent opposition within the Vienna Circle:

> On the way home after one session in which Schlick, Hahn, Neurath and Waismann had talked about language but during which neither Gödel nor I had uttered a word, I said, "Today we managed to out-Wittgenstein these Wittgensteinians once again: We didn't say anything." Gödel answered, "The more I think about language, the more I am surprised that people have ever understood each other at all."[123].

Because of the style and diction of the manifesto issued to honor of Schlick's remaining in Vienna, Menger no longer wished to be identified as a member but rather only as "associated with the Vienna Circle".[124] He decided to found and lead his own "Mathematical Colloquium" with students and foreign guests. He also published a series called Findings of a Mathematical Colloquium (1931–1937). In addition, he was instrumental, along with Hans Hahn,

---

[121] Gödel is characterized as "reserved" and "difficult" in numerous sources, including in the Popper interview in [520] Stadler, Kreis (1997)

[122] „Sein Interesse bekundete er lediglich durch leichtes Kopfschütteln – zustimmend oder skeptisch ablehnend ...; ich berichtete auf Schlicks Wunsch in zwei Sitzungen über den Inhalt meiner ersten drei „Bemerkungen über die Grundlagenfragen". Dabei fiel mir auf, dass Gödel der einzige im Kreise war, der meine Ausführungen, [aus denen das von Carnap sogenannte Toleranzprinzip hervorging], mit lebhaft zustimmendem Kopfnicken aufnahm." [377] Menger, Erinnerungen (1981), p. 1, quoted from [520] Stadler, Kreis (1997), p. 237

[123] „Nach einer Sitzung, in der Schlick, Hahn, Neurath und Waismann über Sprache diskutiert hatten, während jedoch weder Gödel noch ich ein Wort äußerte, sagte ich auf dem Heimweg: «Wir haben heute wieder einmal diese Wittgensteinianer überwittgensteinert: Wir haben geschwiegen.» «Je mehr ich über die Sprache nachdenke», antwortete Gödel, «umso mehr wundert es mich, dass die Menschen einander überhaupt je verstehen»" [377] Menger, Erinnerungen (1981), p. 9, quoted from [520] Stadler, Kreis (1997), p. 237

[124] This was a sentiment shared by English mathematician Frank Plumpton Ramsey (1903–1930), German mathematician and philosopher Hans Reichenbach (1891–1953) and Austrian philosopher Edgar Zilsel (1891–1944)

2.2 Karl Menger: "From the Criterion of Meaning to the Principle of Tolerance"

Hans Thirring and Hermann Mark[125], in the organization of lectures and their publication. This cooperation resulted in

- *Krise und Neuaufbau in den exakten Wissenschaften (Crisis and Reconstruction in the Exact Sciences)* (1933)
- *Alte Probleme – Neue Lösungen in den exakten Wissenschaften (Old Problems - New Solutions in the Exact Sciences)* (1934)
- *Neuere Fortschritte in den exakten Wissenschaften (Newer Advancements in the Exact Sciences)* (1936).

Menger's Mathematical Colloquium was a forum that gave new impetus to communication with scientists from abroad. In particular, established connections with those researchers who had had no previous contact with Schlick circle members or about whom the Vienna Circle knew nothing. Reminiscing in 1993, this impression was shared by Menger's student Franz Alt:

> Menger, though the youngest, was the magnet which pulled the foreign visitors on their pilgrimages to Vienna. Polish, ..., Americans, ... French and Japanese, too, but relatively few from Germany (besides Nöbeling), possibly because there was less interest there in the areas that were at the forefront in Vienna - logic, basic principles, set theoretical topology, etc.[126]

2.2.2 Alfred Tarski's Support

The Vienna Circle first became known in Poland from a small anonymous notice[127] in 1929 describing it as an informal philosophical group that was concerned with a scientific understanding of the world and that sought contacts with people representing similar views.[128] Since no Polish names were

---

[125] Hermann Mark (1895–1992), Austrian-American chemist. Mark was a university professor in Vienna 1932–1938, then emigrated to the US. He founded the Polymer Institute in New York. Mark was the inventor of polymerization - a prerequisite for the production of plastics

[126] „Menger, obwohl der jüngste, war der Magnet, zu dem ausländische Besucher nach Wien pilgerten. Polen, ... Amerikaner, ... auch Franzosen, Japaner, aber verhältnismäßig wenige aus Deutschland (außer Nöbeling), vielleicht weil es dort weniger Interesse gab an den Gebieten, die in Wien führend waren – Logik, Grundlagen, mengentheoretische Topologie usw." Letter to Reinhard Sigmund-Schultze dated 12 July 1993, quoted in [513] Sigmund-Schultze, Mathematiker (1998), p. 251. Sigmund-Schultze noted that in addition to Georg August Nöbeling (born 1907), who earned his doctorate under Menger in 1931 and worked in Erlangen from 1935, Kurt Reidemeister should be mentioned. Reidemeister, who was from Braunschweig, held a professorship from 1922 to 1925, which was taken by Menger upon his return from Amsterdam when Reidemeister himself departed for Königsberg

[127] [571] Wolenski, School (1989), p. 443

[128] Jan Wolenski cites as the source of this notice: *Ruch Filozoficzny*, XI, 1928–29, p. 196

Fig. 2.9. A. Tarski

mentioned, it is assumed that the members of the Vienna Circle, at least at the time, had not heard much about the Lemberg-Warsaw school of logic! In the summer of 1929, mathematicians at the University of Warsaw invited Menger to deliver a guest lecture the following autumn[129] and he then traveled to Warsaw in September. There he met the Polish logician Leon Chwistek[130] near Krakow. A short time later, he met with Alfred Tarski [131], Adolf Lindenbaum[132] and some of their students, whose logical investigations and findings as well as their treatment of philosophical problems with logical techniques had impressed him. He resolved to introduce the body of thought espoused in the Polish school of logic to the members of his Mathematical Colloquium and the Vienna Circle, and he invited Alfred Tarski to Vienna. Tarski accepted

---

[129] Menger had had contact with the topologists in Warsaw since 1923, when he had sent special reprints of his first paper on dimension theory to Warsaw. Works were exchanged regularly from then on. In about 1920, mathematicians in Warsaw had founded the journal *Fundamenta Mathematicae*. It was unique in its orientation toward set theory, point sets and the fundamentals of mathematics! Menger was in contact with several contributors, including Bronislaw Knaste (1893–1990), who became a professor in Lviv (then known as Lemberg) in 1939 and in Wroclaw in 1945, and Kazimierz Kuratowski (1896–1980), who became a professor in Lviv in 1933 and was an *ordinaries* at the University of Warsaw in 1934–35. Like the mathematicians of Warsaw, the logicians there were surrounded by a superb scientific atmosphere, although until this point they had been strangely isolated

[130] Leon Chwistek (1884–1944), Polish logician and philosopher from the Lemberg-Warsaw school

[131] Alfred Tarski (1902–1983), Polish logician and philosopher, emigrated to the US in 1939. He arrived at the University of California at Berkeley in 1942, where he became a professor of mathematics in 1949. Following his retirement he remained active there until his death. Lotfi Zadeh was not in close contact with Tarski, however: "I met him but I did not have any conversation." [705] R. S. interview with L. A. Zadeh (1999)

[132] Adolf Lindenbaum (1904–1941 or 1942), Polish logician

2.2 Karl Menger: "From the Criterion of Meaning to the Principle of Tolerance"

immediately.[133] And so "the first ever presentation on Polish logic in the West"[134] was arranged.[135]

Mathematically, Tarski owed much to the algebraic approach of the point set theory by Banach[136] and Sierpinski[137], and in the area of logic he was also influenced by the algebraic approach by Boole[138], Schröder[139] and Skolem.[140] Additionally, he belonged to the strongly Aristotelian Lemberg-Warsaw school of logicians Jan Łukasiewicz[141] and Stanislaw Lesniewski.[142] He worked on a formulation of Hilbert's proof theory as metamathematics. He also separated the language in which the axioms of logic were derived and formulated from that language of logic itself. Tarski delivered his lectures before Menger's Mathematical Colloquium in February of 1930; his subjects were:

- "Set Theory",
- "Some Fundamental Concepts of Mathematics" and
- "Studies in the Calculus of Propositions".

The three-valued or multi-valued logic Tarski talked about in the third lecture was likely still completely unfamiliar to the Viennese. Nevertheless,

---

[133] Tarski delivered three lectures in the Mathematical Colloquium (February 19, 20 and 21, 1930); the Vienna Circle was also invited to two of them, which were to be of a logical character

[134] [520] Stadler, Kreis (1997), p. 442f

[135] [520] Stadler, Kreis (1997), states that Tarski's trip to Vienna (which he repeated in 1935) led to the integration of the Polish logician and philosopher of science into the Unity of Science movement

[136] Stefan Banach (1892–1945), Polish mathematician, earned his PhD at Jan Kazimierz University in Lviv in 1922, where he qualified and became an extraordinary professor the same year. He became an ordinary professor in 1927

[137] Waclaw Sierpinski (1882–1964), Polish mathematician, studied mathematics and physics at the university in Warsaw until 1908 and became professor of mathematics in Lviv in 1910

[138] George Boole (1815–1864), British mathematician and philosopher. Boole became professor of mathematics at Queen's College in Cork, Ireland in 1848, where he remained until his death

[139] Ernst Friedrich Wilhelm Karl Schröder (1841–1902), German logician, qualified as a professor in Zurich in 1865, became professor of mathematics at the Technical University of Darmstadt in 1874 and at the Technical University in Karlsruhe in 1876. In 1890 and 1891, he committed to paper his lectures on algebra and logic in which he systematized George Boole's findings

[140] Thoralf Albert Skolem (1887–1963), Norwegian mathematician, logician and philosopher, qualified in 1926 at the University of Oslo and was an ordinary professor there from 1938 to 1957

[141] Jan Łukasiewicz (1878–1956), Polish philosopher and logician of the Lemberg-Warsaw school. Between 1915 and 1939, he was a professor at the universities in Lviv and Warsaw. During the Nazi occupation, he was involved with the underground university of Warsaw; he moved to Dublin in 1949

[142] Stanislaw Lesniewski (1886–1939), Polish philosopher and logician of the Lemberg-Warsaw school

the discussions concentrated from the outset on Tarski's thoughts on metamathematics; as can be seen in Carnap's diary, the members met before the first lecture date:

> Afternoon to Hahn's. Menger, Neurathin [= Olga Hahn] and Tarski there, his speaking is reserved, humble and very clear. About metamathematics. Gödel's [!] problem of formalizability of mathematics.[143]

Meetings between the lectures and in more private settings were naturally also used for professional discussions; metamathematics gave Carnap a lot to think about and his diary includes this note:

> 6:45 Tarski comes to my house [...] about an axiomatic. [...] different terms need to be defined metamathematically instead of mathematically [...] Gödel comes in evening. Until 11:30.[144]

Fig. 2.10. Alfred Tarski and Kurt Gödel in Vienna, 1935

Tarski accomplished what Menger had not been able to do: He managed to make the "Wittgenstein faction" unsure of themselves.[145] A hierarchy of languages and logics was now considered in earnest and was no longer rejected, as had been the case following Menger's lectures a year earlier. Tarski's visit had

---

[143] „Nachmittags zu Hahn. Dort Menger, Neurathin [= Olga Hahn] und Tarski, er spricht zurückhaltend, bescheiden und sehr klar. Über Metamathematik. Gödels [!] Problem der Formalisierbarkeit der Mathematik. Rudolf Carnap, Diary, 16 February 1930." Quoted in: [287] Köhler, Gödel (1991), p. 142

[144] „3/4 7h kommt Tarski zu mir [...] über eine Axiomatik. [...] verschiedene Begriffe müsste[n] anstatt mathematisch metamathematisch definiert werden [...] Abends kommt Gödel. Bis 1/2 12h." [520] Stadler, Kreis (1997), p. 441

[145] [520] Stadler, Kreis (1997), p. 441

2.2 Karl Menger: "From the Criterion of Meaning to the Principle of Tolerance"

been enormously important[146] and fruitful for Kurt Gödel, too, and his proof of the law of incompleteness, which he produced the same year, then finally prompted Carnap to abandon his "General Axiomatics" and to follow Tarski's idea of introducing different stages of languages, which he then fleshed out in order to complete his *Logical Syntax of Language* (1934). Alongside Menger, Carnap had now become the second "pluralist of logic and language" in the Vienna Circle![147]

2.2.3 The Principle of Logical Tolerance

Karl Menger maintained his standpoint between intuitionism and logicism and continued to develop his ideas. In his paper on *Intuitionism*[148], he was emphatic when he said that it absolutized the concept of constructivity, but also that it can be interpreted narrowly or broadly.

> Intuitionism acknowledges constructively proven laws and rejects the other findings of mathematics. Before we go into the dogmatics within this program, let us first of all note that no satisfactorily precise definition has yet been provided for the basic concept of constructivity that is inherent in the program. The only thing that is certain is that the requirements of constructivity set by various mathematicians are different from one another. For example, Brouwer rejects the law of the excluded middle but recognizes uncountable sets, while Borel has raised no objections to the law of the excluded middle but rejects uncountable sets for being insufficiently defined. On this matter, the author of this article has argued the position that if it is possible at all to provide a precise definition of constructivity (which has heretofore never been satisfactorily specified), then this more precise statement should at least include different types and should in particular *vary gradually*. Even in the most well-understood areas of geometry, there is no notion and no word that does not inevitably require a certain degree of specification and would not be capable of several different specifications, and the same is true, although certainly in an even greater measure, for the ambiguous word "constructivity", which one can surely define more broadly or less broadly, if at all. *For each of these different concepts of constructivity, an associated deductive mathematics can be developed.* It might be possible, for example, to define a concept of constructivity so strictly that only infinite sets can

---

[146] After Tarski's lectures in the Schlick circle, Gödel asked Menger to arrange a meeting with Tarski so that he could speak to him about the completeness of first-order logic. Tarski expressed great interest in Gödel's findings

[147] Stadler writes that Neurath was likely also with them. [520] Stadler, Wiener Kreis (1997), p. 142

[148] [384] Menger, Intuitionismus (1930), p. 324f

be controlled constructively, or a somewhat *weaker* concept that included *countable* sets, or one even weaker in which *analytical* sets were permitted, or one very general that introduced *arbitrary* sets of real numbers into mathematics. The requirement of freedom of contradiction can in this sense be seen as the broadest possible requirement of constructivity. What the attempts at intuitionism have done so far has been for each of them to *commit* dogmatically to a particular concept of constructivity (usually not clearly circumscribed, as mentioned previously) and to characterize the related developments as meaningful and those more advanced as meaningless. In the opinion of the present author, however, a statement such as this *does not have the slightest epistemological content*. For mathematics and logic are not concerned with what one assumes but rather with what one *derives* from them or by using them. Whether mathematician A declares the axiom of choice to be "admissible", "believes" in it and applies it, and whether mathematician B rejects it as "unconstructive" or "because he can't connect any meaning with it" – these facts may be interesting in the biographies of mathematicians A and B, and possibly historically, but in no way are they important for mathematics and logic. They merely have to do with what follows from the axiom of choice. They seek to discover the statements into which particular statements can be transformed with the aid of particular transformation rules. If one justifies his acceptance or rejection of statements or transformation rules by appealing to intuition, then in the end he will be saying nothing at all but empty words.[149]

---

[149] „Der Intuitionismus anerkennt konstruktiv bewiesene Sätze und lehnt die anderen Ergebnisse der Mathematik ab. Ehe wir auf die in diesem Programm gelegene Dogmatik eingehen, bemerken wir vor allem, dass der in das Programm eingehende Grundbegriff der Konstruktivität bisher nicht befriedigend präzisiert wurde. Fest steht vielmehr bloß, dass die Konstruktivitätsforderungen verschiedener Mathematiker voneinander abweichen. Beispielsweise lehnt Brouwer den Satz vom abgeschlossenen Dritten ab und anerkennt überabzählbare Mengen, während Borel gegen den Satz vom ausgeschlossenen Dritten keine Einwände erhoben hat, aber überabzählbare Mengen als ungenügend definiert ablehnt. Der Verfasser dieses Aufsatzes hat diesbezüglich die Ansicht ausgesprochen, dass der (bisher noch nie befriedigend präzisierte) Begriff der Konstruktivität wenn überhaupt, so jedenfalls auf verschiedene Arten und insbesondere *graduell verschieden präzisierbar* sein dürfte. Sogar in den anschaulichsten Teilen der Geometrie gibt es keine Vorstellung und kein Wort, das zwangsläufig eine bestimmte Präzisierung fordern würde und nicht mehrerer verschiedener Präzisierungen fähig wäre, und ebenso, nur sicher in noch höherem Maß, steht es zweifellos mit dem verschwommenen Worte „Konstruktivität" das man, wenn überhaupt, so sicherlich mehr oder weniger weit fassen kann. *Zu jedem dieser unterschiedlichen Konstruktivitätsbegriffe ließe sich eine zugehörige deduktive Mathematik entwickeln.* Es ist beispielsweise vielleicht möglich, einen so strengen Konstruktivitätsbegriff zu definieren, dass nur endliche Mengen konstruktiv beherrschbar werden, oder einen

2.2 Karl Menger: "From the Criterion of Meaning to the Principle of Tolerance"

It was these arguments put forth by Menger that shook Carnap's previously held convictions to their foundation. Carnap's change of heart had far-reaching consequences, for in 1934 he postulated in his book on *the Logical Syntax of Language*.

> ... the *principle of tolerance: we do not wish to establish prohibitions but to arrive at conclusions*. Some of the previous prohibitions have the historic legacy of having made people keenly aware of important differences. Yet such prohibitions can be replaced by a definitional differentiation. In some cases, this happens by examining language forms of different types side by side (similar to the systems of Euclidean and non-Euclidean geometries) such as a definite language and an indefinite language or a language that allows the law of the excluded middle and a language that does not. Occasionally, a prohibition can be substituted by considering the intended difference within a particular form of language in terms of a suitable classification of expressions and investigating the different types. Thus [Carnap's language[150]] I, for example, differentiates between descriptive and logical predicates, while Wittgenstein and Kaufmann reject logical or arithmetic properties; [Carnap's language] II differentiates between definite and indefinite predicates and establishes their various properties; moreover, we differentiate in II between limitedly universal sentences and analytical unlimitedly universal sentences, while Wittgenstein, Kaufmann and

etwas *schwächeren*, der *abzählbare* Mengen umfasst, oder einen noch schwächeren, demzufolge *analytische* Mengen zulässig werden, oder einen sehr allgemeinen, der *beliebige* Mengen reeller Zahlen in die Mathematik einführt. Die Forderung nach Widerspruchsfreiheit kann in diesem Sinne als die weitestmögliche Konstruktivitätsforderung aufgefasst werden.

Was nun die bisherigen intuitionistischen Versuche taten, war dass jeder von ihnen sich dogmatisch auf einen bestimmten (meist, wie gesagt, gar nicht klar umschriebenen) Konstruktivitätsbegriff *festlegte* und die zugehörigen Entwicklungen als sinnvoll, die weitergehenden als sinnlos bezeichnete. Nach Ansicht des Verfassers dieses Aufsatzes hat aber eine derartige Aussage *nicht den mindesten Erkenntnisgehalt*. Denn es handelt sich in der Mathematik und Logik nicht darum, welche Axiome und Schlussprinzipien man *annimmt*, sondern darum, was man aus ihnen bzw. mit ihrer Hilfe *herleitet*. Ob der Mathematiker A das Auswahlaxiom als „zulässig" erklärt, an dasselbe „glaubt" und es anwendet und ob der Mathematiker B es als „unkonstruktiv" oder „weil er keinen Sinn damit verbinden kann" ablehnt, – diese Tatsachen sind für die Biographie der Mathematiker A und B von Interesse, eventuell für die Geschichte, keinesfalls aber für die Mathematik und Logik. Ihnen handelt es sich lediglich darum, was aus dem Auswahlaxiom folgt. Es handelt sich ihnen darum, in welche Aussagen sich gewisse Aussagen mit Hilfe gewisser Transformationsregeln transformieren lassen. Begründet man die Annahme oder Ablehnung von Aussagen oder Transformationsvorschriften durch einen Appell an die Intuition, so wird damit letzten Endes überhaupt nichts gesagt als leere Worte", [384] Menger, Intuitionismus (1930), pp. 323–325

[150] For information on Carnap's languages I and II, see footnote 155 in Chap. 3

Schlick want to exclude the laws of the third type (laws of nature) from language because they cannot be completely verified. *There is no moral in logic.* Each person can construct his logic, i.e. his form of language, as he wishes. Yet he must indicate clearly, if he wants a discussion with us, how he wants to do it – provide syntactical rules instead of philosophical arguments.[151]

Unfortunately, Carnap completely failed to mention in his *Intellectual Autobiography*[152] that he had incorporated "his" principle of tolerance into his *Logical Syntax of Language* on the basis of Menger's ideas from 1927 to 1929.[153]

---

[151] „... das *Toleranzprinzip: wir wollen nicht Verbote aufstellen, sondern Festsetzungen treffen*. Einige der bisherigen Verbote haben das historische Verdienst, dass sie auf wichtige Unterschiede nachdrücklich aufmerksam gemacht haben. Aber solche Verbote können durch eine definitorische Unterscheidung ersetzt werden. In manchen Fällen geschieht das dadurch, dass Sprachformen verschiedener Arten nebeneinander untersucht werden (analog den Systemen euklidischer und nichteuklidischer Geometrie), z. B. eine definite Sprache und eine indefinite Sprache, eine Sprache ohne und eine Sprache mit Satz vom ausgeschlossenen Dritten. Zuweilen ist ein Verbot dadurch zu ersetzen, dass der gemeinte Unterschied innerhalb einer bestimmten Sprachform durch eine geeignete Einteilung der Ausdrücke und Untersuchung der verschiedenen Arten berücksichtigt wird. So werden z. B. in [Carnaps] Sprache I deskriptive und logische Prädikate unterschieden, während Wittgenstein und Kaufmann logische oder arithmetische Eigenschaften ablehnen; in [Carnaps Sprache] II werden definite und indefinite Prädikate unterschieden und ihre verschiedenen Eigenschaften festgestellt; ferner unterscheiden wir in II beschränkt allgemeine Sätze, analytische unbeschränkt allgemeine Sätze, während Wittgenstein, Kaufmann und Schlick die Sätze der dritten Art (Naturgesetze) aus der Sprache ausschalten wollen, weil sie nicht vollständig verifizierbar sind. *In der Logik gibt es keine Moral.* Jeder mag seine Logik, d. h. seine Sprachform, aufbauen wie er will. Nur muss er, wenn er mit uns diskutieren will, deutlich angeben, wie er es machen will, syntaktische Bestimmungen geben anstatt philosophische Erörterungen." Only after this, and in smaller print, did Carnap then explain: "The tolerant position intended here, in terms of specific mathematical calculi, should be familiar to most mathematicians even though they may not usually say them explicitly. In the conflict over the logical basis of mathematics, this position has been supported with particular enthusiasm (and apparently before anyone else) by [384] Menger, Intuitionism (1930), p. 324f. Menger indicates that the concept of constructivity which intuitionism absolutizes can be understood narrowly or broadly – How important it is in the explanation of philosophical pseudo-problems to relate the tolerant position to the form of the language as a whole will become clear later on (cf. § 78)." [114] Carnap, Unity (1934), p. 44f

[152] [116] Carnap, Unity (1934)

[153] Carnap did not remember much about Menger's contributions at all; he was also no longer aware of the fact that Menger had worked for Brouwer in Amsterdam for two and a half years instead of just one and that he had qualified as a professor there

2.2 Karl Menger: "From the Criterion of Meaning to the Principle of Tolerance"

I was powerfully drawn to the constructionist approach. In my book *Logical Syntax*, I constructed a language called "Language I", which satisfies the essential requirements of constructionism and seemed to be quite superior to Brouwer's language form. In this book, however, I also constructed another language, which was comprehensive enough for the formulation of classic mathematics. According to my principle of tolerance, I explained that however important it might be to make distinctions between constructivist and non-constructivist definitions and proofs, it is nevertheless advisable not to prohibit particular approaches, but instead to examine all practicable approaches. It is true that certain approaches, such as those allowed by constructivism or intuitionism, are more admissible than others. Therefore, it would be well to use these methods to the greatest extent possible. However, there are other processes and methods that are less certain because they cannot provide any proof of their freedom of contradiction but that are nevertheless practically indispensable in physics. This is why it does not seem particularly reasonable to me to prohibit these approaches, at least as long as no contradictions have been discovered.[154]

Instead, Carnap wrote in 1937 in the English edition of his book *The Logical Syntax of Language*:

> ... the earlier position of the Vienna Circle, which was in essentials that of Wittgenstein. On that view it was a question of ‚*the* language' in an absolute sense; it was thought possible to reject both concepts and sentences if they did not fit into *the* language.[155,156]

---

[154] „Ich fühlte mich stark zum konstruktivistischen Ansatz hingezogen. In meinem Buch *Logische Syntax* konstruierte ich eine «Sprache I» genannte Sprache, welche die wesentlichen Erfordernisse des Konstruktivismus erfüllt und der Brouwerschen Sprachform einiges voraus zu haben schien. In diesem Buch konstruierte ich aber noch eine andere Sprache, die umfassend genug zur Formulierung der klassischen Mathematik war. Gemäß meinem Toleranzprinzip erklärte ich, so wichtig es auch sein möge, Unterschiede zwischen konstruktivistischen und nichtkonstruktivistischen Definitionen und Beweisen zu machen, so sei es doch ratsam, bestimmte Verfahrensweisen nicht zu verbieten, sondern alle praktisch brauchbaren Verfahrensweisen zu prüfen. Es stimmt, dass gewisse Verfahrensweisen, etwa solche, die der Konstruktivismus oder Intuitionismus zulässt, zuverlässiger als andere sind. Deshalb tut man gut daran, diese Verfahrensweisen so weit es geht zu verwenden. Es gibt jedoch andere Verfahren und Methoden, die nicht so sicher sind, weil sie keinen Beweis für ihre Widerspruchsfreiheit erbringen können, die aber dennoch in der Physik praktisch unentbehrlich sind. Darum scheint es mir nicht sonderlich vernünftig, diese Verfahrensweisen zu verbieten, jedenfalls solange keine Widersprüche entdeckt worden sind." [117] Carnap, Weg (1963), p. 76

[155] According to McGuinness, the German edition differs greatly. [367] McGuinness, Hahn (1988), p. 142, footnote 1

[156] *Testability and Meaning*, published the same year, states explicitly: "But I was wrong in thinking that the language I dealt with was the language, i. e. the only

That there was resentment among the members of the Vienna Circle is confirmed by the following report by Gustav Bergmann:

> However, the Boltzmanngasse Collective had already reached his apex, I think, in 1927/28, kept its momentum for a few more years, was already starting to show signs of serious divisions and, as a result, decline in 1931/32. When Carnap departed, those representing a consistently physical scientific theory, the "logical construct of the world" and the associated general radical-rational view lost the personality who could have served as the epicenter of an academic colloquium; but the Wittgenstein esoterics, with their aversion to "talking things to death" and to discussion with people who lacked an "eye for the essential", the necessary intuition, and under whose influence Schlick continued to fall, were not displeased to see the disintegration of the old circle, from whose traditional tenor they themselves had become more and more estranged.[157]

2.3 Probabilistic or Statistical Metrics and *Ensembles Flous*

Menger spent 1930 and 1931 as a visiting lecturer at Harvard University and at the Rice Institute, and this sojourn in the United States would serve him well six years later. He continued working with the concept of constructivity

---

legitimate language – as Wittgenstein, Schlick and Lewis likewise seem to think concerning the forms of language accepted by them." [115] Carnap, Testability (1937), p. 20. These passages are in contrast to the statements in his *Intellectual Autobiography* ([117] Carnap, Weg (1963)), although he didn't write that until 25 years later. Carnap stresses there that the idea of the freedom to choose a language from a variation of languages had already been one of his principle ideas in the days before Vienna. The development of the idea of a plurality of languages and logics is one of the facts that may have been completely wrong in Carnap's memory in his later years. Clarence Irving Lewis (1883–1964), American logician and philosopher

[157] „Das Kollektiv Boltzmanngasse hatte aber, wie ich glaube, 1927/28 seinen Höhepunkt bereits erreicht, hielt sich dann noch einige Jahre in Schwung, zeigte 1931/32 bereits deutliche Spaltungs-, und, in deren Gefolge Verfallserscheinungen. Die Vertreter einer konsequent physikalistischen Wissenschaftstheorie, des «logischen Aufbaus der Welt» und der damit verbundenen radikalrationalen Allgemeinhaltung hatten nämlich mit dem Abgang Carnaps die Persönlichkeit, die geeignet gewesen wäre, den Mittelpunkt eines akademischen Colloquiums zu bilden, verloren; die Wittgensteinschen Esoteriker aber, unter deren Einfluß Schlick immer mehr geriet, sahen in ihrer Abneigung gegen das ‹Zerreden› und gegen die Diskussionen mit Leuten, denen der ‹Blick für das Wesentliche›, die nötige Intuition, abging, den Zerfall des alten Kreises, dessen traditioneller Grundhaltung sie selbst sich immer mehr entfremdet hatten, nicht ungern." Gustav Bergmann, Erinnerungen an den Wiener Kreis (Memories of the Vienna Circle). Letter to Otto Neurath. In [519] Stadler, Kontinuität (1988), pp. 171–180

there, as is shown by the communications he sent back to Hans Hahn, who, as a corresponding member of the Academy of Sciences in Vienna, submitted them to the meetings of the mathematical-natural sciences class that took place December 4 to 18, 1930.[158]

Menger did not participate in any more meetings of the Vienna Circle after Moritz Schlick's violent death on June 22, 1936.[159] The German-Austrian Agreement between Adolf Hitler and Kurt von Schuschnigg came on June 11, 1936, Rudolf Carnap emigrated to the USA the same year and Menger also managed to anticipate the impending consequences of the March 1938 *Anschluss* by the German Reich. He described his deteriorating situation in Vienna for friends and acquaintances at the International Congress of Mathematicians in Oslo in 1936. A short time later he was offered a professorship at the University of Notre Dame in Indiana, an offer he accepted.[160] He and his family arrived in South Bend in 1937, and he was naturalized the same year.[161] He held the position of mathematics professor at Notre Dame from 1937 to 1946. There he resumed his *Reports of a Mathematical Colloquium, 2nd Series*, and he later also published the *Notre Dame Mathematical Lectures*.[162] Menger moved to Chicago in 1946, where Rudolf Carnap was also

---

[158] *Über die sogenannte Konstruktivität bei arithmetischen Definitionen (On So-Called Constructivity in Arithmetic Definitions)* by Karl Menger, presently in Cambridge (Mass., USA). Gazette of the Academy of Sciences in Vienna. *Mathematical-Natural Sciences Class*, 1930, No. 25, p. 257f. and *Über den Konstruktivitätsbegriff (On the Concept of Constructivity), Second Communication*. By Karl Menger, presently at Harvard University, Cambridge (Mass.), Gazette of the Academy of Sciences in Vienna. *Mathematical-Natural Sciences Class*, 1931, No. 1, pp. 7–9

[159] It remains unclear to this day whether there was a National Socialist connection to this murder

[160] Norbert Wiener recalled in his autobiography: "Menger wrote to me from Vienna to ask whether he might find refuge in the United States before all hell broke loose in Austria, and we procured an invitation from Notre Dame University in Indiana; later he went to the Illinois Institute of Technology." [567] Wiener, Leben (1962), p. 142

[161] After the 1938 *Anschluss*, he abandoned his professorship in Vienna, and it would not be until the 1960s that he returned to Vienna. Stadler has written that an attempt to reinstall Menger as a professor in Vienna failed miserably ([521] Stadler, Vernunft (1988), p. 121) while ex-Nazis enjoyed "paradigmatic careers". ([513] Sigmund-Schultze, Mathematiker (1998), p. 279.) Menger himself claimed that Vienna University never made any such attempt: "No! And it would have been no use. But they never even tried at all," he added. [223] Haider, Menger (1978), p. 5

[162] This small Catholic institution did not become important in the mathematics community until the arrival of the immigrants! In his memoirs, Menger stresses his efforts around 1937 "to put Notre Dame on the mathematical map". See [202] [201] Golland et al., Menger (1994), p. 216, quoted from: [513] Sigmund-Schultze, Mathematiker (1998), p. 234

living and working. He became a professor at Chicago's Illinois Institute of Technology[163] and there he remained until his retirement in 1971.[164] He died in Chicago in 1985. In his obituary, Seymour Kass[165] gave an account of Menger's situation following his emigration:

> European intellectuals who fled Europe for refuge in America were sometimes uncomfortable with American ways and unaccustomed to teaching elementary courses. Though central European in dress, manner, and style, Menger felt at home in America and enjoyed teaching undergraduates, believing that when properly done, it stimulated research. During the 1960s he lectured to high school students on the subject "What is x?".[166]

The mathematician and philosopher Karl Menger had been interested in the fundamental problems of mathematics since his student days in Vienna. In the Vienna Circle, the discussions revolved around the relationships between logic, language, mathematics and reality. Menger had been taught to think clearly and precisely, and he couldn't help criticizing the American textbooks for things like not adequately explaining the difference between a function f and its value $f(x)$ at a point x. He wrote a number of short articles on this subject in which he expressed his criticism[167], and in 1952 he produced a textbook of his own: *Calculus – A Modern Approach*.[168]

> Written in characteristically vigorous style, it was a radical revision of textbooks of the period, scrapping some traditional notation. It received a lengthy, thoughtful, and cautiously favourable review in the Monthly by H. E. Bray but was never accorded serious attention. He sent a copy to Einstein, who replied that he liked it and recognized the need for some clarity in notation, but advised against attempting too much "housecleaning". That his book was ignored saddened Menger's later years. When Menger addressed the foundations of dimension theory, topology, projective n-space, or differential geometry, attention

---

[163] Oddly, Viktor Kraft writes ([297] Kraft, Kreis (1968), p. 179, footnote 1) that Menger had been a math professor at MIT in Massachusetts! This is probably just a mix-up: as Wiener's quote in the above footnote shows, Menger went to the Illinois Institute of Technology (IIT)

[164] In 1951, Menger was a visiting lecturer at the Sorbonne in Paris and in 1961 a guest professor at various European universities. In 1964, he taught as a guest professor at the University of Arizona, the Ford Institute and again in Vienna, and in 1968 at the Middle East Technical University in Ankara

[165] Seymour Kass is an American mathematician and professor at the University of Massachusetts, Boston

[166] [273] Kass, Menger (1996), p. 560

[167] [374] Menger, Calculus (1953), [396] Menger, Variables (1954), [397] Menger, Variables (1955)

[168] [374] Menger, Calculus (1953)

2.3 Probabilistic or Statistical Metrics and *Ensembles Flous*   101

was paid by the best mathematical minds of his generation: Hahn, Brouwer, von Neumann, Gödel. The failure of the calculus endeavour strained his relations with the mathematical community.[169]

The problems that were so important to Menger can be illustrated here using one brief example: Menger regarded a "scientific" variable to be a pair consisting of an act of reading and the read value. If T is the temperature of a gas, for instance, then T is considered a pair consisting of an act of reading the thermometer and the value that is read. Another example of a scientific variable is gas pressure p. For this one, Menger stipulated that only a gas of the same volume would be considered. In that case, one object of observation is the fact that when two readings of gas pressure p are equal, the read values of gas temperature T are also equal and *vice versa*. Physicists were talking about this very same fact when they said that T is a function of p and p is a function of T. However, according to Menger, this concept of function was much different from a "pure" function because, while it was possible to substitute one pure function with another pure function or else a scientific variable, nothing else can be replaced in a scientific variable:

> A pure function f, in general, permits substitutions of scientific variables and of pure functions [...]. Into a scientific variable neither a variable nor a pure function can be substituted. These differences between the two categories would exist even if both were called functions or both were called variables.[170]

Menger stressed the necessity of categorically differentiating between the entities of pure and applied mathematics. "Originally I had planned to eliminate variables from analysis as much as possible."[171] Even Kurt Gödel, whom he visited at Princeton a few times, had a hard time following: "It is as though you wanted to speak only about mankind, never simply about a man."[172] Menger pursued his program and sought new designations to express what he meant. There were numeric variables in numeric theory[173] and there were function variables in analysis.[174] The fact that the numeric variables x and y existed everywhere within analysis, he said, could be traced back to a fundamental gap in the symbolism of functions, which had arisen during the Renaissance: There are no designations for identity, the nth power and constant functions, which are traditionally identified by their values for x "(the functions $x, x^n, 3, a')$".[175] For the sake of uniformity, this designation was expanded to include logarithm, cosine as well as other functions and function

---

[169] [273] Kass, Menger (1996), p. 560f
[170] [374] Menger, Calculus (1953), p. 958
[171] [202] Golland et al., Menger (1994), p. 227
[172] [202] Golland et al., Menger (1994), p. 227
[173] Menger used the term *numeric variables* here for letters that stand for numbers
[174] Menger used the term *function variables* here for letters that stand for functions
[175] [202] Golland et al., Menger (1994), p. 228

variables: "the functions $\log x, \cos x, f(x)$".[176] However, if this gap were filled by designations such as j, j_n, c_3, c_a analog to log, cos and f, then there would be a completely clear and consistent system of notation for analysis and this was absolutely indispensable to the development of an algebra of functions in which identity plays a role comparable to that of 0 and 1 in arithmetic. Menger and Gödel agreed that those so-called "scientific variables" ought to be carefully distinguished from these number and function variables: pressure, volume, temperature, bridged distance, elapsed time - usually represented by the letters p, V, T, s and t, respectively. Gödel called them *variable quantities*; Menger suggested the term *fluents*:

> A fluent is a certain pairing of numbers to objects such as rolling balls or expanding squares. Is a fluent thus a function? ... A fluent has a nonmathematical domain and, therefore, is an extramathematical concept, limited to a special part or aspect of the world, and of interest only to special branches of science.[177]

Menger emphasized that these conceptual differences were to be taken into account when analysis was applied in the sciences. He wrote again about this differentiation between pure and applied mathematics more than ten years later. At a symposium of the American Association for the Advancement of Science organized in 1966 to commemorate the 50th anniversary of Ernst Mach's death, Menger spoke about *Positivistic Geometry*.[178] Ernst Mach 's thinking had made a lasting impression on the Vienna Circle and Menger tied his comments into Mach's statements in the *Principles of Thermodynamics*.[179] In the chapter entitled *The Continuum*, Mach had begun by characterizing the continuum as a system of members that possessed one or more properties A in varying measures and such that an infinite number of members can fit between each pair of members that differ from each other in a finite way with respect to A. These consecutive members then display just one infinitely small difference from each other. Mach determined that "there was no reason to object to this fiction or the random conceptual construction of such a system". Yet the natural scientist does not deal only in pure mathematics. So he has to ask himself whether there is anything in nature that corresponds to such a fiction! At the end of his chapter on the continuum, Menger quotes Ernst Mach:

> Everything that *appears* to be a continuum could probably consist of *discrete* elements if these were only small enough and numerous enough with respect to our smallest practically applied measures. Anywhere we think we will find a continuum, the fact is that we are still employing analogous considerations to the smallest perceivable parts

---

[176] [202] Golland et al., Menger (1994), p. 228
[177] [202] Golland et al., Menger (1994), p. 228
[178] [380] Menger, Geometry (1970)
[179] [344] Mach, Wärmelehre (1923)

2.3 Probabilistic or Statistical Metrics and *Ensembles Flous*

of the system in question and can observe a behavior similar to those of larger ones. Insofar as *experience* has raised no objections, we can maintain the fiction of continuum, which is by no means harmful but is merely *comfortable*. In *this sense*, we also refer to the system of *thermal conditions* as a *continuum*.[180]

Menger then transitioned to Henri Poincaré, the other great scientist who had substantially influenced the thinking of the Vienna Circle and who had analyzed more aggressively than anyone else the problems with the concept of the physical continuum. In all of his books[181], Poincaré had characterized the physical continuum as follows:

$$A = B, \qquad B = C, \qquad A \neq C.$$

In doing so, he was trying to symbolize the fact that when an element from the physical continuum can be differentiated from two others, these two other elements can nevertheless easily differ from each other.

In Poincaré's view, the mathematical continuum was constructed with the primary intention of overcoming the above formula, which he called "contradictory or at least inconsistent". Only in the mathematical continuum did the equations $A = B$ and $B = C$ imply equation $A = C$. In the observable physical continuum, on the other hand, where "equal" meant "indistinguishable", the equations $A = B$ and $B = C$ did not imply the equation $A = C$ at all. The raw result of this finding must rather be expressed by the following relation and must be considered a formula for the physical continuum:

$$A = B, \qquad B = C, \qquad A < C.$$

In the 1940s and '50s, Menger had expressed a number of thoughts on the concept of physical equality based on this distinction between the mathematical and physical continuum. With regard to Poincaré's above-mentioned argument, he described the difference as a non-transitive relation, and his other examinations led him to an even more radical conclusion:

---

[180] „Alles, was als Continuum *erscheint*, könnte wohl aus *diskreten* Elementen bestehen, wenn dieselben nur unsern kleinsten praktisch angewendeten Maassen gegenüber hinreichend klein, beziehungsweise hinreichend zahlreich wären.
Überall, wo wir ein Continuum vorzufinden glauben, heisst das nur, dass wir an den kleinsten wahrnehmbaren Theilen des betreffenden Systems noch analoge Betrachtungen anstellen und ein analoges Verhalten bemerken können wie an grösseren. So weit die *Erfahrung* noch keine Einsprache erhoben hat, können wir die in keiner Weise schädliche, sondern nur *bequeme* Fiktion des Continuums aufrecht halten. In *diesem Sinne* nennen wir auch das System der *Wärmezustände* ein *Continuum*." [344] Mach, Wärmelehre (1923), p. 76. Menger cited this work in English without providing any further source reference

[181] [447] Poincaré, Science (1902)

A closer examination of the physical continuum suggests that in describing our observations we should sacrifice more than the transitivity of equality. We should give up the assumption that equality is a relation.[182]

Fig. 2.11. Computer-generated likeness of Karl Menger

A simple experiment showed, for example, that irritating two points A and B on the skin at the same time sometimes resulted in just *one* sensation, but sometimes in two sensations, as well. The fact that two different sensations could occasionally be artificially equated with one another occurred merely because we have faith in the majority of the impressions we experience and that we perform averaging processes.

We can obtain a realistic description of the equality of two elements when we associate a number with A and B, namely the *probability* that A and

---

[182] [391] Menger, Relations (1951), p. 178

2.3 Probabilistic or Statistical Metrics and *Ensembles Flous*

B are indistinguishable. In applications, this number can be represented by relative frequencies of the instances in which A and B cannot be distinguished from one another. This idea, Menger further argued, canceled out Poincaré's paradox:

> For if it is only very likely that A and B are equal, and very likely that B and C are equal, why should it not be less likely that A and C are equal? In fact, why should the equality of A and C not be less likely than the inequality of A and C?[183]

For a probability $E(a,b)$ that a and b are equal, Menger then postulated the following properties[184]:

(1) $E(a,a) = 1$ for all a;
(2) $E(a,b) = E(b,a)$,   for all a and b;
(3) $E(a,b) \cdot E(b,c) \leq E(a,c)$,   for all a, b, c.

Menger proposed calling a and b *certainly-equal* if $E(a,b) = 1$. This led to an equality relation, since all elements that are certainly-equal to a can be combined into an *equality set* A and each pair of these sets is either disjoint or identical. He defined $E(A, B)$ as the probability that every element of A is equal to every element of B. This number is not dependent upon the selection of the two elements in each case.

Menger had already been involved with the theory of metric spaces in Vienna, that is, spaces for whose elements a distance function is defined that assigns each pair (p, q) of points the distance $d(p, q)$ between them in the form of a real number.[185] Menger thus used probabilities to define a distance function of this type, as is seen in this following equation:

$$- \log E(A, B) = d(A, B),$$

which fulfills the following properties:

($1_{a'}$) $d(A, A) = 0$;
($1_{b'}$) $d(A, B) \geq 0$;
($1_{c'}$) $d(A, B) \neq 0$, when $A \neq B$;
($2'$) $d(A, B) = d(B, A)$ for all a and b;
($3'$) $d(A, B) + d(B, C) \neq d(A, C)$.

---

[183] [391] Menger, Relations (1951), p. 178
[184] Menger commented that conditions (1) and (2) correspond to reflexivity and symmetry for the equality relation, while condition (3) "expressed a minimum of transitivity". [391] Menger, Relations (1951), p. 178
[185] The first abstract formulation of distance in mathematics was created by the French mathematician Maurice René Fréchet (1878–1973) in the year 1906. In 1914, the German mathematician Felix Hausdorff (1868–1942) then introduced the name "metric space"

These are Maurice Fréchet's postulates for the distance in a metric space.[186]

So if disjoint sets $A, B, \ldots$ form a metric space with the distance function $d(A, B)$ and we define the expression

$$E(A, B) = E(a, b) = e^{-d(A,B)},$$

for all elements a from A and b from B, then $E(a, b)$ satisfies postulates (1), (2), (3) of a probability of equality.

> The systems of probabilities of equality in a set S are thus identical with the systems of negative antilogarithms of the distance for the various possible metrizations of S.[187]

With these studies, Menger had tied into his research in topology and geometry soon after immigrating to the US. He had worked with so-called *convex metric spaces*[188] back in Vienna and had also been engaged in various generalizations. In a short article for the *Proceedings of the National Academy of Sciences* in December 1942, he published a generalization of this theory of metric spaces based on probability theory.[189]

He called a set S a *statistical metric* if a probability function $\Pi(x; p, q)$ that is linked with two of its elements p and q satisfies the following conditions[190]:

1. $\Pi(0; p, p) = 1$,   (The probability is 1 that the distance between p and q is 0.)
2. If $p \neq q$, then $\Pi(0; p, q) < 1$,
3. $\Pi(x; p, q) = \Pi(x; q, p)$,
4. $T[\Pi(x; p, q), \Pi(y; q, r)] \leq \Pi(x + y; p, r)$. (This is a triangle inequality.)

Additionally, $T[\alpha, \beta]$ is a function defined for $0 \leq \alpha \leq 1$ and $0 \leq \beta \leq 1$ such that:

(a) $0 \leq T[\alpha, \beta] \leq 1$.
(b) T is non-decreasing in every variable.
(c) $T[\alpha, \beta] = T[\beta, \alpha]$.
(d) $T[1, 1] = 1$.
(e) If $\alpha > 0$, then $T[\alpha, 1] > 0$.

---

[186] In particular, (3') is a triangle inequality

[187] Menger gave several examples: If S is a straight line, then $E(a, b) = e^{-|b-a|}$ is the probability of equality corresponding to the Euclidean metrization of S. Another example is $E(a, b) = e^{\sqrt{-|b-a|}}$, while the smooth function $e^{-(b-a)^2}$ is not considered, since the associated distance $d(a, b) = (b - a)^2$ does not satisfy the triangle inequality. [391] Menger, Relations (1951), p. 179

[188] "A partial set of a metric space is called convex if it contains an intermediate point for each two of its points" is Menger's definition in [379] Menger, Geometrie (1931), p. 377

[189] [392] Menger, Statistical (1942)

[190] Menger defined a *probability function* here as a non-decreasing function that is defined for all non-negative values of x, is continuous on the right-hand side, assumes the values between 0 and 1 and converges toward 1 when x increases beyond all limits

2.3 Probabilistic or Statistical Metrics and *Ensembles Flous*

Menger called $\Pi(x; p, q)$ the *distance function* of p and q and interpreted it as the *probability* that the distance between points p and $q \leq x$. The "triangle inequality" implies that the following is true for all points q and all numbers x between 0 and z:

$$\Pi(z; p, r) \geq \text{Max } T \, [\Pi(x; p, q), \Pi(z - x; q, r)].$$

Menger called the function T a *triangular norm* or *t-norm*.

In his address on the 50th anniversary of Ernst Mach's death, Menger proposed applying his findings on statistical metrics to positivistic geometry, in particular for the problem of the "physical continuum". In his *Dernières Pensées*, Poincaré had suggested defining the physical continuum as a system S of elements, each pair of said elements being linked together by a chain of elements from S such that every element of the chain is indistinguishable from the following element.

Menger remembered an example of a definition of distances using such chains from Erwin Schrödinger's *Colorimetry* or *Color Metrics* from the 1920s in Vienna (Fig. 2.12).[191] At the time, Schrödinger set the distance between two colors C and C' equal to 1 when they are "clearly distinguishable"[192] and equal to a natural number n when a series of elements $C_0 = C, C_1, C_2, ..., C_{n-1}, C_n = C'$ exists, such that each pair of consecutive elements C_{i-1} and C_i is "clearly distinguishable" and there is no shorter chain of this type.

Schrödinger represented each color as a three-dimensional vector. Every C is then the center of an ellipsoid having internal points that cannot be distinguished from C, while the points on its surface are clearly distinguishable from C. Such "well-defined" neighborhoods (environments) did not exist in the color cone, in Menger's view, and so distance 1 first had to be defined by means of probabilities (fig. 2.13).

Menger saw the most difficult problem for microgeometry in the individual identification of elements of the space. He published a suggestion for it in a French text in 1951, which he penned during his guest residency at the Sorbonne. In addition to studies of well-defined sets, he called for a theory to

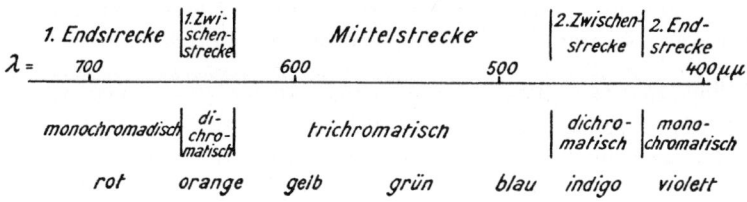

Fig. 2.12. Color Spectrum

---

[191] [497] Schrödinger, Farbenmetrik (1920)
[192] Menger wrote "barely indistinguishable"

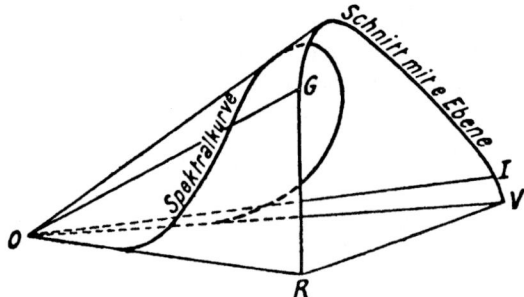

Fig. 2.13. Schrödinger's Color Cone

be developed in which the relationship between elements and sets is replaced by *the probability that an element belongs to a set*:

> Une relation monaire au sens classique est un sous-ensemble $\mathcal{F}$ de l'univers. Au sens probabiliste, c'est une fonction $\Pi_{\mathcal{F}}$ définie pour tout $x \in U$. Nous appellerons cette fonction même un ensemble flou et nous interpréterons $\Pi_{\mathcal{F}(x)}$ comme la probabilité que x appartienne à cet ensemble.[193]

In English, Menger later replaced the term *ensemble flou* with the expression *hazy set* and to elucidate the contrast he referred to conventional sets as *rigid sets*.

Menger clearly recognized that the difficulty of using his term *ensemble flou* in microgeometry, a term which was after all defined by means of probabilities and thus represented a *probabilistic set*, lay in the fact that the individual elements had to be identified. Since this was simply not possible, though, he suggested combining the *ensemble flou* with a geometry of "lumps", for lumps were easier to identify and to differentiate from one another than points. Lumps could assume a position between *indistinguishability* and *apartness*, which would be the condition of overlapping. It was irrelevant whether the primitive (i.e. undefined) concepts of a theory were characterized as *points* and *probably indistinguishable* or as *lumps* and *probably overlapping*. Of course, all of this depended on the conditions that these simple concepts had to fulfill, but the properties stipulated in the two cases could not be identical.

> I believe that the ultimate solution of problems of microgeometry may well lie in a probabilistic theory of hazy lumps. The essential feature of this theory would be that lumps would not be point sets; nor would they reflect circumscribed figures such as ellipsoids. They would rather be in mutual probabilistic relations of overlapping and apartness, from which a metric would have to be developed.[194]

---

[193] [378] Menger, Flous (1951), p. 2002
[194] [380] Menger, Geometry (1970), p. 233

2.3 Probabilistic or Statistical Metrics and *Ensembles Flous*

After Menger had also mentioned the term *ensemble flou* in his speech for the symposium in honor of the 50th anniversary of the death of Ernst Mach, he established a link in a subordinate clause to – as he saw it – a very similar development:

> In a slightly different terminology, this idea was recently expressed by Bellman, Kalaba and Zadeh under the name fuzzy set. (These authors speak of the degree rather than the probability of an element belonging to a set).[195]

It is obvious that Menger was mentioning the commonalities and differences between his *ensembles flous* and Zadeh's *fuzzy sets* by name. Both scientists had been confronted with the problem of representing distances having points in abstract spaces to each other. Menger sought generalizations of the existing theory of metric spaces that could be applied in the sciences. As a mathematician, he started with concepts of probability theory to represent unknown quantities. In the 1950s, Lotfi Zadeh was considering an application problem: How can signals received by a communication system be identified and separated? To do so, he represented these signals as abstract vectors in a vector space and then likewise tried to find a definition of the unknown distance in that kind of space. Zadeh then also considered various metrics.[196] Menger developed a theory for the probability of equality and defined for this statistical metric the term *triangular norm (T-norm)*, and later also the *s-norm*[197], which today are integral elements of the tools in fuzzy theory. Lotfi Zadeh had the fundamental idea in the development of fuzzy theory when he attempted to characterize the relationship between elements and sets without using concepts from probability theory. This story is related in Chap. 5. Before that, Chaps. 3 and 4 will deal with the developments that led to it: the advent and evolution of General System Theory, cybernetics (Chap. 3), automata theory and artificial neuronal networks (Chap. 4) and a further system theory that originated in communication technology.

The subjects revolving around the relationships between language, logic and reality, subjects only briefly sketched out here but discussed extensively in the Vienna Circle, were important to the later establishment of Fuzzy Set Theory in two respects. Thanks to the lasting principle of logical tolerance, which was introduced first by Menger and in turn by Carnap, those logical systems that were not based on classic two-valued logic were not precluded from being studied scientifically. Although the Lemberg-Warsaw school of logic had already developed multi-valued systems of logic at that time, this work – as

---

[195] [380] Menger, Geometry (1970), p. 232

[196] See Chap. 1

[197] Today the t-norms and s-norms (often also called t-conorms) are essential elements of Fuzzy Set Theory. They are used to represent the links between fuzzy sets in a general form. See e.g. [203] Gottwald, Logik (1999), pp. 185–204, [283] Klement et al., Logic (1999), pp. 205–225 or [145] Driankov et al., Introduction (1993), pp. 55–61

mentioned earlier – did not become known to a larger circle outside of Poland until Tarski was given the opportunity to speak about it in the Vienna Circle. It was Tarski's work on metamathematics that prompted Carnap to consider that science does not require a single language and a single logic, but rather that a pluralism of such systems offered much greater implications for science as a whole. The principle of logical tolerance was a basic pre-condition for deviant theories – and Fuzzy Set Theory was considered a "deviant theory" for a long time – to earn their right to belong to the system of sciences.[198]

The second reason to give the Vienna Circle such prominent recognition in the story of the creation of Zadeh's Fuzzy Set Theory is its pursuit of unified science. This program, which later became known as *unity of science*, was a milestone in the history of General System Theory, which will be covered in the next chapter. The influences of General System Theory on Zadeh's Fuzzy Set Theory will be examined in Chaps. 4 and 5.

---

[198] See [221] Haack, Deviant (1974)

3
General Systems Theory and Cybernetics

In February of 1930, a "study group for scientific cooperation" led by Rudolf Carnap was founded under the auspices of the Ernst Mach Society and organized in the Vienna Chamber of Labor. Members of this group included biologists Ludwig von Bertalanffy[1] and Wilhelm Marinelli[2], sociologist Hans Zeisel[3], economists Karl Polanyi[4] and Richard Strigl[5] as well as the

---

[1] Carl Ludwig von Bertalanffy (1901–1972) transferred from Innsbruck to the University of Vienna after one year of study. There he concluded his studies in philosophy and art history and turned his attention to the natural sciences, particularly biology. He earned his doctorate in 1926 with a paper about the physicist and philosopher Gustav Theodor Fechner (1801–1887) and qualified as a professor in 1934 while working with the Vienna Circle philosophers Robert Reiniger (1869–1955) and Moritz Schlick (1882–1936) and with Jan Vesluys (1873–1939) of the Netherlands, who became a professor of zoology in Vienna in 1925. The content of this paper became the first of two volumes which appeared as *Theoretical Biology* [68] Bertalanffy, Theorie (1928), the publication of which was encouraged by Vienna botanist and university professor Richard von Wettstein (1863–1931). Bertalanffy founded a "theoretical biology", which was a popular university subject in the US, and he developed "organismic biology" synthesized from mechanistic and vitalistic ideas into a "General System Studies", which was first published in 1949. He then expanded this to the "General System Theory", the principles of which he then expanded from biology to also include psychology, sociology and anthropology. Bertalanffy died in Buffalo, New York in 1972

[2] Wilhelm Marinelli (1894–1973), zoologist and professor in Vienna, president of the Society for the Prevention of Cruelty to Animals, the Institute for Science and Art and the Ludwig Boltzmann Society

[3] Hans Zeisel (1938–1992), American social scientist and jurist of Austrian origin. Between 1930 and 1932, he worked alongside the Austrian-British social psychologist and sociologist Marie Jahoda (1907–2001) and her husband (until 1934), the Austrian-American sociologist Paul Felix Lazarsfeld (1901–1976), on the "Marienthal study" by the "Economic Psychology Research Center" in Vienna, of which he was a co-founder

[4] Karl Polanyi (1886–1964), Hungarian-born economist and economic historian, lived most of his life in exile, first in Vienna, then in England and finally in the USA. In 1944, he published his manuscript *The Great Transformation*

[5] Richard Strigl (1891–1942), Austrian economist

psychologists and psychoanalysts Egon Brunswik[6], Else Frenkel-Brunswik[7] and Wilhelm Reich.[8] The study group[9] was intended "to bring about a harmonization of the special branches of science and a clarification of their place within the framework of science as a whole ... by means of reports and discussions, particularly about the newer methods, problems, concept formations in the various specialist fields".[10]

The Vienna Circle had shaped the concept of unified science, according to which there can only be pragmatic reasons for the separation of scientific disciplines, for they believed that unified scientific language and logic made it possible for everything to be described using this language. Carnap elucidated this view in 1937, three years before he published "his" principle of logical tolerance in a paper on *The Physical Language as a Universal Language of Science*. Here he contrasted the "generally widespread view" that the sciences "differ fundamentally in terms of their objects, their sources of perception, their methods" with the belief held by the "Wittgenstein faction" in the Vienna Circle:

> By contrast, the view shall be expressed here that science forms one unit: All sentences can be expressed in one language, all facts are of one type, recognizable by one method.[11]

When the Vienna Circle disintegrated after Austria's annexation into the German Reich, it did not spell the end of the unified science project. Following difficult negotiations[12], *Erkenntnis* was continued after 1937 as *The Journal of Unified Science* primarily by Carnap, who had moved from Prague to Chicago,

---

[6] Egon Brunswik (1903–1956), Hungarian-American psychologist and philosopher, first studied mathematics and physics at the Technical University in Vienna, but had also enrolled simultaneously at the university (under a slightly different name – double majors were still forbidden at the time) to study psychology and philosophy. Brunswik wrote his dissertation *Structure and Physics* under Karl Bühler. After the *"Anschluss"* by Germany, he went to the University of California, Berkeley. He committed suicide in 1956

[7] Else Frenkel-Brunswik (1908–1958), Austrian-American psychologist, studied at the university in Vienna and earned her doctorate under Karl Bühler with a paper entitled *Atomism and Mechanism in Association Psychology*

[8] Wilhelm Reich (1897–1957), Austrian psychoanalyst, later professor of medical psychology in the US (New York)

[9] [520] Stadler, Kreis (1997), p. 382, quoted from *Erkenntnis*, 1, 1930–31, p. 79. See also [225] Haller, Neopositivmus (1993), p. 74

[10] [520] Stadler, Kreis (1997), p. 382, quoted from *Erkenntnis*, 1, 1930–31, p. 79. See also [225] Haller, Neopositivmus (1993), p. 74

[11] „Demgegenüber soll hier die Ansicht vertreten werden, dass die Wissenschaft eine Einheit bildet: Alle Sätze sind in einer Sprache ausdrückbar, alle Sachverhalte sind von einer Art, nach einer Methode erkennbar."[113] Carnap, Sprache (1931), p. 432

[12] See [235] Hegselmann et al., Erkenntnis (1991)

and Hans Reichenbach, who had by that time emigrated to Turkey[13], and the series *Schriften zur wissenschaftlichen Weltauffassung und Einheitswissenschaft*, which had heretofore been published at the Springer Verlag in Berlin by Schlick and Frank and then by Neurath, Frank and Charles W. Morris, now appeared as the *International Encyclopedia of Unified Science*.[14]

3.1 General Systems Theory

Carnap's "turn" to the principle of tolerance apparently came at the same time as the publication of his book *The Unity of Science*, which was translated into English by Max Black[15], since it featured both his physical thesis, namely that all meaningful sentences can be translated into physical sentences, and the thesis that all events could be explained with physical laws. Even the laws of biology could be traced back to the laws of physics – a thesis he admittedly could no longer substantiate later on, let alone prove.[16]

Ludwig von Bertalanffy, who emigrated from Austria to Canada after the war, likely also no longer expected this when he called *this* view of a unity of the sciences in the mid-1950s "unrealistic". In his article *General Systems Theory* for the journal *Main Currents in Modern Thought*, he closed with some observations on the unity of science:

> From our point of view, unity of science gains a more realistic aspect. A unitary conception of the world may be based, not upon the possibly futile and certainly far-fetched hope finally to reduce all levels of reality to the level of physics, but rather on the isomorphy of laws in different fields.[17]

Bertalanffy was not pleading for a reductionism but for a "perspectivism":

> We can not reduce the biological, behavioral and social levels to the lowest level, that of the constructs and laws of physics. We can, however, find constructs and possibly laws within the individual levels.[18]

---

[13] The title *Erkenntnis* was kept and appeared in parentheses! Philipp Frank, Joergen Jörgensen (Copenhagen), Charles W. Morris (Chicago), Otto Neurath (The Hague), Louis Rougier (Besançon and Cairo) and Lizzie Susan Stebbing (London) functioned as "associate editors"

[14] Until the German invasion of the Netherlands in 1940, these series were published by Van Stockum & Zoon

[15] Max Black (1909–1988), Azerbaijani-American philosopher. Black studied in Cambridge, Göttingen and London, where he earned his doctorate in 1939. In 1940 he emigrated to the USA. He taught philosophy at the University of Illinois and became professor of philosophy at Cornell University in New York in 1946

[16] [114] Carnap, Unity (1934), quoted from [73] Bertalanffy, System (1968), p. 87

[17] [72] Bertalanffy, System (1954), p. 8

[18] [72] Bertalanffy, System (1954), p. 8

Fig. 3.1. L. v. Bertalanffy

The world is like Neapolitan ice cream – as Aldous Huxley[19] once put it – and its physical, biological, sociological and moral spheres correspond to the layers of chocolate, strawberry and vanilla: "We cannot reduce strawberry to chocolate – the most we can say is that possibly in the last resort, all is vanilla, all mind or spirit."[20] The unified principle was more the organizational structure of the individual areas which join together to form a whole, and this principle aimed to find it. A world model of overlapping organizations like this might play a role in strengthening the sense of reverence for life that seemed to have been lost during the bloody decade of the 1940s (Fig. 3.2).

Bertalanffy also stressed later that unity in science lay in the correspondingly structured organization of its sub-disciplines and not in what seemed to him a "utopian" reductionism:

> ...we are certainly able to establish scientific laws for the different levels or strata of reality. And here we find, speaking of the formal

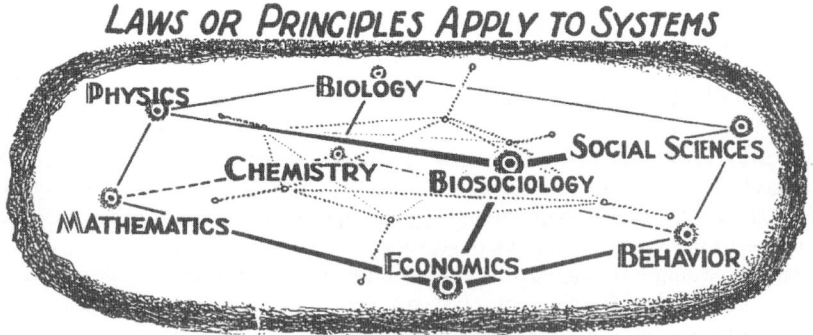

Fig. 3.2. Illustration of Bertalanffy's General Systems Theory

---

[19] Aldous Huxley (1894–1963), British writer, critic of the dehumanization observable in scientific progress, author of the dystopian novel *Brave New World*, which was published in 1932

[20] [72] Bertalanffy, System (1954), p. 8

mode (Carnap), a correspondence or isomorphy of laws and conceptual schemes in different fields, granting the Unity of Science. Speaking in "material" language, this means that the world (i. e. the total of observable phenomena) shows a structural uniformity, manifesting itself by isomorphic traces of order in its different levels of realms.[21]

His philosophical observations on science and nature prompted him to proclaim a science that is superordinated to the sciences, with a system of sciences as its sub-systems. Structures like this which encompassed all of the sciences were known in natural philosophy in the sense of "world views", but the ever-increasing differentiation and specialization made it difficult to devise an integrative system such as this in which every detail of even the most distant sub-discipline of a subject retains its classification. Bertalanffy was not calling for a new "world view," however. He wanted instead to turn attention to the interdisciplinary and transdisciplinary system structures of scientific principles. To this end, he referred to a phenomenon within scientific development that he had observed alongside the ever-increasing fragmentation:

> In contrast to the game of chance played by the atoms according to the mechanistic world view of the previous century, modern physics is shot through with all manner of questions of organization, of wholeness, of dynamic interaction, which amount to nothing less than the center of the problems it addresses today, be it the structure of the atom, the composition of protein bodies or interaction phenomena in modern thermodynamics.[22]

As a natural philosopher and biologist, he was familiar both with the mechanistic view, according to which living things were fragmented into their individual parts and life processes were considered in terms of their sub-processes, as well as with the newer philosophy of organism, which stressed the principles of organization and order by which they were joined as a whole.[23] The parts of the whole behave differently in isolation than in the dynamic of their

---

[21] [73] Bertalanffy, System (1968), p. 87

[22] [74] Bertalanffy, Systemtheorie (1957), p. 8

[23] "The whole is greater than the sum of its parts." This statement by the Greek philosopher Aristotle (384–322 B.C.) was interpreted by the Vitalists in the late 19th and early 20th century to mean that analytical-mechanistic examinations of life processes were not enough to explain them. Instead, everything organic was ruled by a "holistic causality" regulated by an "entelechy" that could be seen as a factor similar to the soul. This entelechy navigated life processes toward its goal. However, a nearly complete renunciation of the vitalist biologists was then made by British physiologist John Scott Haldane (1860–1936) in 1931. Mechanicism could not in turn explain the coordination in living systems. Only holism or organism – both terms were used in the 1920s – replaced vitalism as the "new paradigm". Organism was a term that had already been used in the social sciences by Auguste Coupte (1798–1857), however. See [359] Mayr, Biologie (1997), p. 40 and Chap. 1, remark 14

context.[24] He stated similar views with regard to psychology, medicine and the social sciences[25] and "that these developments occurred independently of one another and usually without any knowledge of the parallel developments in other fields".[26] Bertalanffy saw this "parallelism" even in the "structurally similar models and sets of laws in completely different areas". He cited the exponential law of growth and certain differential equations as examples.[27] Despite these sporadically occurring mathematically expressed rules, it was generally not possible to account for the circumstances of living systems using the traditional tools of physics-oriented mathematics. "So we need to expand our theoretical schemata in order to attain an exact set of laws in those fields where applying physical laws is not adequate or not even possible."[28]

In order to identify "structural similarities", "isomorphies", "principles" that are true "absolutely" or "in general", in other words, to derive properties of "the systems of the most disparate kinds, be they mechanical, caloric, chemical or whatever," Bertalanffy required a "general systems theory".[29]

In this "generalized science" that is referred to as *General Systems Theory*, the cooperation of sciences in different disciplines should be facilitated and fostered and multidisciplinary sets of laws should be sought, found and examined. This "über"-science was initially referred to in German as *General Systems Studies* or *Theory*[30], and Bertalanffy himself also called it a "new

---

[24] It is evident that these formulations trace back to the philosophy of Georg Wilhelm Friedrich Hegel (1770–1831). See e. g. [443] Philips, System (1969)

[25] [74] Bertalanffy, Systemtheorie (1957), p. 8

[26] [74] Bertalanffy, Systemtheorie (1957), p. 8

[27] He does not mention the Lotka-Volterra equation by name here, but at another point in this article he cites Alfred James Lotka's (1880–1949) *Physical Biology* alongside Wolfgang Köhler's (1887–1967) *Theorie der physischen Gestalten* as forerunners of his theory: [74] Bertalanffy, Systemtheorie (1957), p. 9. Vito Volterra (1860–1940), mathematician and physicist born in the Papal State (now Italy). In 1882, he graduated in Pisa with a Ph.D. in physics and became a professor of mechanics there a year later. Volterra was later a professor in Turin and Rome

[28] [74] Bertalanffy, Systemtheorie (1957), p. 8f

[29] [74] Bertalanffy, Systemtheorie (1957), p. 9

[30] The Hungarian philosopher and system theorist Erwin Laszlo (born 1932) writes in his foreword to [70] Bertalanffy, Perspectives (1975), that Bertalanffy used both terms in German at first – *Systemtheorie* and *Systemlehre* – but that the translation of the word "Lehre" (studies) was misleading: "Thus when Bertalanffy spoke of Allgemeine Systemtheorie it was consistent with his view that he was proposing a new perspective, a new way of doing science." Lazslo also notes that Bertalanffy was not interested in a "theory of general systems". Bertalanffy was responsible for far more than an individual theory, which, as we know, can always be falsified and sometimes short-lived: "He created a new paradigm for the development of theories. These theories are and will be system-theories, for they deal with systemic phenomena – organisms, populations, ecologies, groups, societies, and the like. An Allgemeine Theorie that can integrate these in-themselves different

mathesis universalis," with which he emphasized both its universal claim and its formal character as well as positioning himself – it may be – as a successor to Leibniz[31]

> Leibniz dreamt of a time of mathesis universalis, a universal science that would encompass all individual sciences. It may be that a general systems theory represents a step in this direction.[32]

He himself "first formulated the requirement for a general systems theory in 1938 in lectures at the University of Chicago. However, because of the war and other circumstances, my first work was not published until after 1945".[33] He was able to spend this year in the United States thanks to a Rockefeller scholarship, and after first working in Chicago with the mathematician Nicolas Rashevsky[34], he traveled the country and delivered a number of lectures.[35] In another passage, Bertalanffy recalled that the lecture on his ideas about a general systems theory was met with resistance in Charles W. Morris's[36] philosophy seminar in Chicago. This was also true of his theses on a "theoretical biology"[37], although it later found success as a university subject in America.[38]

---

phenomena with the rigor associated with the English concept of "scientific theory" does not exist." Loc cit., p. 12

[31] This thought also appealed to A. Locker, who wrote in the *Österreichische Hochschulzeitung* on September 15, 1966 in honor of Bertalanffy's 65th birthday: "The General Systems Theory is a formal (a priori) universal science, yes, it may be the realization of the mathesis universalis that Leibniz called for"

[32] [74] Bertalanffy, Systemtheorie (1957), p. 12

[33] [74] Bertalanffy, Systemtheorie (1957), p. 9

[34] Nicolas Rashevsky (1899–1972) was born in Chernikov, Russia and after completing his studies in mathematics and physics emigrated to the US in 1924. At the University of Pittsburgh he pursued the idea of a mathematical biology modeled after mathematical physics. He founded the *Bulletin of Mathematical Biophysics* and developed a mathematical theory of neuronal networks before the backdrop of Pavlovian conditioning. Rashevsky died in the Netherlands

[35] Bertalanffy spent the second half of his stay in America at the Marine Biological Laboratory in Woods Hole, Massachusetts. See [425] Nierhaus, Bertalanffy (1981), p. 145

[36] Charles William Morris (1901–1979), American philosopher, was a professor at the University of Chicago from 1931 to 1947 and in Gainesville, Florida from 1958 to 1971

[37] [69] Bertalanffy, Menschen (1970), p. 14

[38] From 1934 to 1948, Bertalanffy was an extraordinary professor in Vienna. Because of the poor living and working conditions there after the war, he moved to Ottawa, Canada, where he became director of the Biological Institute. In 1954–55, he was a senior fellow at the Center for Advanced Study at Stanford, and later the director of research and guest professor of psychology at the University of Southern California in Los Angeles. He then accepted a position at the Menninger Foundation in Topeka, Kansas. In 1961, he was given the chair for theoretical

Bertalanffy's *Modern Theory of Development* appeared in *Schaxels Abhandlungen* in 1928.[39] His search for a more general theory could already be seen here:

> Since the fundamental character of a life form lies in its organization, the usual examination of individual components and individual processes cannot provide a complete explanation of life phenomena. Instead, the laws of living systems must be examined at all levels of organization. We call this interpretation, when considered as research maxims, organismic biology and, as an attempt at explaining it, the *systems theory* of the organism.[40]

The *Modern Theory of Development* proceeded from the tenet that the whole is greater than the sum of its parts, and Bertalanffy's concepts of "open systems", "dynamic equilibrium" and "feedback" can also be found here. Bertalanffy was apparently unaware that the latter had not originally been founded in cybernetics.[41] He employed these three interrelated fundamental concepts again in the 1950s to argue for General Systems Theory: First he separated the "open" from the "closed" systems and introduced his concept of "dynamic equilibrium":

> Conventional physics is concerned with closed systems, that is, those that do not exchange any material with their environment. ... However, we find systems in nature that, by their essence and their definition, are not closed systems. Every living organism is an open system, for it maintains itself in a constant inflow and outflow, constant building up and breaking down of its components. As long as the organism lives, it is never in a state of rest, of chemical and thermodynamic equilibrium. Rather, it supports itself in a so-called dynamic equilibrium, i.e. a balance of import and export that is far removed from actual equilibrium. This is the essence of that basic phenomenon of life which we call metabolism.

---

biology in the department of zoology and psychology in Edmonton, Canada. He ended his employment there at the age of 65 and accepted a research and teaching position at the State University of New York at Buffalo

[39] [68] Bertalanffy, Theorie (1928)

[40] This is probably – as Bertalanffy himself also wrote – one of the earliest places in which the term "systems theory" appears. Quoted from [76] Bertalanffy, Vorläufer (1972), p. 20. Bertalanffy cites himself here "from the 1920s", but unfortunately does not list a more precise source

[41] The concept of feedback originated with the Munich physiologist Richard Wagner (1893–1970), who first published his findings from the 1920s on the control mechanisms in living creatures as a monograph about 30 years later: [545] Wagner, Regelung (1954). A letter from Wagner to Bertalanffy explaining this dates from 1968; it was in the possession of Mrs. Maria von Bertalanffy. A mention of this letter can be found in: [425] Nierhaus, Bertalanffy (1981), footnote 14. See also the section on Richard Wagner in: [138] DBE (1998–1999)

Bertalanffy explained in conclusion that even the target-oriented behavior of open systems did not represent a problem for General Systems Theory, since there were models available for this which he attributed to the latest developments in electrical engineering:

> ... such as the control mechanism or so-called feedback, which underlies the theory of cybernetics. Models with target-oriented behavior were repeatedly constructed as electric apparatus, like Ashby's so-called homeostat and Walter's electric turtle.[42]

After the war, when developments in Norbert Wiener's cybernetics began to fulfill his call for a general systems theory, Bertalanffy renewed his efforts to realize his ideas. He was now receiving much more support than before the war – it seems many scientists from the various disciplines had begun to see a transdisciplinary system theory as a desideratum. The political economist Kenneth E. Boulding[43], for example, wrote Bertalanffy in 1953:

> I seem to have come to much the same conclusions as you have reached, though approaching it from the direction of economics and the social sciences rather than from biology – that there is a body of what I have been calling "general empirical theory" or "general system theory" in your excellent terminology, which is of wide applicability in many different disciplines. I am sure there are many people all over the world who have come to essentially the same position that we have, but we are widely scattered and so not know each other, so difficult is it to cross the boundaries of the disciplines.[44]

When the American Association for the Advancement of Science met in Berkeley, California in 1954, Bertalanffy, Kenneth Boulding, the neurophysiologist Ralph W. Gerard[45] and biomathematician Anatol Rapoport[46] founded

---

[42] [74] Bertalanffy, Systemtheorie (1957), p. 12. William Ross Ashby (1903–1972), British psychiatrist, was director of research at the Barnwood House Hospital, Gloucester, from 1947 to 1949 and director of the Burden Neurological Institute in the department of electrical engineering at the University of Illinois, Urbana from 1961 to 1970. In 1971, he became a fellow of the Royal College of Psychiatry. William Grey Walter (1910–1977), British neurophysiologist and robotics pioneer. Using electric components and switching elements, Walter constructed two robots which resembled turtles; he named them Elmer and Elsie. These devices, also known as "machine speculatrix", were capable of recharging their batteries themselves. They possessed photocells as "eyes" and were designed to move toward light

[43] Kenneth Ewert Boulding (1910–1993), American economist

[44] Quoted from [73] Bertalanffy, System (1968), p. 14. All but the last sentence of the quote can also be found in [70] Bertalanffy, Perspectives (1975), p. 155

[45] Ralph Waldo Gerard (1900–1974), American physiologist

[46] Anatol Rapoport (born 1911), Russian mathematician and biologist

the Society for General Systems Research[47] as "Group L" of the association. This was a plan they had devised shortly before during a meeting at the newly established Center for Advanced Study in the Behavioral Sciences in Palo Alto, California.[48] The first edition of Rapoport's annual *General Systems*[49] was published two years later, and the journal *Mathematical System Theory* was also published beginning in 1967.

3.2 Cybernetics

In 1948, a book came onto the market in Paris, Cambridge (Mass.) and New York that would shortly become a scientific best seller. It bore the title *Cybernetics or Control and Communication in the Animal and the Machine* and had been written for the most part in Mexico City by MIT professor Norbert Wiener.[50] It contained research findings and interpretations developed over a number of years by Wiener and his colleagues: mainly MIT engineer Julian Bigelow[51] and Mexican neurophysiologist Arturo Rosenblueth[52].

> We have decided to call the entire field of control and communication theory, whether in the machine or in the animal, by the name *Cybernetics*, which we form from the Greek word for "steersman".[53]

---

[47] The group was first called the Society for General System Theory
[48] The Center for Advanced Study in the Behavioral Sciences (CASBS) was founded by the Ford Foundation in 1954
[49] The first contribution was a reprint of Bertalanffy's article *General System Theory* ([72] Bertalanffy, System (1954)) from two years earlier. The only change for this reprint was an expansion of the footnotes. The annual *General Systems* was published from 1956 to 1976
[50] Norbert Wiener (1894–1964) was the son of the distinguished Slavic scholar Leo Wiener, who taught French and German as a professor of modern languages at the University of Missouri-Columbia. He was considered a "Wunderkind". He first attended Tufts College until the final exams in 1909. He went to Harvard and studied biology for a short time before switching to philosophy, and earned his doctorate at Harvard in 1912 with a paper on formal logic. Following this, he studied with Bertrand Russell in Cambridge, then went to Göttingen to study mathematics with David Hilbert (1862–1943) and Edmund Landau (1877–1938) and philosophy with Edmund Husserl (1859–1938). In 1915, back in the USA, he completed his studies in Cambridge and at Columbia University in New York and soon became an assistant professor of philosophy and logic at Harvard and then mathematics lecturer at the University of Maine in 1916. He resigned from this position when the United States entered the First World War. He then worked in the turbine department of General Electric and later for Encyclopedia Americana in Albany. From 1929 until the end of his life, he taught in the mathematics department at MIT
[51] Julian Bigelow (1913–2003), American mathematician and engineer
[52] Arturo Rosenblueth (1900–1970), Mexican physiologist
[53] [558] Wiener, Cybernetics (1948), p. 19

This ancient Greek word is derived from the name of the feast days celebrated in antiquity to honor two pilots, the "kybernetes".[54] André-Marie Ampère[55] had borrowed the word from Plato in 1834 as *cybernétique* to denote the *study of procedures in government,* in other words, the methods for steering toward a political goal.[56] Some 30 years later, James Clerk Maxwell published a paper in which he discussed third-order linear differential equations for a closed control circuit. He titled this paper *On Governors* and in it he defines:

> A "governor" is a part of a machine by means of which the velocity of the machine is kept nearly uniform, notwithstanding variations in the driving-power or the resistance.[57]

Wiener, whose book *Extrapolation, Interpolation, and Smoothing of Stationary Time Series* was also called "the yellow peril" because of its mathematically very challenging presentation – or as John Robinson Pierce[58] put it, "because of the headaches it caused" – [59] hoped that with *Cybernetics* he could:

> ... give an account of the new information theory which was being developed by Shannon and myself, and of the new prediction theory which had its roots in the prewar work of Kolmogorov and in my researches concerning anti-aircraft predictors. I wished to bring to the attention of a larger public than had been able to read my "yellow peril" the relations between these ideas and show it a new approach to communication engineering which would be primarily statistical. I also wished to alert this larger public to the long series of analogies

---

[54] Plato used the word "$\kappa \upsilon \beta \varepsilon \rho \nu \eta \tau \eta \varsigma$" in comparing the state to a ship with different parties fighting over its rudder. Of course, Plato condemned those who gain control of the ship and "want to steer, although they know nothing of the art; and they have a theory that it cannot be learned. If the helm is refused them, they drug the captain's posset, bind him hand and foot, and take possession of the ship. He who joins in the mutiny is termed a good pilot and what not; they have no conception that the true pilot must observe the winds and the stars, and must be their master, whether they like it or not; – such a one would be called by them fool, prater, star-gazer." Quoted from [252] Ilgauds, Wiener (1980), p. 58

[55] André-Marie Ampère (1775–1836), French mathematician and natural scientist. In 1805, he became a professor at the École Polytechnique and at the Collège de France. Ampère became a member of the Académie des Sciences in 1814

[56] [20] Ampère, Essai (1934)

[57] Originally, "governor" was the term for the centrifugal force regulator in the steam engine invented by James Watt (1736–1819). The word can also be derived from the Latin "gubernare", and this comes directly from the Greek *kybernetes,* i. e. the art of the steersman. Quoted from [252] Ilgauds, Wiener (1980), p. 59

[58] John Robinson Pierce (born 1910), American radio engineer

[59] Of course, this name also refers to the yellow cover in which Wiener's book was published. See [445] Pierce, Information (1961), p. 41

between the human nervous system and the computation and control machine which had inspired the joint work of Rosenblueth and me. However, I could not undertake this multiform task without an intellectual inventory of my resources. It became clear to me almost at the very beginning that these new concepts of communication and control involved a new interpretation of man, of man's knowledge of the universe, and of society.[60]

Fig. 3.3. N. Wiener

Wiener had worked with anti-aircraft weaponry during the First World War. He had been denied entry into the armed forces because of his poor eyesight and had gone to work in the civilian sector instead. He then received a letter from Professor (and now Major) Oswald Veblen[61] with the offer to become "computer at the US Army Proving Grounds in Aberdeen, Maryland,"[62] and thus got the chance "to work with a colorful group of civilian and military mathematicians".[63] This position primarily involved drawing up and calculating ballistic tables for the artillery. Wiener was active there first as a civilian, later as a soldier, for over half a year before being honorably discharged in 1919.

It was a period in which all the armies of the world were making the transition between the rough of formal ballistic to the point-by-point solution of differential equations, and we Americans were behind neither our enemies nor our allies. In fact, in the matter of interpolation and the computation of the correction of the primary ballistic tables,

---

[60] [567] Wiener, Mathematician (1962), p. 325
[61] Oswald Veblen (1880–1960), American mathematician. From 1932, he was a professor at the Institute for Advanced Study in Princeton, which he was instrumental in founding
[62] [353] Masani, Wiener (1990), p. 68
[63] [567] Wiener, Mathematician, (1962), p. 29

Professor Bliss of Chicago made a brilliant use of the new theory of functionals. Thus the public became aware for the first time that we mathematicians had a function to perform in the world.[64]

Even if Wiener had also been able to mirror the success he had enjoyed in his scientific career thus far in a study of philosophy, his future was now laid out for him, for he felt great enthusiasm for mathematics. "Whatever we did, we always talked mathematics", he recalled later, and "... I am sure that this opportunity to live for a protracted period with mathematics and mathematicians greatly contributed to the devotion of all of us to our science."[65]

Fig. 3.4. J. Bigelow

During the Second World War, he had actually wanted to develop mathematical and mechanical decoding methods, but when the appropriate projects failed to materialize for him, he found a field of scientific activity that was just as vital to the war effort, namely the development of fire control systems and anti-aircraft weaponry. At that time, anti-aircraft guns were still extraordinarily difficult to handle and extremely inaccurate and the crews were fully occupied with trying to operate them by consulting tables and making adjustments by hand.

At the beginning of the war the only known method for tracking an airplane with an anti-aircraft gun was for the gummer to hold it in his sights by a humanly regulated process. Later on in the war, as radar became perfected, this process was mechanized. It became possible to couple directly to the gun the radar apparatus by which the plane is localized, and thus to eliminate the human element in gun pointing.

---

[64] [559] Wiener, Ex-Prodigy (1953), p. 256, quoted from [353] Masani, Wiener (1990), p. 68
[65] [559] Wiener, Ex-Prodigy (1953), p. 258, quoted from [353] Masani, Wiener (1990), p. 68

However, it does not seem even remotely possible to eliminate the human element as far as it shows itself in enemy behavior.[66]

An attempt was made to incorporate the brand new developments in servo mechanisms into the anti-aircraft guns that were to be controlled by analog computers.[67] Wiener and Bigelow were involved in related research at MIT to investigate how to exploit the opportunities provided by Vannevar Bush's Differential Analyzer. It was clear to Wiener that the computations needed for the control apparatus on the anti-aircraft battery had to be built into this apparatus, since airplanes had become so fast and maneuverable by the start of World War II that classical methods of directing fire had been rendered obsolete. In his autobiography, he recalled:

> These were rendered much more difficult by the fact that, unlike all previously encountered targets, an airplane has a velocity which is a very appreciable part of the velocity of the missile used to bring it down. Accordingly, it is exceedingly important to shoot the missile, not at the target, but in such a way that missile and target may come together in space at some time in the future. We must hence find some method of predicting the future position of the plane.[68]

Since they did not think they could solve the problem psychologically, Wiener and Bigelow attempted to think both like a gunner and like a pilot and to imitate their behavior mechanically. They recognized that both the pilot and the gunner were aware of particular patterns of behavior and were trying to avoid making the same errors in the future. This resembled the control method of the controllers that they knew from electric circuits and that were used in servo mechanisms, the so-called feedback principle.

> However, this improvement in behavior does not continue indefinitely, for after a certain stage, with a large measure of amplification in the feedback, the apparatus will go into spontaneous oscillation and behave in such a wild way that we have decreased rather than increased the load independence of the apparatus. We expected that if human control also were to depend on feedback, there would be certain pathological conditions of very great feedback, under which the human system, instead of acting effectively as a control system, would go into

---

[66] [567] Wiener, Mathematician (1962), p. 251

[67] The first electric fire-control system was developed by David B. Parkinson at Bell Telephone Laboratories in 1940. Department D-2, Fire-Control of the National Defense Research Committee then awarded Bell Laboratories a contract to develop a prototype, and Western Electric received the contract for its production. See [465] Roch, Maus (2003), Axel Roch: Die Maus. Von der elektrischen zur taktischen Feuerleitung. $http://mikro.org/Events/19991006/roch.htm$. See also [169] Fagen, History (1975), p. 148

[68] [558] Wiener, Cybernetics (1948), p. 11

wilder and wilder oscillations until it should break down or at least until its fundamental method of behavior should be greatly changed.[69]

It was therefore also possible to assume that a medical problem existed. Wiener's friend, the Mexican neurophysiologist Arturo Rosenblueth, was conducting research at Harvard Medical School at the time[70] and could easily be consulted with this problem:

> The specific question we put was: Are there any known nervous disorders in which the patient shows no tremor at rest, but in which the attempt to perform such an act as picking up a glass of water makes him swing wider and wider until the performance is frustrated, and (for example) the water is spilled? Dr. Rosenblueth's answer was that such pathological conditions are well known, and are termed "intention tremors"; and that very often the seat of the disorder lies in the cerebellum, which controls our organized muscular activity and the level on which it takes place. Thus, our suspicions that feedback plays a large role in human control were confirmed by the well-established fact that the pathology of feedback bears a close resemblance to a recognized form of the pathology of orderly and organized human behavior.[71]

In August of 1940, the German air force had begun the first phase of severe attacks on Great Britain. British anti-aircraft defenses were no match for the Luftwaffe and losses were very high.

Fig. 3.5. A. Rosenblueth

---

[69] [567] Wiener, Mathematician (1962), p. 253

[70] In the early '40s, Rosenblueth was professor of physiology at Harvard Medical School, but was hired by the Instituto Cardiologia in Mexico in 1944. See [558] Wiener, Cybernetics (1948), p. 14

[71] [567] Wiener, Mathematician (1962), p. 253

The United States had entered the world war following the Japanese bombing raid on Pearl Harbor, and to protect England from total collapse the newly installed anti-aircraft systems, after initial tests during the second Battle of Britain, were used to shoot down most of the V1 flying bombs launched by the German military. The flak guns outfitted with the new technology proved easier to operate and the kill rate was optimized. In these new apparatus, analog computers used the electric signals that were provided by optical enemy tracking and entered by hand to extrapolate the anticipated flight path of the object to be intercepted. The shot could then be fired to a position which the enemy object was likely to reach at the same time.[72] This method of extrapolating the enemy flight path into the immediate future was based on Wiener's mathematical-statistical prediction theory.

Working together, Wiener, Bigelow and Rosenblueth authored the article *Behavior, Purpose and Teleology*, in which they characterized their studies as a "behavioristic approach":

> Given any object, relatively abstracted from its surroundings for study, the behavioristic approach consists in the examination of the output of the object and of the relations of this output to the input. By output is meant any change produced in the surroundings by the object. By input, conversely, is meant any event external to the object that modifies this object in any manner.[73]

They concluded in this text that deliberate actions in humans, like those in machines, could be explained with the aid of the feedback principle used in the engineering sciences. The fact that machines, such as target-seeking torpedoes, function in this way did not require any further explanation. To illustrate analogous mechanisms in humans, the authors offered the following example:

> When we perform a voluntary action what we select voluntarily is a specific purpose, not a specific movement. Thus, if we decide to take a glass containing water and carry it to our mouth we do not command certain muscles to contract to a certain degree and in a certain sequence; we merely trip the purpose and the reaction follows automatically.[74]

In contrast to the concept of *positive* feedback, which engineers used to indicate that a portion of the output energy is returned to the apparatus as

---

[72] See [465] Roch, Maus (2003), Axel Roch: Die Maus. Von der elektrischen zur taktischen Feuerleitung. $http://mikro.org/Events/19991006/roch.htm$
[73] [472] Rosenblueth et al., Behavior (1943), p. 18
[74] [472] Rosenblueth et al., Behavior (1943), p. 19

input[75], Wiener, Bigelow and Rosenblueth here employed the restrictive concept of *negative* feedback to denote that the behavior of the object is controlled by the incorrect position it assumes with respect to the target to be hit at the respective time of observation.[76] In *Cybernetics*, Wiener expressed it this way:

> Now, suppose that I pick up a lead pencil. To do this, I have to move certain muscles. However, for all of us but a few expert anatomists, we do not know what these muscles are; and even among the anatomists, there are few, if any, who can perform the act by a conscious willing in succession of the contraction of each muscle concerned. On the contrary, what we will is *to pick the pencil up*. Once we have determined on this, our motion proceeds in such a way that we may say roughly that the amount by which the pencil is not yet picked up is decreased at each stage. This part of the action is not in full consciousness.[77]

The interdisciplinary approach taken by Wiener, Bigelow and Rosenblueth culminated in their hypothesis that the behavioral mechanisms in machines and in living organisms were – at least roughly – the same, although it was acknowledged that particularities might occur in one way or another.[78] There were naturally functional differences, as well, between living things and machines; if an engineer were to design a robot that was supposed to act like an animal, he would not be likely to build it out of proteins and other colloids, but would instead probably use metallic implements, a few dielectrics and a lot of vacuum tubes.[79]

Wiener noted in his *Cybernetics* that the views they had developed differed greatly from the beliefs prevalent among physicians:

---

[75] As an example, the authors describe an electric amplifier with feedback: "The feed-back is in these cases positive – the fraction of the output which re-enters the object, has the same sign as the original input signal. Positive feed-back adds to the input signals, it does not correct them." [472] Rosenblueth et al., Behavior (1943), p. 19

[76] "The feed-back is then negative, that is, the signals from the goal are used to restrict outputs which would otherwise go beyond the goal." [472] Rosenblueth et al., Behavior (1943), p. 19

[77] [558] Wiener, Cybernetics (1948), p. 14

[78] "Thus, no machine is available yet that can write a Sanscrit-Mandari dictionary. Thus, also, no living organism is known that rolls on wheels – imagine what the result would have been if engineers had insisted on copying living organisms and had therefore put legs and feet in their locomotives, instead of wheels." [472] Rosenblueth et al., Behavior (1943), p. 22

[79] While the movements of a robot can sometimes be faster and stronger than those of the organic original, its memory and ability to learn are by contrast more rudimentary. A lot of catching up can be done in the future, however, and the robot could be very similar to the mammal in behavior and structure: "The ultimate model of a cat is of course another cat, whether it be born of still another cat or synthesized in a laboratory." [472] Rosenblueth et al., Behavior (1943), p. 23

> The central nervous system no longer appears as a self-contained organ, receiving inputs from the senses and discharging into the muscles. On the contrary, some of its most characteristic activities are explicable only as circular processes, most emerging from the nervous system into the muscles, and re-entering the nervous system through the sense organs, whether they be proprioceptors or organs of the special senses.[80]

Not content with the fact that these ideas contained enough dynamite to flatten the disciplinary bulwarks between the various scientific disciplines, Wiener managed to build a bridge between his research results and his scientific knowledge, a step that can also credited to Claude E. Shannon and Andrei N. Kolmogorov[81]: The concept of information was central to the idea of feedback as a principle of control in machines and animals; Wiener understood feedback processes to be processes of information manipulation and decision making:

> On the communication engineering plane, it had already become clear to J. H. Bigelow and myself that the problems of control engineering and of communication engineering were inseparable, and that they centered not around the technique of electrical engineering but around the much more fundamental notion of the message, whether this should be transmitted by electrical, mechanical, or nervous means.[82]

As discussed in Chap. 1, Wiener examined time series, that is, series of measurable events – whether discrete or constantly recurring. In order to calculate the expected future development of such time series, Wiener had developed his prediction theory, in which an operator – a "predictor" – was applied in each case to the preceding element of the time series. In theory, this predictor corresponds to a mathematical calculation scheme; in practice, such as predicting the flight path of an enemy object, it was realized by a technical apparatus – and with the assistance of Bigelow's engineering knowledge. Rosenblueth's confirmation of an analogy to muscular trembling turned out to be the icing on the cake:

---

[80] [558], Cybernetics (1948), p. 15

[81] Andrei Nikolaevich Kolmogorov (1903–1987), Russian mathematician, was publishing papers on set theory, Fourier analysis and probability theory while he was still completing his mathematics degree. In 1929, he finished his studies with Nikolai Nikolaevich Lusin (1883–1950). His study trips in 1930 and 1931 led him to Göttingen, Munich and Paris; he became a professor at the university in Moscow in 1931. In 1933, he published the book *Fundamentals of Probability Calculus*, in which he established axioms on probability theory. He was promoted in mathematics and physics in 1934, and in 1939 became a member of the Russian Academy of Sciences. In 1941 he was awarded the Stalin Prize. He became a member of London's Royal Society in 1964 and a member of the Académie des Sciences in 1968

[82] [558] Wiener, Cybernetics (1948), pp. 15–16

The better the apparatus was for smooth waves, the more it would be set into oscillation by small departures from smoothness, and the longer it would be before such oscillations would die out. Thus the good prediction of a smooth wave seemed to require a more delicate and sensitive apparatus than the best possible prediction of a rough curve, and the choice of the particular apparatus to be used in a specific case was dependent on the statistical nature of the phenomenon to be predicted.[83]

3.3 Communication Theory

Applying statistical prediction theory to communication technology seemed the completely obvious course to Wiener, especially since he had worked with the closely related problems of noise and communication filtering.[84]

> What is not generally realized is that the rapidly changing sequences of voltages in a telephone line or a television circuit or a piece of radar apparatus belong just as truly to the field of statistics and time series, although the apparatus by means of which they are combined and modified must in general be very rapid in its action, and in fact must be able to put out results *pari passu* with the very rapid alterations of input.[85]

Conveying information by means of communication technology meant, then as now, transmitting alternatives; each alternative is based on two possibilities. Deciding in favor of one alternative means choosing: "heads" or "tails", "yes" or "no", "zero" or "one". Transmitting information thus also connotes transmitting decisions from among two possibilities (which might have probabilities associated with them). Binary decisions of this type helped Wiener come up with a definition of the amount of information:

> The least amount of information a message gives is whether it is sent or not. This minimal message is, however, entirely abnormal and, if sent, need occupy no long time. A continued message, sent for example by telephone, telegraph or wigwag, presents a sequence of successive decisions. It is fair to state that the amount of information transmitted is proportional to the number of decisions made.[86]

Wiener defined the amount of information that a system transmits on the basis of the number of decisions necessary to receive the information; here he was assuming – in a first approach – equally probable possibilities of each

---
[83] [558] Wiener, Cybernetics (1948), p. 16
[84] See also Chap. 1
[85] [558] Wiener, Cybernetics (1948), p. 74
[86] [557] Wiener, Communication (1949), p. 74

alternative. If n different decisions must be made, then this corresponds to a division of the basic set into 2^n equally probable sets, and the probability of each is thus $1/(2^n)$. The logarithm of this value to base 2 is $-n$.

Wiener concluded from this line of argumentation that the amount of information n is numerically proportional to the negative logarithm (to base 2) of the probability of the subdivision sets as determined by the decisions.[87]

Wiener had arrived at this definition from processes used in statistical mechanics; this field uses the logarithm of a probability to define the quantity *entropy*, which functions as a measure of the disorder of a statistical state. He was not bothered by the negative sign in front of his analogously formed quantity of the amount of information:

> Its negative, which we are here calling an amount of information, is thus a measure of order. This identification of order and information is entirely natural.[88]

Fig. 3.6. A. Kolmogorov

As a generalization for those cases in which it was not possible to say for certain whether something belonged to a set but only to give certain probabilities for whether it belonged to certain subsets – due to imprecise measurement results, for example – Wiener combined the information with the corresponding probability distribution, and so with regard to the aforementioned decisions, the original probability distribution was then overlaid by a second probability distribution that depended upon the exactness of our observation. Here, too, he proceeded in a manner analog to the entropy formula: If there is a subdivision with respective probabilities $P_1, ..., P_n$, then the result is the entropy of the whole to

$$\sum P_n \log P_n + C, \text{ where } C \text{ is a constant.}$$

---

[87] See [555] Wiener, Communication (1949), p. 75 and [558] Wiener, Cybernetics (1948), p. 86ff. Wiener oriented his argumentation to the statistical mechanics devised by American mathematician and physicist Josiah Willard Gibbs (1839–1903)

[88] [557] Wiener, Communication (1949), p. 75

Taking into account the negative sign, the result for the amount of information in the case transmitted in this way is as follows:

$$-\sum P_n \log P_n - C.$$

If the probabilities for the subdivision are then designated with p_n, then the amount of information can be calculated thusly:

$$\sum P_n \log P_n - p_n \log P_n.$$

Very similar definitions for the concept of information, an item of information, its value or its content were being proposed at the same time by Claude E. Shannon[89], Ronald A. Fisher[90] and Andrei N. Kolmogorov, a fact which Wiener readily acknowledged.[91] *The Mathematical Theory of Communication* by Shannon and Wiener's *Cybernetics* had appeared simultaneously, but in both cases the publication was delayed by the war. However, Wiener mentioned the fact that, since Shannon was a Bell employee, his research projects were geared toward realizing and marketing his findings as quickly as possible, whereas Wiener, as a college professor, was able to approach his research freely. He had "found the new realm of communication ideas a fertile source of new concepts not only in communication theory, but in the study of the

---

[89] Claude Elwood Shannon (1916–2001) studied mathematics first at Gaylord High School in Michigan and, starting in 1932, at the University of Michigan, where he also majored in electrical engineering. In 1936, he became Vannevar Bush's research assistant in department of electrical engineering at MIT and worked with him on the Differential Analyzer. He submitted his paper on realizing Boolean expressions by means of electric circuits under the title *A Symbolic Analysis of Relay and Switching Circuits* in 1937; in 1938 it was published in the *AIII Transactions*: [508] Shannon, Relay (1938). At MIT he earned his Master's degree and his doctorate in mathematics. He then spent a year as a national research fellow at the Institute for Advanced Study at Princeton, where he worked with the German-American mathematician and physicist Hermann Weyl (1885–1955). In 1941, Shannon went to Bell Telephone Laboratories, where he worked in the field of message transmission

[90] Sir Ronald Aylmer Fisher (1890–1962), British statistician, evolutionary theorist and geneticist, became a professor of eugenics at University College in London in 1933 and was appointed the Balfour Professor of Genetics at Cambridge in 1943

[91] "This idea occurred at about the same time to several writers, among them the statistician R. A. Fisher, Dr. Shannon of the Bell Telephone Laboratories, and the author. Fisher's motive in studying this subject is to be found in classical statistical theory; that of Shannon in the problem of coding information; and that of the author in the problem of noise and message in electrical filters. Let it be remarked parenthetically that some of my speculations in this direction attach themselves to the earlier work of Kolmogoroff in Russia, although a considerable part of my work was done before my attention was called to the work of the Russian school." [558] Wiener, Cybernetics (1948), p. 18. Wiener then cites [294] Kolmogoroff, Interpolation (1941)

living organism and in many related problems".[92] In the manuscript for his book *Invention: The Care and Feeding of Ideas*[93], which he never finished, Wiener wrote in 1957:

> Once I had alerted myself and the public in general to the statistical element in communication theory, conformation began to flow in from all sides. At the Bell Telephone Laboratories there was, and is, a young mathematical physicist by the name of Claude Shannon. He had already applied mathematical logic to the design of switching systems, and throughout all his work he has shown a love for the discrete problems, the problems with a small number of variable quantities, which come up in switching theory. I am inclined to believe that from the very start, a large part of his ideas in communication theory and its statistical basis were independent of mine, but whether they were or not, each of us appreciated the significance of the work of the other. The whole subject of communication began to assume a new statistical form, both for his sorts of problems and for mine. This is not the point for me to give the genealogy of every single piece of apparatus which this new statistical communication theory has fostered, but I can say that the impact of the work has gone from one end of communication theory to the other, until now there is scarcely a recent communication invention which has not been touched by statistical considerations. Thus, this whole wide spreading branch of science represents a subtle working out of concepts implicit in Gibbs and in the Lebesque-Borel team, but if I may say so, implicitly implicit, so that until some forty years had passed, no one could have seen the direction in which the earlier thought was bound to lead. This is, in my mind, a key example of a change in intellectual climate, and of the effect it has had both in discovery and in invention.[94]

Today Shannon's name is associated almost unanimously with mathematical information and communication theory – these terms are often considered synonyms – although people had spoken at first of the "Wiener-Shannon Communication Theory". The fact that Wiener's name is no longer mentioned here could be due to the simple and succinct communication pattern with which Shannon illustrated his text. This "communication model" became widely known in the technical sciences but also in other social sciences, such as psychology.[95] Another reason might be seen in the minus sign which Wiener

---

[92] Wiener in conversation with Bello (technology editor of *Fortune*) on 13 October 1953, Box 4.179, MC22, Wiener Papers, MIT. Quoted from [275] Kay, Leben (2000), p. 138

[93] [565] Wiener, Invention (1993). Steve Joshua Heims published the book fragment in 1993

[94] [565] Wiener, Invention (1993), p. 22f

[95] See e. g. [29] Badura, Sprachbarrieren (1971). Today, however, these approaches are considered at the very least to be in need of supplementation. See also [347] Maletzke, Kommunikationswissenschaft (1998), p. 108

Fig. 3.7. C. Shannon

had put in his definition of information as negative entropy. Shannon on the other hand had used the usual entropy formula and thus devised a "positive and powerful formulation that endured".[96]

Fig. 3.8. H. Nyquist

Claude E. Shannon had arrived at Bell Telephone Laboratories from MIT in 1941. There he was occupied simultaneously with questions involving the transmission of communication and with secret cryptology projects which involved, among other things, the digital coding of continuous sound waves. While the latter subject can be tied in to his famous Master's thesis, the former was part of a long tradition at Bell, for when the American Telephone and Telegraph Company (AT&T) was consolidated with its development

---

[96] [275] Kay, Leben (2000), p. 138. Lily E. Kay has summarized this very nicely. For more on this and the irritation caused by the significant similarities between these two theories, see all of Chap. 3 in [275] ("Production of Discourse: Cybernetics, Information, Life"), particularly pp. 137–140. Kay indicates that the related set of concepts in the two "theories of information" could be traced back to the fact that the MIT professor Wiener was involved in the education of the MIT student Shannon

departments in 1924 to form Bell Telephone Laboratories, Harry Nyquist[97] and Ralph Vinton Lyon Hartley[98] were working there on matters of information transmission. Nyquist had publicized a telegraph theory in February of 1924 in which he compared the efficiency of various telegraphing and coding systems. He was also the first to differentiate between signal, letter and message transmission speed and he showed the relationships between signal speed and bandwidth.[99] In 1928 he presented a self-contained "Theory of Telegraphy",[100] in which he also provided details about the criteria for distortion-free signal transmission.

In 1928, Hartley established a first comprehensive, abstract and self-contained concept with which the efficiency of different transmission techniques could be compared to one another, and it was here that the term "theory of information" was first used in communication technology, as researchers had heretofore spoken of "intelligence", "amount of intelligence" and "transmission of intelligence".

A system's ability to transmit any given sequence of symbols seemed to depend only upon whether the selections made on the transmission side could be recorded on the receiver side. Yet this was completely independent of the symbols' meaning. Also for the first time in 1928, Hartley introduced into this theory of information a quantitative measurement of the set of information, which he found in the number of possible messages. He derived the logarithmic law for the transmission of :

$$H = K \log s^n.$$

In this formula, H is the amount of information, K is a constant, n is the number of symbols in a message and s is the set of symbols. Finally, s^n indicates the number of symbol sequences of a specified length n. [101]

As mentioned, Shannon was engaged in further developing Hartley's theory. Even when he went to Princeton in the autumn of 1940 for a four-month National Research Fellowship at the Institute for Advanced Study, he continued working toward a *General Theory of Transmission and Transformation of Information*. The parallel work on secret binary codes and ways to improve them obviously led him, as it had Wiener, to employ statistical considerations:

---

[97] Harry Nyquist (1889–1976), Swedish-American electrical engineer, immigrated to the US in 1907 and worked 1917–1934 for the American Telephone and Telegraph Company (AT&T) and 1934–1954 at Bell Telephone Laboratories

[98] Ralph Vinton Lyon Hartley (1888–1970), American radio engineer, worked in the Research Laboratory of the Western Electric Company and, after WWI, in the Bell Telephone Laboratories

[99] Midwinter Convention of the A.I.E.E. in Philadelphia, February 4–8, 1924. [428] Nyquist, Factors (1924), p. 124. See also [429] Nyquist, Topics (1928)

[100] [429] Nyquist, Topics (1928), p. 617ff

[101] In September 1927, he presented his findings at the International Congress of Telegraphy and Telephony at Lake Como under the title *Transmission of Information* [230] Hartley, Transmission (1928)

Fig. 3.9. R. Hartley

A source selects from a set of symbols with particular probabilities. The information that is unique to a conveyed symbol increases if its probability of occurrence increases. This assumption led to the following generalization of Hartley's law:

$$H_n = -\sum p_i \log p_i$$

with H_n as the amount of information and with the probabilities p_i for each selection of one of the n symbols ($i = 1, ..., n$).

Shannon's assumption of the logarithm to base 2 can be explained by the Master's thesis he had published ten years earlier.[102] The electric relay switches had two possible positions, and binary decisions could therefore be realized technically and the amount of information contained in a message consisting of n binary-coded symbols was:

$$H_n = -\sum p_i \log_2 p_i.$$

The unit for this "measurement of information" was termed a "bit".[103]

So with the exception of the minus sign, Shannon and Wiener had arrived at the same result by different paths. They had thus "made of communication engineering design a statistical science, a branch of statistical mechanics".[104] For Wiener this was a very satisfying step toward the technology platform of *communication engineering*. Until now, a student of this discipline had been lacking many fundamentals of communication theory. "He studies communication circuits as they are affected by sinusoidal and information-carrying inputs."[105]

---

[102] See Chap. 1

[103] The suggestion to use the term "bit" – binary digit – came from John Wilder Tukey, a statistician at Princeton University who worked with John von Neumann at the Institute for Advanced Studies (IAS) after the war and, starting in 1946, simultaneously at Bell Telephone Laboratories. See [280] Kittler et al., Shannon (2000), footnote 4

[104] [558] Wiener, Cybernetics (1948), p. 17

[105] [557] Wiener, Communication (1949), p. 74

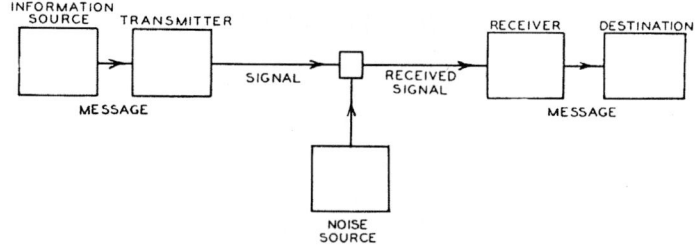

Fig. 3.10. Shannon's Communication Model

For the publication of his *Mathematical Theory of Communication*[106], Shannon had drawn a diagram of a general communication model (Fig. 3.10):

- An *information source* produces a series of messages that are to be delivered to the receiver side. Transmission can occur via a telegraph or teletypewriter system, in which case it is a series of letters. Transmission can also occur via a telephone or radio system, in which case it is a function of time $f(t)$, or else it is a function $f(x, y, t)$, such as in a black and white television system, or it consists of complicated functions.
- The *transmitter* transforms the message in some way so that it can produce signals that it can transmit via the channel. In telegraphy, these are dot-dash codes; in telephony, acoustic pressure is converted into an electric current.
- The *channel* is the medium used for transmission. Channels can be wires, light beams and other options.
- The *receiver* must perform the opposite operation to that of the transmitter and in this way reconstructs the original message from the transmitted signal.
- The *destination* is the person or entity that should receive the message.

According to this, Shannon conceived of communication purely as the transmission of messages – completely detached from the meaning of the symbols! Warren Weaver[107] stressed this in 1949 when he wrote a "recent contribution to the mathematical theory of communication" in *Scientific American*, which was then also published in book form along with the two parts of Shannon's text[108], possibly also in order to afford Shannon's disquisition a degree of renown comparable to that enjoyed by Wiener's cybernetics.[109] In his text, which was placed before Shannon's (as an introduction), Weaver divided the communication problem into three levels::

---

[106] [507] Shannon, Communication (1948)
[107] Warren Weaver (1894–1978), American mathematician, Director of Natural Sciences at the Rockefeller Foundation (1932–1955), later also vice-president of the Alfred P. Sloan Foundation
[108] See [505] Shannon et al., Communication (1976)
[109] See [275] Kay, Leben (2000), p. 142

- *Level A* contains the purely technical problem involving the exactness with which the symbols can be transmitted,
- *Level B* contains the semantic problem that inquires as to the precision with which the transmitted signal transports the desired meaning,
- *Level C* contains the pragmatic problem pertaining to the effect of the symbol on the destination side: What influence does it exert?

Weaver underscored very clearly the fact that Shannon's theory did not even touch upon any of the problems contained in levels B and C, that the concept of information therefore must not be identified with the "meaning" of the symbols:

> In fact, two messages, one of which is heavily loaded with meaning and the other of which is pure nonsense, can be exactly equivalent, from the present viewpoint, as regards information.[110]

Weaver also mentioned another, already well-known problem:

> In the process of being transmitted, it is unfortunately characteristic that certain things are added to the signal which were not intended by the information source. These unwanted additions may be distortions of sound (in telephony, for example) or static (in radio), or distortions in shape or shading of picture (television), or errors in transmission (telegraphy or facsimile), etc. All of these changes in the transmitted signal are called *noise*.[111]

It was this noise that had formed the starting point for Wiener's papers on electrical filtering.

3.4 "The Cybernetics Group"[112]

Wiener's *Cybernetics* had been an enormously successful book, but he reached thousands more readers when he followed it up a short time later with one that was intended for the general public: *The Human Use of Human Beings: Cybernetics and Society*.[113] An oft-quoted line from *Cybernetics* goes: "Information is information, not matter or energy."[114] The sentence that follows, which concludes the chapter, is usually omitted, since it leaves no doubt that Wiener intended his comments to be understood much more comprehensively. It says: "No materialism which does not admit this can survive at the present day."[115]

---

[110] [548] Weaver, Communication (1976), p. 18
[111] [548] Weaver, Communication (1976), p. 17
[112] Title of a book by Steve Joshua Heims: [236] Heims, Cybernetics (1991)
[113] [561] Wiener, Human (1950)
[114] [558] Wiener, Cybernetics (1948), p. 132
[115] [558] Wiener, Cybernetics (1948), p. 132

In *The Human Use of Human Beings*, in the chapter entitled "Organization As the Message", he wrote about the essence of the individuality of living things. This individuality was not, as had been assumed, material in nature but was essentially information – stored information – that was transmitted during cell division. Since the amount of information contained in a germ cell was comparable to all of the volumes of the *Encyclopedia Britannica*, he did not see any reason why humans should not one day be able to transport this information via technical communication channels. It was thus possible in principle to transport all of the information that makes up a human. The most serious problem would probably remain that of keeping the person alive in the process.[116] Living organisms could therefore be interpreted as items of information, Wiener wrote, and in so doing he anticipated much of the present euphoria about views of cyberspace in the future.[117]

The visions of the future called up by the success of the cybernetics books flanked the resurging calls for a General Systems Theory: Bertalanffy had never denied that the two inter- and transdisciplinary movements were very similar; he had even stressed that the feedback principle could explicitly be found in Wiener's theory and could be presented as one of the principles that were found spanning the sciences. At the time, he even used Wiener's example, though without quoting it directly, saying it was "well known that a new discipline, called Cybernetics, was introduced by Norbert Wiener to deal with these phenomena":

> Furthermore, feedback systems comparable to the servomechanisms of technology exist in the animal and human body for the regulation of actions. If we want to pick up a pencil, report is made to the central nervous system of the amount of which we have failed the pencil in the first instance; this information then is fed back to the central nervous system so that the motion is controlled till it reaches its aim.[118]

Bertalanffy also tied Shannon's *Mathematical Theory of Communication* closely into the network of General System Theory:

> It has often been said that energy is the currency of physics, just as economic values can be expressed in dollars or pounds. There are, however, certain fields of physics and technology where this currency is not readily acceptable. This is the case in the field of communication which, due to the development of telephones, radio, radar, calculation

---

[116] Wiener emphasized that he did not want to get tangled up in science fiction. The problem in question was solved in later science fiction stories, such as the American TV series *Star Trek* in the 1960s, by what was called "beaming"!

[117] The statements about this passage rely heavily upon [275] Kay, Leben (2000), p. 131f

[118] [72] Bertalanffy, System (1954), p. 5f

machines, servomechanisms and other devices, had led to the rise of a new mathematical field.[119]

Bertalanffy had seen General Systems Theory "as a logical-mathematical area, the task of which is the formulation and derivation of those general principles that apply absolutely to 'systems'. In this way, it is possible to make exact formulations of system properties, such as entirety and sum, differentiation, progressive mechanization, centralization, hierarchical order, finality and equifinality etc., in other words, characteristics which appear in all sciences that are concerned with systems and determine their homology".[120]

First of all, this idea differed from that of cybernetics, and the original approaches of the two "meta-sciences" also stemmed from completely separate fields, biology and technology. Additionally, General Systems Theory sought general dynamic interactions while cybernetics elevated feedback to its central principle and expanded into other fields (Fig. 3.11).

Bertalanffy therefore deserves the credit for a more comprehensive approach, yet war-related developments had not been good for his program. Following the Second World War, General Systems Theory and the much more popular cybernetics cross-pollinated each other's efforts, and new and similar scientific conceptions were added. So Bertalanffy did not claim that his idea held a monopoly on interdisciplinary movement within the system of sciences. Instead he presented it as *one* member of a "group of modern currents, which also included the theory of information, cybernetics, game theory, operational research theory and others".[121]

The enormously successful development of this entire "group of currents" in the second half of the 20th century need not be discussed any further here.[122] In the mid-1970s, the editors of an anthology on *System Theory and System Technology* found that it was "not sensible to attempt here and now

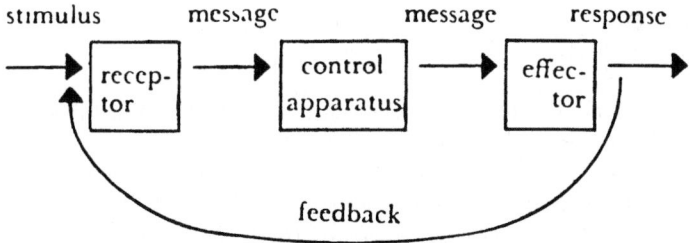

Fig. 3.11. Bertalanffy's Stimulus-Reaction Diagram

---

[119] [72] Bertalanffy, System (1954), p. 5f
[120] [76] Bertalanffy, Vorläufer (1972), p. 21
[121] [72] Bertalanffy, System (1954), p. 9
[122] See for example: [284] Klir, Approach (1969), [444] Pichler, Systemtheorie (1975), [266] Kalmann et al., Topics (1969), [399] Mesarovic, Views (1964), [398] Mesarovic et al., Systems (1975), [196] Gigch, Systems (1974), [228] Händle et

to define precise delimitations between the individual schools, whose findings, axioms and instruments can be used imaginatively as a great box of tools".[123] However, the "large field of system theory/cybernetics"[124] – a scientific tool box! – made the scientists see very plainly the ivory towers in which they had primarily been sitting and continued to sit, as George Klir summed up in his 1991 overview of the "facets" of the "system sciences". Generally speaking, the arguments put forth for decades had made the scientists increasingly sensitive to the boundaries of their disciplines, and they were becoming more aware of the fact that the important real-world problems could be understood only if they transcended the boundaries of the individual sciences to study those problems.[125] Two examples of the effects of this shift in consciousness are given here:

- The study *The Limits to Growth*, published in 1972 as a report to the Club of Rome[126] by an international team of scientists at MIT led by Dennis Meadows, inspired a great deal of uproar and consternation due to its pessimistic prognosis for the world but also because the scientists had used a procedure that was based on a system theory approach.
- The International Institute for Advanced Systems Analysis (IIASA) was founded at Laxenburg Palace near Vienna in 1972 by twelve scientific organizations in the east and west after US President Lyndon B. Johnson[127] suggested in 1966 that scientists from the USA and USSR might work together on non-military problems.[128]

---

al., Systemtheorie (1974), [319] Lenk, Systemtheorie (1978), [466] Rogers et al., Networks (1981)

[123] [228] Händle et al., Systemtheorie (1974), p. 14
[124] [228] Händle et al., Systemtheorie (1974), p. 15
[125] [285] Klir, Facets (1991), p. 175f
[126] The Club of Rome, an informal gathering of scientists, industrialists, economists and humanists founded by Italian industrialist Aurelio Peccei (1908–1984), had commissioned the study in 1970. Can the Earth physically cope with growth rates? How many people can live on it, with what degree of prosperity and for how long? These were the questions examined by the team led by American economist Dennis Meadows (born 1942) at MIT. Using a system dynamic method by the American computer pioneer and cyberneticist Jay Wright Forrester (born 1918), they represented the interactions in complex systems and related time delays in a computer-simulated model called "World B". The parameters Meadows chose as the most important factors for the developments of humanity were population growth, food production, industrialization, pollution and the exploitation of raw materials. According to the data material the group compiled, all five factors were growing exponentially in the 20th century. [371] Meadows et al., Growth (1973)
[127] Lyndon Baines Johnson (1908–1973), American politician, was Vice-President under John Fitzgerald Kennedy (1917–1963) and became the 36th President of the United States (1963 to 1969) upon Kennedy's assassination
[128] See: $http://www.iiasa.ac.at/docs/history.html?sb = 3$

To return to the topic from the beginning of this chapter, a brief intermediate summary is in order. The "unified science project" developed in ways completely different from how its creators in the Vienna Circle had ever intended or expected. Emigration, war and the inter- and transdisciplinary concepts that were subsequently favored had all had a hand in combining the scientific disciplines – from mathematics through the natural sciences, humanities, social and economic sciences all the way to engineering sciences– and linked them all to an open network by which the scientists of the day could find easy access to the like-minded from other institutions.

Among the foundations supporting interdisciplinary research in the 1940s and '50s were the *Macy Foundation*, the *Ford Foundation* and the *Sloan Foundation*.

The Josiah Macy Jr. Foundation[129] organized a number of group conferences, designating for each group a small number of scientists as its "core". Care was taken to invite a representative from each scientific discipline involved with the subject, and guests were also occasionally admitted. At the meetings, which were scheduled over two days, there were two or three presentations, but they were not supposed to take on the character of a formal lecture; the atmosphere of the presentations and discussions was expected to be as unceremonious as possible.

In May 1942, Rosenblueth delivered a presentation at such a meeting in New York on the theme of *Problems of Central-Inhibition in the Nervous System*; he spoke about the hypothesis he had developed with Wiener and Bigelow, which was to be published the following year as the article *Behavior, Purpose and Teleology*. Among the attendees were Gregory Bateson[130], Lawrence Schlesinger Kubie,[131] Warren McCulloch[132], Margaret Mead[133] and the medical director of the foundation, Frank Fremont-Smith[134], all of whom were so taken with this approach that they asked the foundation to form a

---

[129] Kate Macy Ladd founded the Josiah Macy Jr. foundation in 1930 in memory of her father, who died young. The foundation supported interdisciplinary research in the field of medicine, especially psychosomatic medicine. See: $http://www.josiahmacyfoundation.org$

[130] Gregory Bateson (1904–1980), British-American anthropologist and social scientist

[131] Lawrence Schlesinger Kubie (1896–1973), American psychologist and psychiatrist

[132] Warren Sturgis McCulloch (1898–1968), American physician and physiologist. McCulloch was first an assistant professor at Yale, then professor of psychiatry at the University of Illinois College of Medicine (1941–1948) and finally at the Electronics Research Laboratory at MIT

[133] Margaret Mead (1901–1978), American ethnologist and anthropologist. From 1926 to 1969, she was curator of the American Museum of Natural History in New York and in 1954 became professor of ethnology at Columbia University. In 1969, she founded the Institute for Intercultural Studies at the Museum of Natural History

[134] Frank Fremont-Smith (1895–1974), American physician, led the medical department of the Josiah Macy Jr. Foundation in the 1940s. In 1964, he became the

separate group just for this subject. Chairman of this new group was the neurophysiologist Warren McCulloch; its core was chosen from members of this conference on "cerebral inhibition" as well as some of the mathematicians, physiologists and engineers who had attended a meeting organized at Princeton by Norbert Wiener and John von Neumann in the winter of 1943–44. In addition to Warren S. McCulloch, Walter H. Pitts[135] and Rafael Lorente de No[136], this meeting was also attended by designers of the most modern computer systems.[137]

The other Macy conferences held by the new group took place between 1944 and 1953.[138] They were initially entitled *Circular Causal and Feedback Mechanisms in Biological and Social Systems* and were conducted by McCulloch. After Heinz von Foerster[139], who had just emigrated from Vienna, had given a lecture at a conference in 1949, the group appointed him secretary of the conference proceedings and editor of its publications.[140]

---

director of the Interdisciplinary Communications Program (ICP) at the New York Academy of Sciences

[135] Walter H. Pitts (born 1924) never completed college. He went to Chicago in 1937, where he worked in Nicolas Rashevsky's group on Mathematical Biology. Here he met Warren S. McCulloch, with whom he researched the theory of artificial neurons. In 1943, he became Norbert Wiener's research assistant at MIT. He suffered a nervous breakdown in 1951, destroyed all evidence of his past life and disappeared. He probably died in the 1960s

[136] Rafael Lorente de No (1902–1990), Spanish-American physician and neuroanatomist, worked in Madrid with Santiago Ramón y Cajal (1852–1934) during the latter's final years. Lorente De No later worked at the Rockefeller Institute

[137] See [526] Stewart, Origins (1959)

[138] All but the last of these conferences took place in New York City, at the Beekman Hotel, 575 Park Avenue. Out of respect for John von Neumann, the final conference was held at the Nassau Inn in Princeton, New Jersey

[139] Heinz von Foerster (1911–2002), Austrian-American physicist, studied engineering at the Technical University in Vienna until 1935 and was promoted in physics at the University of Breslau in 1944. He moved to the US in 1949, where he was a professor of communication engineering at the University of Illinois, Urbana from 1951 to 1975. He was also a professor of biophysics from 1962 to 1975

[140] Heinz von Foerster recalled the story of how the conference got its name as follows: "As a guest at the 6th Macy conference on March 24th and 25th, 1949, I was excluded from the business meeting that took place in the evening. When they asked me to come back in, though, the chairman Warren McCulloch announced that, because of my poor command of English, they had been trying to find a way for me to acquire this language as soon and as thoroughly as possible. And, they told me, they had found such a possibility. I was assigned to write the minutes of the conference, which were supposed to be published as soon as possible. I was completely floored! After I had gathered myself, I stated that the title of the conference "Circular, Causal Feedback Mechanisms in Biological and Social Systems" seemed too clumsy to me, and that I had wondered whether it might be possible to call the conference simply "Cybernetics" and to use the present name as a sub-title. When this suggestion was welcomed with immediate applause and

The meetings of the group newly renamed *Cybernetics* became known in the history of AI research[141] as the Macy meetings.[142]

At least as important as the "three-man paper" *Behavior, Purpose and Teleology* was for cybernetics – it has also been called "the birth certificate of cybernetics"[143] – was a "two-man paper" which was likewise published in 1943. It was called *A Logical Calculus of the Ideas Immanent in Nervous Activity* and its authors were McCulloch and the 19-year-old math student Walter H. Pitts. The paper appeared in the *Bulletin of Mathematical Biophysics*, which Nicolas Rashevsky had started just four years earlier.[144]

Walter Pitts had worked with Rashevky's bio-mathematical group, which may be the reason this journal was chosen. On this matter, McCulloch said later:

> Thanks to Rashevky's defense of logical and mathematical ideas in biology, this paper was published in his journal, where, so far as biology is concerned, it might have remained unknown.[145]

However, in the next sentence he noted the reason this text had become so significant:

> But John von Neumann picked it up and used it in teaching the theory of computing machines.[146]

---

laughter from all present, Norbert Wiener left the room with tears in his eyes in order to hide his emotions." [175] Foerster, KybernEthik (1993)

[141] AI = artificial intelligence

[142] Steve J. Heims dedicated an entire chapter of his book *The Cybernetics Group* ([236] Heims, Cybernetics (1991), Chap. ??) to the Macy Foundation and the Macy meetings

[143] [526] Stewart, Origins (1959), Sect. 4; Stewart is citing [307] Latil de, Pensée (1953)

[144] Rashevsky had hosted Bertalanffy in 1938. The fact that the article by McCulloch and Pitts appeared in this publication can be attributed to Pitts's acquaintance with the publisher. Wiener wrote in *Cybernetics*: "Another young migrant from the field of mathematical logic to cybernetics is Walter Pitts. He had been a student of Carnap at Chicago and had also been in contact with Professor Rashevsky and his school of biophysicists. Let it be remarked in passing that this group contributed much to directing the attention of the mathematically minded to the possibilities of the biological sciences, although it may seem to some of us that they are too dominated by problems of energy and potential and the methods of classical physics to do the best possible work in the study of systems like the nervous system, which are very far from being closed energetically." [558] Wiener, Cybernetics (1948), p. 21. The journal was published in Chicago, Illinois: University of Chicago Press, 1939–1972

[145] [366] McCulloch, Number (1961), p. 12, quoted from: [364] McCulloch, Geist (2000), p. 15

[146] [366] McCulloch, Number (1961), p. 12

Fig. 3.12. Members of the "Cybernetic Group" at a meeting in Paris. Left to right: W. Ross Ashby, Warren McCulloch, Grey Walter and Norbert Wiener

The text linked the activities of a network of nerve cells (neurons) with a complete logical calculus for time-dependent signals in electric circuits. Time is measured here as a synaptic delay. With that, McCulloch and Pitts had shown that a system of "artificial neurons" like this could calculate every logical consequence of their input, for:

> The "all-or-none" law of nervous activity is sufficient to insure that the activity of any neuron may be represented as a proposition. Psychological relations existing among nervous activities correspond, of course, to relations among the propositions; and the utility of the representation depends upon the identity of these relations with those of the logic of propositions. To each reaction of any neuron there is a corresponding assertion of a simple proposition.[147]

This paper by McCulloch and Pitts presented the first model of what were later called "artificial neuronal networks" (ANN), a model that, just like Shannon's paper six years earlier, was based on analogy. Following his studies, McCulloch, a psychologist working in the field of psychiatry, had one life-long research objective: to understand the physiology of the nervous system. In addition, he had been interested in logic since 1919 and occasionally "sunk his teeth into" an attempt to "manufacture a logic of transitive verbs".[148] After he had abandoned these attempts, he examined and recognized the possibilities that the logic of propositions afforded him[149]

---

[147] [361] McCulloch et al., Calculus (1943), p. 117
[148] [366] McCulloch, Number (1961), p. 8
[149] In his evaluation of the field of logic at that time, which he admittedly did not publish until 1961, McCulloch also mentioned that "the Polish school was well on

My object, as a psychologist, was to invent a kind of least psychic event, or 'psychon', that would have the following properties. First, it was to be so simple an event that it either happened or else it did not happen. Second, it was to happen only if its bound cause had happened – shades of Duns Scotus! – that is, it was to imply its temporal antecedent. Third, it was to propose this to subsequent psychons. Fourth, these were to be compounded to produce the equivalents of more complicated propositions concerning their antecedents.[150]

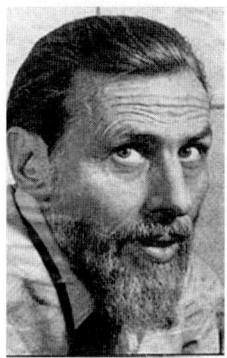

Fig. 3.13. W. McCulloch

Since 1929, McCulloch had been walking around with the subliminal notion that these "psychons" might be regarded as the "all-or-none impulses of neurons, combined by convergence upon the next neuron to yield complexes of propositional events"[151], but it was only while working with the mathematician Pitts that he was able to develop a corresponding model. Two difficulties were associated with the network connections of the neurons: 1) *Facilitation and extinction*: If a neuron was active and sending impulses into a region of a network, this affected that region's ability to react to new stimulus for a certain amount of time. 2) *Learning*: Stimuli that occurred at the same time could lead to lasting changes, and so stimuli that had earlier been inadequate to produce an effect were now effective. Both problems could be overcome by making appropriate substitutions in the model using "artificial" neurons with permanent connections and thresholds:

> But one point must be made clear: neither of us conceives the formal equivalence to be a factual explanation. *Per contra!* – we regard

---

its way to glory". It can thus be assumed with some certainty that he was also familiar with Tarski's work
[150] [366] McCulloch, Number (1961), p. 11
[151] [366] McCulloch, Number (1961), p. 12

facilitation and extinction as dependent upon continuous changes in threshold related to electrical and chemical variables, such as afterpotentials and ionic concentrations; and learning as an enduring change which can survive sleep, anaesthesia, convulsions and coma. The importance of the formal equivalence lies in this: that the alterations actually underlying facilitation, extinction and learning in no way affect the conclusions which follow from the formal treatment of the activity of nervous nets, and the relations of the corresponding propositions remain those of the logic of propositions.[152]

By modeling neurons after electric on-off switches, which according to Shannon's 1938 paper can be interconnected such that each Boolean statement can be realized, McCulloch and Pitts now "realized" the entire logical calculus of propositions by "neuron nets".[153] They presented it in "*Language II of R. Carnap (1938)*"[154] and they arrived at the following assumptions:

1. The activity of the neurons is an "all-or-none" process.
2. A certain fixed number of synapses must be excited within the period of latent addition in order to excite a neuron at any time, and this number is independent of previous activity and position on the neuron.
3. The only significant delay within the nervous system is synaptic delay.
4. The activity of any inhibitory synapse absolutely prevents excitation of the neuron at that time.
5. The structure of the net does not change with time.[155]

Every McCulloch-Pitts neuron is a threshold element: If the threshold value is exceeded, the neuron becomes active and "fires". By "firing" or "not firing", each neuron represents the logical truth values "true" or "false". Appropriately linked neurons thus carry out the logical operations like conjunction, disjunction, etc. In this paper, the psychologist McCulloch recognized great progress in his field and in neurophysiology, although he gave his co-author most of the credit:

---

[152] [361] McCulloch et al., Calculus (1943), p. 117

[153] McCulloch and Pitts actually spoke of "realizability" here

[154] In his book *The Logical Syntax of Language* [114], Carnap had differentiated formal Language I from formal Language II. Language I is a limited language of first-order predicate logic, with which a few statements of arithmetic can be represented. However, the system of Language I is not decidable and not complete. Language II is of a higher order and can be used to represent all of arithmetic. Language II is an expansion of the axiomatic system of Language I: For each statement that can be derived (or proven) from statements of Language I, the same is true for Language II. Carnap's Language II was expanded to include several characteristics that had been borrowed from B. Russell and A. N. Whitehead (1927), including the use of point notation in the *Principia Mathematica*. [361] McCulloch et al., Calculus (1943), p. 118

[155] [361] McCulloch et al., Calculus (1943), p. 116

But we had done more than this, thanks to Pitts's modulo mathematics. In looking into circuits composed of closed paths of neurons wherein signals could reverberate we had set up a theory of memory – to which every other form of memory is but a surrogate requiring reactivation of a trace. Now a memory is a temporal invariant. Given an event at one time, and its regeneration at later dates, one knows that there was an event that was of the given kind. The logician says: There was some x such than x was a ψ. In the symbols of the *Principia Mathematica*, $(\exists x)(\psi x)$. Given this and the negation, for which inhibition suffices, we can have $(\exists x)(\ \psi x)$, or, if you will, $(x)(\psi x)$. Hence we have the lower-predicate calculus with equality, which has recently been proved to be a sufficient logical framework for all of mathematics.[156]

Fig. 3.14. W. H. Pitts

The work done by McCulloch and Pitts initiated the research program of "Neuronal Information Processing", a collaboration involving psychology and sensory physiology, for which other groups of researchers were soon interested – scientists such as Jerome Y. Lettvin[157] and Humberto Maturana[158] with their work on neurons in frog eyes[159] or David Hunter Hubel and Torsten Nils Wiesel with their research on nerve cells in the brains of cats.[160] The science historian Olaf Breidbach described the situation at the time in this way:

---

[156] [361] McCulloch et al., Calculus (1943), p. 116
[157] Jerome Ysrael Lettvin (born 1920), American neuroscientist. Professor emeritus of electrical and bioengineering and communications physiology at MIT
[158] Humberto Maturana Romesín (born 1928), Chilean biologist
[159] Jerome Y. Lettvin and Humberto R. Maturana, published jointly: [358] Maturana et al., Anatomy (1960)
[160] David Hunter Hubel (born 1926) and Torsten Nils Wiesel (born 1924) received the Nobel Prize in Medicine in 1981 "for their discoveries concerning information processing in the visual system". In studies conducted in area 17 of narcotized cats in 1961, they were able to discern the cell responses with respect to linearity for spatial summation of partial stimuli in the receptive field

The neurophysiologists seemed to be on the right track. Should the brain not be understood to be a sum of simple logical circuits, as McCulloch had claimed? Was it possible, then, to simulate the logical structure and thus the performance of the brain, at least in its very earliest contours, by networking some appropriately constructed relays?[161]

Indeed, all subsequent work on artificial neuronal networks can be traced back to McCulloch and Pitts's deliberations, although several more years would pass before the McCulloch-Pitts neurons could be recreated in a machine.

Even then, though, Breidbach's rhetorical questions couldn't be answered in the affirmative. Bitter setbacks led to much discouragement, and as a result interest was renewed in developing so-called 'Von Neumann computers'. The inception of these machines should be viewed in very close relation to the McCulloch-Pitts paper, and for this reason, that aspect should be touched upon before the artificial neuronal networks are discussed.

3.5 Von Neumann, the Computer and the Brain

It was a coincidence one day in 1944 that John von Neumann[162] and Herman Goldstine[163] wanted to take the train from Aberdeen, Maryland at the same time. Both were coming from the Ballistic Research Laboratory's proving grounds there.[164] Neumann was headed back to Princeton, where he was a professor at the Institute for Advanced Study, and Goldstine was traveling to Philadelphia, where the ENIAC had been created to calculate rocket trajectories and target intercept data for the army.[165] When Goldstine told Neumann

---

[161] [98] Breidbach, Materialisierung (1997), p. 368

[162] John von Neumann (1903–1957), Hungarian-American mathematician and computer pioneer

[163] Herman Heine Goldstine (1913–2004), American mathematician, director of the Electronic Computer Project at the Institute for Advanced Study in Princeton (1946–1957)

[164] Neumann was hired as a part-time advisor by the Ballistic Research Laboratory in Aberdeen. As an expert on theoretical hydrodynamics, he compiled calculations concerning the behavior of the shock waves of explosions as well as ballistics tables. In September 1942, he also got a contract from the Navy Department, Bureau of Ordnance; he was to work for the navy in Washington until the end of 1942 and then in England from January to July of 1943. His task was to discover the patterns produced by the explosions of German underwater mines. His shock wave calculations would also prove to be important when he joined the atom bomb project in Los Alamos in 1943

[165] The ENIAC (Electronic Numerical Integrator And Computer) was the first electronic digital computer. It was built under the direction of electrical engineers John Presper Eckert and John William Mauchly at the Moore school in

3.5 Von Neumann, the Computer and the Brain

of the development of this "electron computer", which was supposed to perform 333 multiplications per second, von Neumann was so interested that just a few days later – probably on August 7, 1944 – he was standing in front of this (not yet completed) computer.

The ENIAC had an *electronic* working memory, so the individual processing operations of the entered data were exceptionally fast. However, each program that was going to be run had to be hard wired, and so reprogramming required several hours of work.[166] Neumann recognized very quickly, though, that this was a major drawback to the huge computer, and he was soon looking for ways to modify it.

Fig. 3.15. J. v. Neumann

Today the novel concept of a central programming unit in which programs are stored in coded form is attributed to John von Neumann. Instead of creating the program by means of the internal wiring of the machine, the program is installed directly in the machine. Basic operations like addition and subtraction remain permanently wired in the machine, but the order and combinations of these basic functions could be varied by means of instructions that were entered into the computer just like the data.

The ENIAC was not rebuilt, however, but was constructed just as it had been designed. Nevertheless, the next project was being planned and designed during this period. The next computer was to be called EDVAC[167] and was not supposed to suffer from the "childhood diseases" that had afflicted the

---

Philadelphia. All told, the ENIAC consisted of 100,000 parts; in addition to the tubes, there were 1,500 relays, 70,000 resistors, 10,000 capacitors and 6,000 toggle switches. It computed 500 times faster than the electromagnetic computers of the day. It could do the work of 75,000 human "computers"

[166] Although there were program patch panels that were interchangeable, and so some of the preliminary work could be performed outside of the system beforehand, this machine was really not very dynamic. Ultimately, though, the ENIAC was only intended for a limited type of task

[167] EDVAC (Electronic Discrete Variable Automatic Computer)

ENIAC. To this end, Neumann's principle of store programming was used, and thus the principle that went down in the history of the computer as "Von Neumann architecture" was realized for the first time.[168]

In the spring of 1945, Neumann was asked to prepare a report on the logical principles of the EDVAC, since the ENIAC had not had any such description and it had been sorely missed.

John von Neumann wrote a text that is recorded in computer history as *First Draft of a Report on the EDVAC*. Neumann identified three fundamental parts of the computer that should be examined separately:

- a central arithmetic part (CA),
- a central control part (CC) and
- a memory (M).

> The three specific parts CA, CC (together C) and M correspond to the associative neurons in the human nervous system. It remains to discuss the equivalents of the sensory or afferent and the motor or efferent neurons. These are the input and the output organs of the device, and we shall now consider them briefly.[169]

The fourth chapter of the text bears the title *Elements, Synchronism Neuron Analogy*. Here Neumann quotes the article by McCulloch and Pitts and, just as they had, he ignores the more complicated aspects of neuronal functions:

> Thresholds, temporal summation, relative inhibition, changes of threshold by after effects of stimulation beyond the synaptic delay, etc. It is however convenient to consider occasionally neurons with fixed thresholds 2 and 3, that is neurons which can be excited only by (simultaneous) stimuli on 2 or 3 excitatory synapses (and none on an inhibitory synapsis).[170]

Finally, Neumann noted that neuronal functions simplified in this way can be emulated by telegraph relays or vacuum tubes.[171]

It is a testament to his legendary lightening-quick powers of perception and implementation that John von Neumann was not only familiar with the vocabulary of the McCulloch and Pitts article, which had been published just a year before, but that he used these words to describe the EDVAC computer, as Heike Stach showed:

---

[168] The EDVAC was first developed exclusively for the army. However, new theoretical considerations kept coming up and so its completion stalled. It took a total of eight years for the EDVAC to be put into operation in 1952

[169] [421] Neumann, EDVAC (1945), Sect. 2.6; John von Neumann: *First Draft of a Report on the EDVAC*. The document can be accessed online, e.g.: http://www.alt.ldv.ei.tum.de/lehre/pent/skript/VonNeumann.pdf

[170] [421] Neumann, EDVAC (1945), Sect. 4.2

[171] [421] Neumann, EDVAC (1945), Sect. 4.2

3.5 Von Neumann, the Computer and the Brain

The idea that the human nervous system and the von Neumann device are very similar permeates the terminology of the "First Draft": The device has organs – 'specialized organs' – for addition, subtraction, multiplication, division and square roots, 'general control organs', 'memory organs' etc. They themselves are made up of sub-organs or networks of 'E,-elements', which von Neumann introduced as 'analogs of human neurons'. The organs, networks and E-elements receive 'stimuli', they are 'excited' or 'quiescent'.[172]

Stach also provides an answer as to why Neumann did this:

> The similarity between neurons and electric switching elements was apparently so clear to him that he did not thoroughly question it. He adopted his neuron model from a paper that explained the brain and nervous system to a logical computer and drew the inverse conclusion: If brain and nervous system are essentially a logical computer, then it must also be possible to describe computers with the help of neuron-like elements.[173]

Everyone involved was naturally aware at the time of how important the simplifications were to McCulloch and Pitts's neuronal network model; in the inverse conclusion, of course, this meant that they were quite far removed from also being able to prove this fundamental similarity between brain and computer for the actual facts any time in the near future. Neumann spent the rest of his life working on the question of how knowledge about the functions of biological brains could be utilized in the design of computers.[174] Nonetheless, he also expressed very negative views, particularly when he talked about the great complexity of the brain compared to the computers that were considered feasible at the time. He was also thinking about this when he spoke up during the discussion after a lecture by McCulloch at the Hixon Symposium held on September 20, 1948 at the California Institute of Technology.[175] The symposium's subject was the parallels between the so-called "electron computer" and the brain. Warren McCulloch, the second speaker, presented a lecture called "Why the Mind is in the Head".[176] This was extremely provocative, since neurophysiologists found the model far too simple and feared an overwhelming mathematization of neurophysiology that might not ever return to the actual nervous system. After all, the McCulloch-Pitts model couldn't even come close

---

[172] [517] Stach, Rechner (2003)
[173] [517] Stach, Rechner (2003)
[174] In the summer of 1956, when von Neumann was already bed-ridden with cancer, he was invited to deliver the Silliman Lectures at Yale. He managed to prepare two lectures, which were later published by his widow as the book *The Computer and the Brain*. [420] Neumann, Brain (1958)
[175] This discussion can be read in [364] McCulloch, Geist (2000), pp. 107–158
[176] [365] McCulloch, Mind (1951): McCulloch, Warren S.: Why the Mind is in the Head. [258] Jeffress, Mechanisms (1951), pp. 42–111

to mapping the immense multitude of connections running between the natural neurons in the human brain via their axons, dendrites and synapses, nor could it account for the influence they have on one another through electrochemically transported signals. That there were neurons of different types was also being ignored. Moreover, the McCulloch-Pitts neurons were not capable of learning, as they were statically linked to each other and so they could not allow for any variation in the influence they had on each other. In the Macy conferences, controversial discussions were now taking place about:

> whether logic (as stated earlier) is indeed sufficient to understand the brain's operations, since logic neglects the brain's distributed qualities. Alternative models and theories were put forth, which were accepted enthusiastically. This enthusiasm was fleeting at first, but remained secretly intact until it could be revived as an important alternative in cognitive science and technology in the 1980s.[177]

As a member of the generation of cognitive scientists that followed the *Cybernetics Group*, Francesco J. Varela[178] stressed a point that could not be denied even then:

> ... that in actual brains there seem to be no rules, no central logical processor, nor does information appear to be stored in precise addresses. Rather, brains can be seen to operate on the basis of massive interconnections in a distributed form, so that the actual connections among ensembles of neurons change as a result of experience. In brief, these ensembles present us with a self-organizing capacity that is nowhere to be found in the paradigm for symbol manipulation.[179]

In the chapter entitled *A Mutual Interest, but ..* of his "double biography" of John von Neumann and Norbert Wiener[180], Steve J. Heims writes extensively about the increasing alienation and squabbles within the *Cybernetics Group*. The group broke apart in the mid-1950s.

A new and important impulse in modeling neuronal networks now came from the study of natural neuronal networks when psychologist Donald O. Hebb[181] sought to identify the learning and the formation of a memory with changes in the strengths of the connections between the neurons. He presented the hypothesis that would become known as the "Hebb rule" in his 1949 book *The Organization of Behavior*[182]: When neurons fire together and simultaneously, the synaptic connection that exists between them is strengthened.

---

[177] [544] Varela, Kognitionswissenschaft (1988), p. 36
[178] Francesco J. Varela (1946–2001), Chilean biologist, close colleague of Humberto Maturana
[179] [544] Varela, Kognitionswissenschaft (1988), p. 54
[180] [237] Heims, Life (1980), pp. 201–220
[181] Donald Olding Hebb (1904–1985), American psychologist
[182] [234] Hebb, Organization (1949)

("Nerve cells that fire together wire together.") The "learning" should take place in this way.[183]

This rule was easy to understand, could be programmed and led to the first simulations of (small) neuronal networks in the 1950s: The switching of two neurons that are connected by a synapse should be changeable and should be proportionally correlated with the activity in front of and behind the synapse. If the synapse strength of a neuron under an input vector $x = (x_1, x_2, ..., x_L)$ is referred to as w_i and its stimulus as $y(x)$ and the size of the learning step is $\varepsilon > 0$, then the change in synapse strength is:

$$\triangle w_i = \varepsilon \cdot y(x) \cdot x_i.$$

3.6 Pattern Recognition with the Perceptron

The classic problem that an artificial neuronal network was supposed to solve was the classification of patterns of features (*pattern classification*), such as handwritten characters. Under the concept of a pattern, objects of reality are usually represented by pixels; frequency patterns that represent a linguistic sign, a sound, can also be characterized as patterns. In 1957 and 1958, Frank Rosenblatt at the Cornell Aeronautical Laboratory at Cornell University in Ithaca, New York developed a first machine for pattern classification.[184] He described this early artificial neuronal network, called Mark I Perceptron, in an essay for the *Psychological Review*.[185] It was the first model of a neuronal network which was capable of learning and in which it could be shown that the proposed learning algorithm was always successful when the problem had a solution at all.

Four hundred photocells formed an imaginary retina in this *perceptron* – a simulation of the retinal tissue of a biological eye – and over 500 other neuron-like units were linked with these photocells by the principle of contingency so they could supply them with impulses that came from stimuli in the imaginary retina. The actual *perceptron* was formed by a third layer of artificial neurons, the processing or response layer. The units in this layer formed a pattern associator.

---

[183] In 1951, the American mathematician and AI pioneer Marvin Lee Minsky (born 1927) had worked with Dean Edmonds in Princeton to develop a first neurocomputer, which consisted of 3,000 tubes and was called SNARC (Stochastic Neural-Analog Reinforcement Computer), in which the weights of neuronal connections could be varied automatically, but SNARC was never practically employed

[184] Frank Rosenblatt (1928–1971), an American neurobiologist, and Charles Wightman developed this first artificial neuronal network for pattern recognition problems. The Mark I Perceptron had a 20x20 pixel image sensor, could recognize simple figures and had 512 motor-driven potentiometers, which were responsible for each of the variable connection weights

[185] [471] Rosenblatt, Perceptron (1958), p. 392

In this classic *perceptron*, cells can be differentiated into three layers: Staying with the analogy of biological vision, the *input layer* with its (photo) cells or "stimulus units" (S cells) corresponds to the retinal tissue, and the middle *association layer* consists of so-called association units (A cells), which are wired with permanent but randomly selected weights to S cells via randomly linked contacts. Each A cell can therefore receive a determined input from the S layer. In this way, the input pattern of the S layer is distributed to the A layer. The mapping of the input pattern from the S layer onto a pattern in the A layer is considered "pre-processing". The *output layer*, which is what actually makes up the perceptron and which is thus also called the "perceptron layer", contains the pattern-processing response units (R cells), which are linked to the A cells. R and A cells are McCulloch-Pitts neurons, but their synapses are variable and are adapted respectively according to the Hebb rule. When the sensors detect a pattern, a group of neurons is activated, which prompts another neuron group to classify the pattern, i.e. to determine the pattern set to which said pattern belongs. A pattern is a point in the $x = (x_1, x_2, ..., x_N)$ N-dimensional vector space, and so it has N components. A pattern $x = (x_1, x_2, ..., x_N)$ such as this also belongs to one of L "pattern classes". This membership occurs in each individual use case. The *perceptron* "learned" these memberships of individual patterns beforehand on the basis of known classification examples it had been provided. After an appropriate training phase, it was then "shown" a new pattern, which it placed in the proper classes based on what it had already "learned". For a classification like this, each unit r of the *perceptron* calculated a binary output value y_r from the input pattern x according to the following equation:

$$y_r = \theta(\sum_{i=1}^{N} w_{ri} x_i).$$

The weightings w_{ri} were adapted by the unit r during the "training phase" in which the perceptron was given classification examples, i.e. pattern vectors with an indication of their respective pattern class C_s, such that an output value $y_r = 1$ occurred only if the input pattern x originated in its class C_r. If an element r delivered the incorrect output value y_r, then its coefficients w_{ri} were modified according to the following formula:

$$\Delta w_{ri} = \varepsilon_r \cdot (\delta_{rs} - y_r) \cdot x_i.$$

In doing so, the postsynaptic activity y_r used in the Hebb rule is replaced by the difference between the correct output value δ_{rs} and the actual output value y_r. These mathematical conditions for the perceptron were not difficult: Patterns are represented as vectors, the similarity and disparity of these patterns can be represented if the vector space is normalized; the dissimilarity of two patterns v_1 and v_2 can then be represented as the distance between these vectors, such as in the following definition:

$$d(v_1 v_2) = \| v_2 - v_1 \|.$$

3.6 Pattern Recognition with the Perceptron

The perceptron appeared to be a universal machine and Rosenblatt had also heralded it as such in his 1961 book *Principles of Neurodynamics – Perceptrons and the Theory of Mind*:

> For the first time, we have a machine which is capable of having original ideas. ... As concept, it would seem that the perceptron has established, beyond doubt, the feasibility and principle of nonhuman systems which may embody human cognitive functions ... The future of information processing devices which operate on statistical, rather than logical, principles seems to be clearly indicated.[186]

The euphoria came to an abrupt halt in 1969, however, when Marvin Minsky and Seymour Papert[187] completed their study of perceptron networks and published their findings in a book.[188] The results of the mathematical analysis to which they had subjected Rosenblatt's perceptron were devastating: Artificial neuronal networks like those in Rosenblatt's perceptron are not able to overcome many different problems! For example, it could not discern whether the pattern presented to it represented a single object or a number of intertwined but unrelated objects. The perceptron could not even determine whether the number of pattern components was odd or even. Yet this should have been a simple classification task that was known as a "parity problem".

The either-or operator of propositional logic, the so-called XOR, presents a special case of the parity problem that thus cannot be solved by Rosenblatt's perceptron. Therefore, the logical calculus realized by this type of neuronal networks was incomplete. Since the question of "linear separability" that was behind this problem was vital to further development – especially to the genesis of *Fuzzy Set Theory* – this matter should be discussed in detail here: The truth table of the logical functor XOR allocates the truth value "0" to the truth values of the two statements x_1, and x_2 when their truth values agree, and the truth value "1" when they have different truth values:

x_1	x_2	x_1 XOR x_2
0	0	0
0	1	1
1	0	1
1	1	0

x_1 and x_2 are components of a vector of the intermediate layer of a perceptron, so they can be interpreted, for example, as the coding of a perception by the retina layer. So $y = x_1 XOR x_2$ is the truth value of the output neuron, which

---

[186] [471] Rosenblatt, Perceptron (1958), quoted from [227] Hamilton, Netze (1993), p. 21f
[187] Seymour Papert (born 1928), American mathematician and AI pioneer
[188] [403] Minsky et al., Perceptrons (1969)

is calculated according to the truth table. The activity of x_1 and x_2 determines this value. It is a special case of the parity problem in this respect: For an even number, i.e. when both neurons are active or both are inactive, the output is 0, while for an odd number, where just one neuron is active, the value is 1.

To illustrate this, the four possible combinations of 0 and 1 are entered into a rectangular coordinate system of x_1 and x_2 and marked with the associated output values. In order to see that, in principle, a perceptron cannot learn to provide the output values demanded by XOR, the sum of the weighted input values is calculated:

$$w_1 x_1 + w_2 x_2.$$

The activity of the output depends on whether this sum is larger or smaller than the threshold value, which results in the plane extending between x_1 and x_2 as follows:

$$\Theta = w_1 x_1 + w_2 x_2, \quad \text{which results in:} \quad x_2 = -\frac{w_1}{w_2} x_1 + \frac{\Theta}{w_2}.$$

This is the equation of a straight line in which, on one side, the sum of the weighted input values is greater than the threshold value ($w_1 x_1 + w_2 x_2 > \Theta$) and the neuron is thus active (fires) but, on the other side, the sum of the weighted input values is smaller than the threshold value ($w_1 x_1 + w_2 x_2 < \Theta$) and the neuron is thus not active (does not fire).

However, the attempt to find precisely those values for the weights w_1 and w_2 where the associated line separates the odd number with (0, 1) and (1, 0) from the even number with (0, 0) and (1, 1) must fail (see Fig. 3.16). The proof is very easy to demonstrate by considering all four cases:

$x_1 = 0, x_2 = 1:$ y should be 1 . $\rightarrow$ $w_1 0 + w_2 1 \geq \Theta$ $\rightarrow$ neuron active!
$x_1 = 1, x_2 = 0:$ y should be 1 . $\rightarrow$ $w_1 1 + w_2 0 \geq \Theta$ $\rightarrow$ neuron active!
$x_1 = 0, x_2 = 0:$ y should be 0 . $\rightarrow$ $w_1 0 + w_2 0 < \Theta$ $\rightarrow$ neuron inactive!
$x_1 = 1, x_2 = 1:$ y should be 0 . $\rightarrow$ $w_1 1 + w_2 1 < \Theta$ $\rightarrow$ neuron inactive!

Adding the first two equations results in: $w_1 + w_2 \geq 2\Theta$.
From the last two equations comes: $\Theta > w_1 + w_2 \geq 2\Theta$, and so $\Theta > 2\Theta$.
This applies only where $\Theta < 0$. This is a contradiction of $w_1 0 + w_2 0 < \Theta$.
Q.E.D

The limits of the Rosenblatt perceptron had thus been demonstrated and they were very narrow, for it was not even able to classify linearly separable patterns.

In their book, Minsky and Papert estimated that more than 100 groups of researchers were working on perceptron networks or similar systems all

3.6 Pattern Recognition with the Perceptron

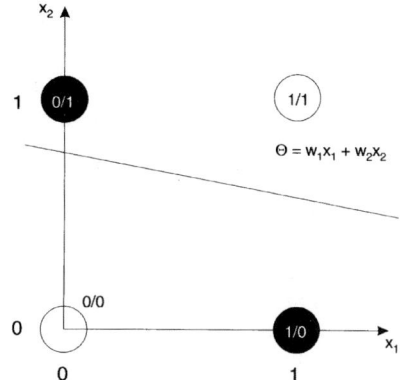

Fig. 3.16. Problem of linear separability

over the world at that time.[189] As a result of their fundamental criticism, many of these projects were shelved or at least modified in the years leading up to 1970.[190] The pattern recognition and learning networks had faltered on elementary questions of logic in which their competitor, the digital computer, had proven itself immensely powerful. Breidbach summed it up this way:

> The logification and the hierarchical-sequential development of a stimulus processing operation accordingly seemed to be the most promising procedure for tracing the phenomenon of the cognitive. Parallel programs – neuronal networks, such as those Hebb proposed as a model of brain functions – did not appear adequate here. The modeling of the brain functions, which also faced the problem of not being able to provide an adequate mathematical description of the interaction processes going on in the corresponding models, was initially considered obsolete and was abandoned.[191]

---

[189] In their paper *Adaptive Switching Circuits*, Bernard Widrow (born 1929) and Marcian Edward Hoff (born 1937) publicized the linear adaptive neuron model ADALINE, an adaptive system that was quick and precise thanks to a more advanced learning process which today is known as the "Delta rule", [556] Widrow et al., Adaptive (1960). In his 1958 paper *Die Lernmatrix*, German physicist Karl Steinbuch (born 1917) introduced a simple technical realization of associative memories, the predecessor of today's neuronal associative memories, [524] Steinbuch, Lernmatrix (1961). In 1959, the paper *Pandemonium* by Oliver Selfridge was published in which dynamic, interactive mechanisms were described that solved the problem of Morse code translation. [502] Selfridge, Pandemonium (1958)

[190] In the 15 years that followed, almost no research grants were approved for projects in the area of artificial neuronal networks, especially not by the US Defense Department for DARPA (Defense Advanced Research Projects Agency)

[191] [98] Breidbach, Materialisierung (1997), p. 369

3.7 Automata

The Hixon Symposium in September 1948 in Pasadena, California was important in the history of computer science for another reason: It was here that John von Neumann first delivered a lecture on his *General and Logical Theory of Automata*.[192] He spoke about analog and digital computers, and particularly about the ENIAC, he compared computing machines to living organisms, dealt with the "formal neural netting" of McCulloch and Pitts and outlined in broad terms a "future logical theory of automata" – he stressed that one did not exist yet. Then he motivated his audience and his readers for his subsequent statements:

> Thus there exists an apparent conflict of plausibility and evidence, if nothing worse. In view of this, it seems worth while to try to see whether there is anything involved here which can be formulated rigorously. So far I have been rather vague and confusing, and not unintentionally at that. It seems to me that it is otherwise impossible to give a fair impression of the situation that exists here. Let me now try to become specific.[193]

Here he turned to *Turing's Theory of Computing Automata*: "The English logician, Turing, about twelve years ago ... wanted to give a general definition of what is meant by a computing automaton."[194] Alan Turing[195] had constructed an abstract machine in 1936 that represented the process of computing on a paper band subdivided into fields. Neumann was interested in

---

[192] [419] Neumann, Automata (1951)
[193] [423] Neumann, Theorie (1967), p. 18
[194] [423] Neumann, Theorie (1967), p. 18
[195] Alan Mathison Turing (1912–1954), British mathematician, studied mathematics at Cambridge from 1931 to 1934. In 1936, he wrote the paper *On Computable Numbers, with an Application to the Entscheidungsproblem* [539], in which he reformulated Kurt Gödel's findings. He replaced Gödel's universal, arithmetic-based, formal language with simple, formal "automata" – known today as the Turing machine. He proved that this machine could solve every conceivable mathematical problem as long as there was an algorithm for it. In 1937 and 1938, Turing studied at Princeton, earning his doctorate there in 1938. During WWII, he was significantly involved in decrypting the German Enigma code at Bletchley Park. He worked 1945–48 on designing the Automatic Computing Engine (ACE) at the National Physical Laboratory, and in 1949 he became assistant director of the computer department at the University of Manchester, where he helped develop the software of the Manchester Mark I. In 1950, he turned his attention to the research field of Artificial Intelligence, and in his article *Computing Machinery and Intelligence* [540] he proposed an "imitation game" with which it could be determined whether or not a computer was a "thinking machine". Following criminal proceedings against him after the secret of his homosexuality became known, Turing is suspected to have poisoned himself

forming a class of calculable functions and he was convinced that this "computing" could also be carried out by a machine that had a finite number of discrete states $0, ..., m$, where $m \geq 0$, a reading-writing head and a scrolling mechanism, by means of which the computing tape, which is (potentially) unlimited in both directions and subdivided linearly into fields, can be moved by one field.

- The starting status is 0,
- The tape field over which the reading-writing head is positioned is called the working field.
- The scrolling mechanism can move the working field either to the right or left by one field.
- The reading-writing head can
  - read,
  - delete
  - and print the letters of the alphabet $\{*, |\}$.
- When the Turing machine stops, a lamp lights up.[196]

Turing's machine was a purely theoretical model, a kind of universal minimal computer, which replicated the idea of an effective computing process (Fig. 3.17); it was inadequate for Neumann's thought processes, however:

> His automata are purely computing machines. Their output is a piece of tape with zeros and ones on it. What is needed for the construction to which I referred is an automaton whose output is other automata.[197]

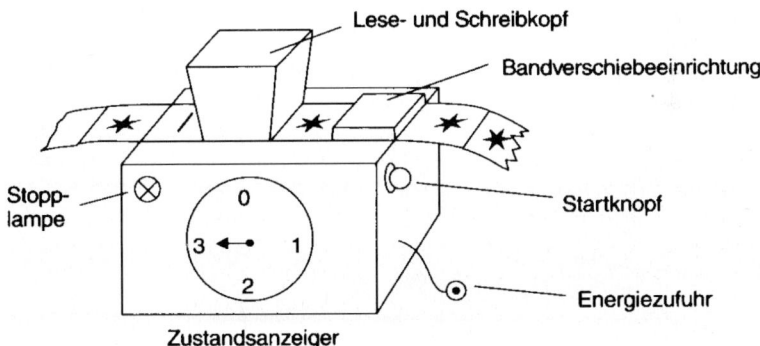

Fig. 3.17. Drawing of a Turing machine by Klaus Mainzer

---

[196] For the sake of comprehensibility, I am using the description of the Turing machine in [346] Mainzer, Computer (1995), p. 79ff
[197] [423] Neumann, Theorie (1967), p. 19

John von Neumann wanted automata that could propagate themselves and construct the other automata, and he also considered what "mutations" might be in this case. He concluded what he called his "very crude steps in the direction of a systematic theory of automata" with the claim that "automata which can reproduce themselves, or even construct higher entities"[198] would also become possible, and he conjectured that this "fact" would be important in the future. With that, a concept was called into being – it only needed to be acted upon – that would become the topic of a conference organized by mathematician John McCarthy[199] at Dartmouth in the summer of 1956: *Artificial Intelligence.*[200]

Working with Marvin Minsky, Nathaniel Rochester and Claude Shannon, John McCarthy (Fig. 3.18), who had just come from Princeton to be a professor at Dartmouth College, suggested discussing this subject in a definitive group:

> We propose that a 2 month, 10 man study of artificial intelligence be carried out during the summer of 1956 at Dartmouth College in Hanover, New Hampshire. The study is to proceed on the basis of the conjecture that every aspect of learning or any other feature of

Fig. 3.18. Marvin L. Minsky und John McCarthy

---

[198] [423] Neumann, Theorie (1967), p. 22
[199] John McCarthy (born 1927), American mathematician
[200] "The proposal for the Dartmouth Summer Research Project on Artificial Intelligence was, so far as I know, the first use of the phrase *Artificial Intelligence*," McCarthy wrote on his website: $http : //www-formal.stanford.edu/jmc/history/dartmouth.html$

intelligence can in principle be so precisely described that a machine can be made to simulate it. An attempt will be made to find how to make machines use language, form abstractions and concepts, solve kinds of problems now reserved for humans, and improve themselves. We think that a significant advance can be made in one or more of these problems if a carefully selected group of scientists work on it together for a summer.[201]

Ten people took part in these Dartmouth workshops: John McCarthy, Marvin L. Minsky, Claude E. Shannon, Nathaniel Rochester, Raymond J. Solomonoff[202], Oliver Selfridge, Trenchard More, Arthur Samuel[203], Herbert A. Simon[204] and Allen Newell.[205] In today's historiography of informatics, this workshop symbolizes a changing of the guard from one generation to the next. The presentation of *Logical Theorist* by Allen Newell and Herbert A. Simon heralded a new era, and at McCarthy's suggestion the new field of research received a new name: *Artificial Intelligence*. John von Neumann's ideas had been built upon the Turing machine, and from them had developed the automata theory, which became the basis for the Theoretical Informatics. People from far outside the circle of specialists were soon seeking access to this theory.[206]

---

[201] J. McCarthy, M. L. Minsky, N. Rochester, C. E. Shannon, *Proposal for the Dartmouth Summer Research Project on Artificial Intelligence*, August 31, 1955, $http://www-formal.stanford.edu/jmc/history/dartmouth/dartmouth.html$

[202] Raymond J. Solomonoff (born 1926)

[203] Arthur Samuel (1901–1990), American electrical engineer, studied (among other places) at MIT, where he earned his Master's degree in 1926. In 1928, he went to Bell Telephone Laboratories and in 1946, Samuel became a professor of electrical engineering at the University of Illinois. In 1949, he worked at IBM on their first stored-program computer, the IBM 701. In 1966, following his retirement from IBM, he went to Stanford University

[204] Herbert Alexander Simon (1916–2001), American social scientist with an education in mathematics and economics. Beginning in 1942, Simon worked as a political scientist at the Illinois Institute of Technology in Chicago. From 1954, he researched decision-making processes by means of computer simulation. He was awarded the Nobel Prize in Economics in 1978

[205] Allen Newell (1927–1992) studied physics at Stanford University, where he obtained a Bachelor's degree in 1949. After a year at Princeton University and while working for the RAND Corporation (1950–1961), he earned a Ph.D. at the Carnegie Institute of Technology in 1957 with a paper about industrial administration. He became a professor there in 1961

[206] The editorial staff of the German journal *Kursbuch* had the impression in March 1967 that mathematics was penetrating into a wide spectrum of modern life and was "embodied" by automata. Since this also "seems familiar to the layperson from the outside" but nevertheless formed part of the basics that "remained hidden to him by an underbrush of symbols, formulae, equations", an attempt was made here to provide clarification. Two already somewhat dated papers were translated and printed: *Can a Machine Think?* by Alan Turing, which had been published in

In 1956, Claude E. Shannon and John McCarthy published a very well-regarded and influential collection of texts entitled *Automata Studies*, to which researchers from the younger and the older generations had contributed. Frank Rosenblatt[207] cited five of these papers in his text about the perceptron and elsewhere. The voluminous tome was split into three sections: *Finite Automata, Turing Machines* and *Synthesis of Automata*, so the new automata theory could be treated from different sides, as the table of contents shows [208]

CONTENTS

Preface

FINITE AUTOMATA

S. C. Kleene:	Representation of Events in Nerve Nets and Finite Automata
J. von Neumann:	Probabilistic Logics and the Synthesis of Reliable Organisms from Unreliable Components
James T. Culbertson:	Some Uneconomical Robots
M. L. Minsky:	Some Universal Elements for Finite Automata
Edward F. Moore:	Gedanken-Experiments on Sequential Machines

TURING MACHINES

Claude E. Shannon:	A Universal Turing Machine with Two Internal States
M. D. Davis:	A Note on Universal Turing Machines
John McCarthy:	The Inversion of Functions Defined by Turing Machines
K. de Leeuw, Edward F. Moore, C. E. Shannon, N. Shapiro:	Computability by Probabilistic Machines

SYNTHESIS OF AUTOMATA

Ross Ashby:	Design for an Intelligence Amplifier
D. M. MacKay:	The Epistemological Problem for Automata
Albert M. Uttley:	Conditional Probability Machines and Conditional Reflexes
Albert M. Uttley:	Temporal and Spatial Patterns in a Conditional Probability Machine

---

the journal *Mind* in 1950, [540] Turing, Machine (1950), and John von Neumann's *The General and Logical Theory of Automata*, which he had delivered in 1948 at the symposium *Cerebral Mechanisms in Behaviour* at the Hixon Foundation at the California Institute of Technology, [423] Neumann, Theorie (1967)

[207] [471] Rosenblatt, Perceptron (1958)
[208] See [504] Shannon et al., Automata (1956)

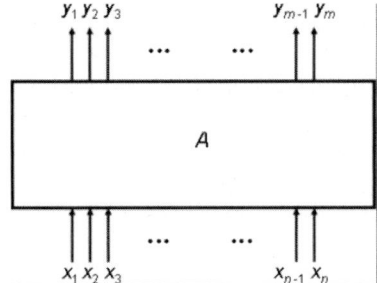

Fig. 3.19. Automaton A with inputs and outputs

According to Neumann's concept, a finite deterministic automaton is a machine with a finite number of inputs through which it receives information, a finite number of outputs through which it relays information and a finite memory in which it processes information.

The machine obtains the information as impulses (*inputs*) via an input channel, it processes the information by modifying its state and it passes it on as impulses (*outputs*) via an output channel. If the automaton receives an external impulse via one of its inputs while it is in a particular state, then it switches to a new state and, if necessary, sends an impulse out via an output. It is this reaction, consisting of a change in state and an output, that allows another impulse input.

A finite, non-deterministic automaton A of this type can be described as a system $A = (U, \Sigma, Y, fg)$. In this description:

U is a finite set of inputs,
Σ is a finite, non-empty set of states,
Y is a set of outputs,
f is a transformation $f : \Sigma \times U \times T \longrightarrow \Sigma$,
g is a transformation $g : \Sigma \times U \times T \longrightarrow Y$,

with T as the set of integers $\{..., -1, 0, 1, ...\}$.

Non-deterministic automata are automata whose behavior cannot be determined with certainty. Rather, it is possible to state only with a certain probability how the change of states occurs from time t to time $t+1$.

The next chapters will show that deterministic and non-deterministic automata were treated as special cases of a technically oriented system theory that had developed in the area of information and communication technology. However, the different orientations of the two system theories did not obscure what they also had in common.

4
From Circuit Theory to System Theory[1]

In March and June of 1956, Claude E. Shannon and Norbert Wiener, the co-editors of the *IRE Transactions on Information Theory*, each wrote an editorial. Under the title *The Bandwagon*, Shannon called for readers to bear in mind that, despite all of the popularity information theory had enjoyed over the previous few years, it was not a "universal remedy" and that they should thus return to serious research and development at the highest scientific levels.[2] The fact that information theory had been applied successfully in so many fields (the concepts of *information, entropy* and *redundancy* were now even being used in psychology, economics and the social sciences) was good news, but it also obscured the abstract meaning of these terms: "Indeed, the hard core of information theory is essentially a branch of mathematics, a strictly deductive system."[3]

Fig. 4.1. Claude Shannon

---

[1] This is the title of an article that is central to the history presented here: [662] Zadeh, System (1962)
[2] [506] Shannon, Bandwagon (1956)
[3] [506] Shannon, Bandwagon (1956)

Fig. 4.2. Norbert Wiener

Shannon was pushing back against the interdisciplinary expansion of his mathematical theory, and thus naturally against Wiener's cybernetics, as well. He concluded by calling upon the future authors of the journal to publish only the clearest and the best efforts: "Research rather than exposition is the keynote, and our critical thresholds should be raised."[4] Three months later, Wiener took the opportunity to respond. His plea was also: "Back to the roots!" Of course, he could not avoid – for the sake of the old argument over priority – referring to the different opinions on the roots of the theory. His editorial was titled with the question *What is Information Theory?* and after the reader learns once again in the first sentence that Shannon and Wiener should be considered the founders of information theory[5], and after the second sentence concurs with the statement in Shannon's earlier foreword that information theory was not a "magic key", this third sentence follows:

> I am pleading in this editorial that Information Theory go back of its slogans and return to the point of view from which it originated: that of the general statistical concept of communication.[6]

With references to the statistical character of the mechanics of Josiah Willard Gibbs[7] and the quantum theory that dominated physics as a whole, Wiener also placed information theory in this overall context:

> What I am here entreating is that communication theory be studied as one item in an entire context of related theories of a statistical nature, and that it should not lose its integrity by becoming a special vested interest attached to a certain set of slogans and clichés.[8]

---

[4] [506] Shannon, Bandwagon (1956)
[5] See Chap. 1
[6] [563] Wiener, Information (1956)
[7] Josiah Willard Gibbs (1839–1903), American mathematician and physicist, was professor of mathematical physics at Yale College in New Haven from 1871. He was one of the founders of modern thermodynamics
[8] [563] Wiener, Information (1956)

Finally, Wiener was clear in his dismissal of Shannon's "purism":

> I hope that these Transactions may encourage this integrated view of communication theory by extending its hospitality to papers which, why they bear on communication theory, cross its boundaries, and have a scope covering the related statistical theories. In my opinion we are in a dangerous age of overspecialization.[9]

Shannon's entreaty to concentrate on the core of information theory painted him into a scientific corner; it was much more desirable for the statistically-based information theory to draw from as many scientific areas as possible.

4.1 Dynamic Programming

A "step in the direction of a broad theory of communication, as contemplated by Wiener"[10] is how the mathematicians Richard Bellman[11] and Robert Kalaba[12] described their article *On the Role of Dynamic Programming in Statistical Communication Theory*, which appeared in the same journal the following year: They made explicit reference to Wiener's editorial, then considered a simple model of just one aspect of the general communication problem and reduced the function of the communication channel to a mathematical transformation: At discrete points in time, source S produces series members of pure signals x with noise r, which can be stochastic or deterministic. The combined signal $x' = f(x, r)$ is the input in the communication channel that is considered a black box and an output signal y is emitted by this black box (Fig. 4.3). Based on the observation of this output, it should be possible to draw a conclusion about the properties of the original pure signal x.

---

[9] [563] Wiener, Information (1956)
[10] [48] Bellman et al., Communication (1957)
[11] Richard Ernest Bellman (1920–1984) received his B.A. at Brooklyn College in 1941, his M.A. at the University of Wisconsin in 1943 and his Ph.D. in mathematics at Princeton University in 1946. At Princeton he was then an instructor and assistant professor before becoming an associate professor at Stanford University in 1948. Bellman took a position at the RAND Corporation in 1952, where he remained until 1965. Following that, he served as a professor of mathematics, electrical engineering and medicine at the University of Southern California until 1984. See the obituary by Lotfi A. Zadeh: In Memoriam Richard Bellman (1920–1984) in: *IEEE Transactions on Automatic control*, Vol. AC-29, No. 11, November 1984, p. 961. Further biographical information on Bellman can be found in [446] Plail, Entwicklung (1998), p. 118ff
[12] Robert E. Kalaba (1926–2004), American mathematician and electrical engineer. Kalaba qualified as a professor in electrical engineering at New York University in 1958. From 1969 until his death, he was a professor in the departments of biomedical engineering, economics and electrical engineering at the University of Southern California

Fig. 4.3. Communication channel as a black box

The transformation T, corresponding to the communication channel, represents the transmission of the input signal x', and so the following is true: $y = T(x') = T(f(x,r))$.

Therefore, the set of all communication systems corresponds to the set of all transformations T, and if an order or metric is introduced to this set, then it could be used to compare communication systems or evaluate their performance.

Fig. 4.4. R. Bellman

Bellman and Kalaba next discussed different methods of evaluating the performance of communications systems, but in this article they were ultimately interested in establishing the method of *dynamic programming* in information theory, a method Bellman had devised and developed since he had arrived at the RAND Corporation[13] in 1952 and had begun working with multi-stage decision processes. This work stemmed from questions from game theory, were based in part on findings by John von Neumann, Oscar

---

[13] The RAND Corporation was founded in 1948 and was initially part of the Douglas Aircraft Company in Santa Monica. With the financial support of the Ford Foundation, RAND became a non-profit organization and was thus formally independent of the US government. In fact, however, RAND receives a majority of the contracts from the Defense Department. See: [446] Plail, Entwicklung (1998), p. 55; [184] Friedman, Corporation (1963), p. 62; [102] Brentano et al., Beschreibung (1966).

Morgenstern[14] and Abraham Wald[15] and had a military motivation: "Examples of applications included the optimal use of guided weapons against moving or stationary enemy targets. The rockets were to be positioned such that the targets would sustain the maximum damage."[16]

Bellman's first publication on this subject had appeared during his first year at the RAND Corporation.[17] To show his description of the problem, I rely here upon the explanation provided by Michael Plail[18]: A set X is partitioned into portions Y and X–Y and the gain from this distribution is $g(Y) + h(X-Y)$. What remains is the portion $aY + b(X-Y)$, with which the partitioning process can be continued. For the maximum gain $f(x)$ obtained by an optimal allocation process in each stage of this process, Bellman devised the following functional equation, which appeared in a RAND report the next year[19]:

$$f(x) = \max_{0 \leq Y \leq X} \{g(X) + h(X - Y) + f(aY + b(X - Y))\}.$$

For an N-stage process, Bellman thus obtained:

$$f_1(x) = \max_{0 \leq Y \leq X} \{g(X) + h(X - Y)\}$$

$$f_N(x) = \max_{0 \leq Y \leq X} \{g(X) + h(X - Y) + f_{N-1}(aY + b(X - Y))\}, N = 2, 3, ...$$

Bellman formulated the general principle of setting up functional equations of this type as an "optimal policy":

> An optimal policy had the property that whatever the initial state and initial decision are, the remaining decision must constitute an optimal policy with regard to the state resulting from the first decision.[20]

From this optimal policy, Bellman constructed the method he called *dynamic programming* in order to emphasize the importance of time and the sequence of operations.[21] Plail outlines Bellman's *dynamic programming* as follows: A system S, described by the state set $p = (p_1, p_2, ..., p_M)$, passes through an n-stage decision process, at each stage of which a decision q_i is made. This

---

[14] Oscar Morgenstern (1902–1977), German-American economist (see footnote 69 in Chap. 2)
[15] Abraham Wald (1902–1950), Hungarian mathematician (see footnote 71 in Chap. 2)
[16] [446] Plail, Entwicklung (1998), p. 121; see also: [49] Bellman, Hurricane (1984)
[17] [57] Bellman, Theory (1983)
[18] [446] Plail, Entwicklung (1998), p. 122
[19] [676] Bellman, Introduction (1953)
[20] [676] Bellman, Introduction (1953); see also [446] Plail, Entwicklung (1998), p. 122
[21] See "Bellman's little story about coming up with the name for dynamic programming" in [446] Plail, Entwicklung (1998), p. 124

decision q_i transfers the system from a state p_i into a state p_{i+1}. The new state $p_{i+1} = T(p_i, q_i)$ can therefore be understood to be the result of a transformation T. $R(p_i, q_i)$ is the contribution to the gain function. The maximum gain $f_n(p)$ of the n-stage process depends merely upon the beginning state p and on n, the number of stages in the process. Therefore, the gain in each stage of the process can be calculated:

$$f_1(p) = \max_q R(p, q)$$

$$f_{n+1}(p) = \max_q \{R(p, q) + f_n(T(p, q))\} \qquad n = 1, 2, 3....$$

In his book about dynamic programming, Bellman made it clear that this method should be incorporated into the definition of each respective problem: "It burst upon the scene."[22] He describes the situation in this way:

a) In each case we have a physical system that is characterized in each stage by a small set of parameters, the so-called random variables.
b) At each stage of each process, we have the choice of a number of decisions.
c) The effect of a decision is a transformation of the state variable.
d) The past of a system is unimportant in the determination of future actions.
e) The purpose of the process is to make a function of the state variables into a maximum.[23]

This description was intentionally left "somewhat imprecise" in order to ensure that none of the "inspiration" was lost; Plail also calls it "unsharp".[24] The great popularity of this method was a result of this unsharp description, and Bellman himself was always looking for new areas where he could apply his optimal policy. "He acted as a skilled salesman of his findings with a flood of publications and applications in the most disparate fields. He published 621 journal articles – though with many, sometimes word-for-word, repetitions – and 40 books."[25] *On the Role of Dynamic Programming in Statistical Communication Theory* would be a further milestone in the triumphal procession of this method.

4.2 Identification Problems and a "Time Out" in Princeton

During his work on filters and general electrical networks, the issue of evaluating their performance was also of particular interest to Zadeh. As was shown at the end of Chap. 1, Zadeh was using his studies of black boxes and multipoles

---

[22] [48] Bellman, Dynamic (1957), p. 81
[23] Quoted from [446] Plail, Entwicklung (1998), p. 125
[24] [446] Plail, Entwicklung (1998), p. 123
[25] [446] Plail, Entwicklung (1998), p. 139

4.2 Identification Problems and a "Time Out" in Princeton 171

to examine electrical circuits and networks with respect to their input and their output. This abstraction of the inner states of systems then also made it possible to inquire completely abstractly as to the input-output relationship of a system and whether this could be "identified" by experimental means. Zadeh's thoughts *On the Identification Problem* appeared in the December 1956 edition of *IRE Transactions on Circuit Theory*.[26] In other scientific disciplines, very similar concepts were known as the *characterization problem, measurement or evaluation problem* or even *Gedanken Experiment*[27], but in the case of the present problem Zadeh found it more appropriate to say that a system should be *identified*: Given 1) a black box B whose input-output relationship is not known a priori and 2) the input space of B, which is the set of all time functions on which the operations with B are defined and 3) a black box class A that contains B, which is known a priori. Based on the observed response behavior of B for various inputs, an element of A should be determined that is equivalent to B inasmuch as its responses to all time functions in the input space of B are identical to those of B. In a certain sense, one can claim to have "identified" B by means of this known element of A.

Of course, this "system identification" can turn out to be arbitrarily difficult to achieve: Only insofar as information about black box B is available can black box set A be determined. If B has a "normal" initial state in which it returns to the same value after every input, such as the resting state of a linear system, then the problem is not complicated. If this condition is not fulfilled, however, then B's response behavior depends on a "not normal" initial state and the attempt to solve the problem gets out of hand very quickly.

Although Zadeh mentions some of the different approaches to solving the problem that had been proposed up to that point as well as the difficulties that had arisen[28], he ultimately considered these developments with little optimism. All of these efforts had been of theoretical interest, but they were not very helpful in practice and, on top of that, many of the suggested solutions did not even work when the "black box set" of possible solutions was very limited.[29] In the course of the article, Zadeh only looks at very specific non-linear systems which are relatively easy to identify by observation as sinus waves with different amplitudes. The identification problem remained unsolved for Zadeh. His resignation is palpable here, and it would soon lead to capitulation. Zadeh was nearing a crossroads.

---

[26] [618] Zadeh, Identification (1956)
[27] See [407] Moore, Gedanken (1956)
[28] Zadeh mentions Norbert Wiener's approach using Laguerre functions and Hermite polynomials as well as series from the '40 and early '50s, which included his own work
[29] For example, when only 2-poles "without memory" are considered. Such structures would then be characterized by a function $y = f(x)$, but it would not have either a Laguerre-Hermite or a Hermite representation

172   4 From Circuit Theory to System Theory

Fig. 4.5. H. Robbins

Herbert E. Robbins[30] was the chairman of Columbia University's department of mathematical statistics at the time. He was good friends with Lotfi Zadeh as well as with Deane Montgomery[31], a member of the Institute for Advanced Study in Princeton. Robbins and Montgomery campaigned for the approval of the IAS guest residency for which Zadeh had applied, even though it was rare for requests by scientists who were neither mathematicians nor theoretical physicists or historians to receive a positive response.[32] Zadeh initially took a half-year sabbatical from Columbia University in 1956. He wanted to write a book about linear systems[33] and he wanted to learn more about logic[34], an interest he had cultivated since 1950, when he predicted that logic,

---

[30] Herbert E. Robbins (1915–2001) was professor of mathematical statistics at the University of North Carolina and beginning in 1953 was a Higgins professor at Columbia University in New York. After his retirement in 1985, he became a professor of statistics at Rutgers University in New Jersey, retiring from there in 1997. Thereafter he lived in Princeton and conducted research 1952–1953 as a Guggenheim Fellow at the Institute for Advanced Study. See ISI-International Statistical Institute, *Newsletter*, Volume 25, No. 3 (75), 2001: $http : //www.cbs.nl/isi/memoriam01 - 3.htm$. [705] R. S. interview with L. A. Zadeh (1999)

[31] Deane Montgomery (1909–2002), American mathematician

[32] [708] R. S. interview with L. A. Zadeh (2001)

[33] As mentioned in Chap. 1, Zadeh published this book with Charles A. Desoer in 1963 with the title *Linear System Theory: The State Space Approach*, [586] Zadeh et al., System (1963). Charles A. Desoer (born 1926), Belgian-American electrical engineer, graduated as a radio engineer from the University of Liège in 1949. In 1953 he earned the Sc.D. in electrical engineering at MIT. Following this, he worked at Bell Telephone Laboratories until he moved to the University of California, Berkeley in 1958. There he held the Miller Institute Research Professorship from 1960 to 1969

[34] "... at that time I was planning to write a book on linear systems. That was my principle objective, to have a chance to write that book, and eventually the book was written together with Desoer in 1963, so it took quite a few years before a

4.2 Identification Problems and a "Time Out" in Princeton

and particularly multi-valued logic, would become increasingly more important to the problems of electrical engineering in the future.[35]

In addition to IAS professor Kurt Gödel[36], who permanently resided there, Zadeh also encountered Willard van Orman Quine[37] and Georg Kreisel[38], from whom he received much food for thought on the subjects of mathematical logic and the philosophy of mathematics.

For Zadeh, Princeton was the "Mecca for mathematicians"[39]; it inspired him very quickly. He attended lectures by Steven Kleene, who had also continued developing the multi-valued logic devised by the Polish school of logic. Kleene became Zadeh's friend and mentor at Princeton: "Steven Kleene was my teacher in logic. Yes, I learned logic from Steven Kleene!"[40] The residency was extended to an entire year and Zadeh started working on his book on linear systems.[41]

Fig. 4.6. S. Kleene

Zadeh found multi-valued logic to be a natural generalization of the conventional logic of just two values into n values, similar to the leap from two-dimensionality to n-dimensionality in mathematics.[42] He was now also toying with the idea of introducing multi-valued logic into automata theory and

---

great project came to fruition. That was one thing. The second objective that I had was to learn more about logic." [706] R. S. interview with L. A. Zadeh (2000)

[35] [667] Zadeh, Thinking (1950), cf. Chap. 1
[36] However, during the year he spent in Princeton, Zadeh did not manage to exchange a single word with Gödel, who was known to be a very reserved person. He did not see him at any institute events although they rode there together in the same car every morning. [707] R. S. interview with L. A. Zadeh (2001). See Chap. 2
[37] Willard van Orman Quine (1908–2000), American philosopher and logician
[38] Georg Kreisel (born 1923), Austrian-American philosopher and mathematician
[39] [706] R. S. interview with L. A. Zadeh (2000)
[40] [705] R. S. interview with L. A. Zadeh (1999)
[41] This book was not finished until a few years later, however, when Zadeh worked with Charles A. Desoer in Berkeley. [639] Zadeh et al., System (1963)
[42] [707] R. S. interview with L. A. Zadeh (2001)

implementing it in electric circuits, and once he had returned to Columbia University in New York he assigned two dissertations that dealt with the subjects of multi-valued logic in the design of transistor circuits[43] and with multi-valued coding[44],[45] "That's why I wanted to know about logics!" Zadeh recalls.[46]

Back at Columbia University from the Institute for Advanced Study in Princeton, Zadeh carried with him many very positive and lasting impressions. This residency had revealed to him some completely new perspectives of scientific life and work. New ways of thinking had come from the mathematics philosophers in Princeton, like Quine and Kreisel. Thanks to the mathematicians there he had learned new mathematical methods from statistics, from game and decision theory. He also experienced new views of system theory and the newly established automata theory[47], and all of the knowledge, impetus and impressions he found at the IAS would have a lasting effect on Zadeh's future endeavors!

Zadeh returned from Princeton with new enthusiasm. The value to that scientific community which he recognized in an institute like IAS in Princeton can also be seen in his dedication to establishing a similar institute for his own scientific community. When he once again composed an editorial for the *IRE Transactions on Information Theory* in 1960, it was entitled *Toward an Institute for Research in Communication Science*.[48] He had in the mean time become a professor at Berkeley and he apparently sometimes longed for the freedom he had enjoyed as a guest scientist in Princeton. He called for an institute for communication science to be founded where scientists could spend a year or two concentrating exclusively on their research without being distracted by teaching and administrative duties, contract negotiations and

---

[43] Oscar Lowenschuss wrote the dissertation *Multi-Valued Logic and Sequential Machines or Non-Binary Switching Theory* the following year: [326] Lowenschuss, Logic (1958). Parts of this paper had been published previously: [326] Lowenschuss, Comment (1958). See also the later publication [328] Lowenschuss, Organs (1959)

[44] Werner Ulrich managed to finish his dissertation *Nonbinary Error Correction Codes* in 1957: [542] Ulrich, Error (1957)

[45] Both dissertations on applications of multi-valued logic in electrical engineering were successfully defended and helped their authors obtain jobs in the industry. Lowenschuss, who was employed by the Sperry Company after his dissertation, also submitted a correspondence article about pattern redundancy in October 1957: [326] Lowenschuss, Comment (1958)

[46] [705] R. S. interview with L. A. Zadeh (1999)

[47] During his stay in Princeton, Zadeh had apparently become familiar with *Automata Studies*, published during this period by Shannon and McCarthy at Princeton University Press. It also contained a contribution by Steven Keene. Zadeh cited the article by Edward F. Moore in *Automata Studies* as early as 1956 in [618] Zadeh, Identification (1956), with the note: "On leave at Institute for Advanced Study, Princeton, N. J." See also Chap. 3

[48] [650] Zadeh, Research (1960)

4.2 Identification Problems and a "Time Out" in Princeton 175

doctoral advising. This was the only way to guarantee a free choice of research topics and scientifically communicative exchange with no outside pressure!

> A number of well-known institutions both in the United States and abroad have these characteristics, but they are embodied perhaps in their purist form in the Institute for Advanced Study, Princeton, N. J., which since its inception in 1930 had played a very significant role in the development of mathematics in their country.[49]

The IAS, Germany's Max-Planck-Institutes or the Institute for Automatics and Telemechanics in the Soviet Union did not necessarily have to serve as models for the institute he hoped to establish.

> Rather it would be designed to meet the specific needs and interests of workers in the fields of information theory, communication theory, system theory, control theory, automata, biological systems, computation, machine translation of languages and related fields. It would be concerned with both theoretical and experimental research in these areas.[50]

After Zadeh had also joined the editorial board of the *IRE Transactions on Automatic Control* in 1962, he wrote *A Critical View of Our Research in Automatic Control*[51] for the April edition, an article in which he repeated his call for the founding of such an institute. Here he suggested establishing an Institute for Control Science and Engineering, then he added: "(or, more broadly, an Institute for Research in Information Sciences) which would serve as a focal point on the national level for research in control theory and its applications as well as in such related fields as system theory, information and communication theories, circuit theory, machine computation, automata theory, bionics, etc."[52] It was only appropriate for the information and communication sciences, he reasoned, to create an institute like the one in Princeton, and it should also rest upon a similarly broad interdisciplinary basis. Here, too, Lotfi Zadeh was following in the tradition of Norbert Wiener.[53]

It's hard to resist the impression that Zadeh thought quite a lot about his year in Princeton:

> Traditionally, it was and is the function of universities rather than industry to support the basic non-profit-orientated scientific research. Our universities are performing this function very well in such fields as

---

[49] [650] Zadeh, Research (1960)
[50] [650] Zadeh, Research (1960)
[51] [603] Zadeh, Critical (1962)
[52] [603] Zadeh, Critical (1962), p. 74
[53] Both Shannon and Wiener influenced Zadeh considerably. It is my impression from conversations with Zadeh, however, that Zadeh was much closer to Wiener and his thinking. [705] R. S. interview with L. A. Zadeh (1999), [711] R. S. interview with L. A. Zadeh (2002)

mathematics and physics, where there is a long tradition of pure and applied research by university professors. The same can not be said of automatic control or, more generally, of engineering. The reason for this is that a typical university professor in theoretical engineering is usually so busy with teaching, textbooks writing, committee work, supervision of M.S. and Ph.D. candidates, etc. that he can devote only a small part of his time and energy to his own research, and even this he can do in a concentrated fashion only during the summer recess. Furthermore, he can not afford to spend much time in the library or the laboratory. As a result, his research output tends to be concerned with somewhat artificial solutions rather than with realistic complex problems which require a concerted and prolonged effort to yield results of theoretical of practical value. There are, of course, exceptions to this rule, but they are relatively few and far between.[54]

The many new impressions, encounters, discussions and insights in Princeton had steered Zadeh's future scientific work in new directions. Some of the texts he produced after this residency at the IAS and after his move from New York to Berkeley in 1959 show this very clearly, and they also prove – as indicated in the above quote – that he was not working on "artificial" problems but on "realistic and complex" ones. Accordingly – as the above passage also shows – his solutions also were not "elegant and publishable".[55] As is demonstrated in the examples delineated below, the fruits of these labors in the first half of the 1960s would lead to success only slowly and belatedly, though as the Theory of Fuzzy Sets and Fuzzy Systems, it was a success that looked very different from what was expected.

4.3 "Optimal Fetishism", Expanded Linearity Concepts and Adaptive Systems

In my view, the development described below should be seen as an intermediate step on Zadeh's path toward Fuzzy Set Theory, for it demonstrates clearly how Zadeh was trying to refine concepts whose definitions were no longer defined sharply enough for the interests of modern technology: "optimal", "linear", "adaptive". These attempts were the result of his intensive preoccupation with the mathematical foundations of electrical systems owing to his work on the textbook as well as the fact that he was becoming more and more aware of the incongruity between these theoretical systems and those

---

[54] [603] Zadeh, Critical (1962), p. 74

[55] The texts by Zadeh discussed here were printed as smaller essays, editorials and correspondence contributions in the respective professional journals. Zadeh surely could not have published them in this form as fully developed journal articles, but he nevertheless had them printed as thought-provoking items! They reveal very clearly the development of the scientific work he was doing at the time!

real systems in the practice of the laboratory, a disparity he had perceived since his student days at MIT.[56] The gap between these two areas of communication technology seemed to be getting wider and Zadeh was attempting to counteract this; he also wanted to place system theory on a new basis and he found a new way to do so in what was termed the "state space approach".

4.3.1 Optimality

When Zadeh penned an editorial for the March edition of the *IRE Transactions on Information Theory*, he pleaded – as Shannon had done two years earlier – for a critical examination of the situation in his own specialist field. He titled his foreword – as Wiener had done two years earlier – with a question: "What Is Optimal?" He asked the readers how reasonable it actually was to insist on optimal solutions. After all, this approach had only established itself since Wiener had publicized his findings on optimal filters and prediction, and proponents were well on their way "to make a fetish of optimality. If a system is not "best" in one sense or another, we do not feel satisfied. Indeed, we are apt to place too much confidence in a system that is, in effect, optimal by definition."[57]

Fig. 4.7. L. Zadeh

Finding an optimal system would mean choosing a performance criterion, then specifying a class of acceptable systems according to various conditions with respect to design, costs, etc. and finally accepting one of these systems from the specified class as the "best" with regard to these criteria. Zadeh now doubted that this method was any more sensible than the "relatively unsophisticated approach of the pre-Wiener era".[58]

The selection of a single performance criterion leaves all of the other criteria that would likewise contribute to performance evaluation unconsidered.

---

[56] See Chap. 1
[57] [637] Zadeh, Optimal (1958)
[58] [637] Zadeh, Optimal (1958)

The problem was oversimplified when scalar loss functions were used; vector-valued functions might possibly be more suitable but also more difficult to manage.

Zadeh similarly criticized the rational selection of decision functions under uncertainty: "What should be done when the probabilities of the "state of nature" characterizing a problem are not known?" Here Zadeh rejects the usual solution methods based on stochastics or game theory:

> At present no completely satisfactory rule for selecting decision functions is available, and it is not very likely that one will be found in the foreseeable future. Perhaps all that we can reasonably expect is a rule which, in a somewhat equivocal manner, would delimit a set of "good" designs for a system. In any case, neither Wiener's theory nor the more sophisticated approaches of decision theory have resolved the basic problem of how to find a "best" or even a "good" system under uncertainety.[59]

Zadeh mentioned the names of mathematicians and statisticians – Lionel Weiss[60], Herman Chernoff[61] and Witold Hurewicz[62] – with whose work he had apparently become familiar in Princeton. Michael Plail characterized this text as a "critique of optimal control, which at that time was just taking on a clear shape, in which, for example, distinct objective functions were provided and one was always trying for more precise calculations".[63] This assessment is not far-reaching enough, in my opinion, because Zadeh was trying to accomplish much more here than simply express criticism within the field of control theory, particularly since this text appeared in the *Transactions on Information Theory!* This plea is early evidence of Zadeh's doubts about traditional mathematics, which had heretofore rarely been challenged as a tool for understanding real systems. Zadeh was still reluctant to turn his back on mathematics altogether, however. As we shall see in the following section, he spent a while longer looking for alternatives, but this was just a rearguard action!

---

[59] [637] Zadeh, Optimal (1958), Editorial
[60] Lionel Weiss (1924–2000), American mathematician
[61] Herman Chernoff (born 1923), American mathematician
[62] Witold Hurewicz (1904–1956) (see footnote 80 in Chap. 2) became Menger's successor as Brouwer's assistant (1928–1936) after Menger was hired as a professor in Vienna (see Chap. 2). In 1936 he visited the IAS in Princeton during a study visit and decided to remain in the USA. He became a professor at the University of North Carolina and then, from 1945 to the end of his life, at MIT
[63] [446] Plail, Entwicklung (1998), p. 224. I am indebted to Michael Plail for alerting me to this Zadeh editorial, which he discovered while doing research for his book. However, Plail erroneously characterized Zadeh in his book (loc. cit.) as "coming from control engineering"

4.3.2 Linearity

Zadeh shook the mathematical foundations of system theory a second time when he discussed the problems associated with the concept of "linear". In the *Proceedings of the IRE* in June of 1961, he suggested expanding the definition of the mathematical concept of linearity. The usual definitions of linearity made sense only to a limited extent because they required that the system be at rest in its initial state. The more general definition he was suggesting did not require this limitation. This criticism was motivated by his knowledge from electrical engineering that time-varying networks are linear in their ground state, but not necessarily in the states they later assume.[64]

Zadeh defines the states of system B as vectors of a finitely or possibly infinitely dimensional state space. The state $s(t)$ of the system at time t describes the inner relationships of system B at time t, and $s(t_0)$ is accordingly its state at time t_0.

An input $u(t_0)$ in system B results in an output $y(t_1)$, but what happens in system B between times t_0 and t_1? To solve this problem, Zadeh introduced a new symbol for the "input segment" between these two points in time:

$$u_{t_0 \leq t \leq t_1}.$$

He then defined the system output $y(t_1)$ as a function F of the state at time t_0, of the time segment between t_0 and t_1 and of these two points in time themselves:

$$F[s(t_0); u_{t_0 \leq t \leq t_1}; t_0, t_1].$$

This is fully characterized for a given function F and a given initial state $s(t_0)$ of the system. The above equation can be used to calculate all possible system outputs that B can assume on the basis of the input segment $u_{t_0 \leq t \leq t_1}$, when $s(t_0)$ passes through all of the possible values. The output time function (over all intervals $[t_0, t_1]$) that results from an input $u_{t_0 \leq t \leq t_1}$ in B in the initial state $s(t_0) = q$ is referred to by Zadeh as $B[s(t_0); u_{t_0 \leq t \leq t_1}]$ or more simply as $B[q; u_{t_0 \leq t \leq t_1}; t_0, t_1]$ or $B[q; u]$.

Zadeh now defined the ground state of the system that is to be differentiated from the initial state as follows: Assuming that the initial state of system B at time t_0 is q and that it receives a zero input $u(t) = 0, t_0 \leq t \leq t_1$. The system state of B at time t_1 is $s(t_1)$; for $t_1 \to \infty$ it can converge to a solid state in the state space of B or not. If it converges to a solid state $s(\infty)$ and if it is independent of the initial state q, then $s(\infty)$ is called the *ground state* of B, which Zadeh symbolized with $\mathcal{O}$.

He defined a system B as *linear with respect to the ground state q* precisely when then following equation was true for all pairs of the input time functions u and v, all values of t_0 and t_1 and all real constants k:

$$k[B(q; u_{t_0 \leq t \leq t_1}) - B(q; v_{t_0 \leq t \leq t_1})] = B[\mathcal{O}; k(u_{t_0 \leq t \leq t_1} - v_{t_0 \leq t \leq t_1})].$$

---

[64] [624] Zadeh, Linearity (1961). See also Zadeh's earlier problematization of the "identification problem", which is outlined in Sect. 4.2 of the present chapter

He defined the *general linearity of systems as linear with respect to all possible initial states*, which means that for all $t_0, t_1, u_{t_0 \le t \le t_1}, v_{t_0 \le t \le t_1}, k$ and all q in the state space of B, the above equation must apply:

$$k[B(q; u_{t_0 \le t \le t_1}) - B(q; v_{t_0 \le t \le t_1})] = B[\mathcal{O}; k(u_{t_0 \le t \le t_1} - v_{t_0 \le t \le t_1})].$$

One month later, Zadeh submitted a reformulation of this definition of system linearity[65], in which he now used the zero state instead of the ground state $\mathcal{O}$. Although these two states were usually identical and the definitions thus equivalent, this was not always the case.[66]

4.3.3 "Non-Inferiority"

In September 1962 another of Zadeh's correspondence contributions appeared in the *IEEE Transactions on Automatic Control*; the title was *Optimality and Non-Scalar-Valued Performance Criteria*.[67] As he had done in *What Is Optimal?*, Zadeh criticized the scalar-valued performance criteria. One of the serious weaknesses of the theories of optimal control at the time, he wrote, was that they were based on the assumption that the performance of a system S could be measured by a single number, the real-valued performance index $P(S)$, and a system S_0 would then be called optimal in a class (set) Σ when the following applied: $P(S_0) \ge P(S)$ for all S in Σ.

However, since there was generally more than one approach to evaluating the performance of a system S and in most cases this approach could not be condensed into a single scalar-valued criterion, the difficulties with defining the concept of optimality began to reveal themselves. It was entirely possible for a system S to be better (*superior*) than a system S' in one respect, but worse (*inferior*) in another respect. The set of all considered systems Σ was therefore partially but not fully ordered.

Zadeh's suggested solution implicitly required that the optimality was differentiated from another system property, which he called *non-inferiority*. A subset (*constraint set*) C of Σ is defined by the restrictions on system S, and a partial order $\ge$ is defined on Σ, by which the following three disjoint constraint sets of Σ can be assigned to each system S in Σ.

1) The constraint set $\Sigma_>(S)$ of all systems that are superior to S.
2) The constraint set $\Sigma_\le(S)$ of all systems that are inferior or equal to S.
3) The constraint set $\Sigma_\sim(S)$ of all systems that are not comparable to S.

Since every system is an element of Σ, every system falls within one of these categories and the combination of $\Sigma_>(S), \Sigma_\le(S)$ and $\Sigma_\sim(S)$ is Σ, the set of all considered systems. With the aid of these constraint sets of Σ. Zadeh defined the system properties of *non-inferiority* and *optimality*.

---

[65] [625] Zadeh, Linearity (1962)
[66] [625] Zadeh, Linearity (1962)
[67] [639] Zadeh, Optimality (1963)

4.3 "Optimal Fetishism", Expanded Linearity Concepts and Adaptive Systems

Definition 1. *A system S_0 is non-inferior in $\mathcal{C}$ when the intersection of $\mathcal{C}$ and $\Sigma_>(S_0)$ is empty.*

$$\mathcal{C} \cap \Sigma_>(S_0) = \emptyset.$$

Therefore, there is no system in $\mathcal{C}$ that is better than S_0. It is also possible to state equivalently: S_0 is not inferior to other systems in $\mathcal{C}$.

Definition 2. *A system S_0 is optimal in $\mathcal{C}$ when $\mathcal{C}$ is retained in $\Sigma_\leq(S_0)$.*

$$\mathcal{C} \subseteq \Sigma_\leq(S).$$

Every system in $\mathcal{C}$ is therefore inferior to S_0 or equal to S_0.

The definitions of the sets $\Sigma_>(S_0)$ and $\Sigma_\leq(S_0)$ illustrate directly that an optimal S_0 is necessarily non-inferior but that the inverse need not apply.

If the set Σ of all considered systems is fully ordered by a scalar criterion, then $\Sigma_\sim(S_0)$ is the empty set while $\Sigma_>(S_0)$ and $\Sigma_\leq(S_0)$ are the complementary sets. If the intersection of $\mathcal{C}$ and $\Sigma_>(S_0)$ is empty, then $\Sigma_\leq(S_0)$ certainly contains the set $\mathcal{C}$. Thus *non-inferiority* and *optimality* are equivalent in this case and the difference between these concepts cannot be discerned for scalar-valued criteria.

Zadeh now proposed – and here he was dealing with a special case of the above definition – taking into account the partial ordering of Σ by a vector-valued performance criterion: The system S is characterized by a vector $x = (x_1, ..., x_n)$, the real-valued components of which are, for example, the values of n variable parameters of the system S, and $\mathcal{C}$ is a partial set of the n-dimensional Euclidean space. The performance of system S is measured by an m-dimensional vector $p(x) = [p_1(x), ..., p_m(x)]$, wherein $p_i(x), i = 1, ..., m$ is a real-valued function of x. Therefore, $S = S'$ applies precisely when $p(x) = p_i(x'), i = 1, ..., m$.

To illustrate a number of statements about non-inferior systems, Zadeh considered the case in which $\Sigma_>(S)$ or equivalently $\Sigma_>(x)$ is a solid cone with its apex at point x and the constraint set $\mathcal{C}$ is a closed restricted partial set in the vector space (Fig. 4.8).

This circumstance can occur, for example, when $p_i(x), i = 1, ..., m$ is in the following form: $p_1(x) = a_i^i x_1 + ... + a_n x_n^i$, where $a_i = (a_i^i, ..., a_n^i)$ is a constant vector, namely the gradient of $p_i(x)$, i.e. $a_i = grad\, p_i(x)$. In this case, $\Sigma_>(x)$ is the polar cone of the cone which is spanned by a^i.

It immediately follows from definition 1 that a *non-inferior* point cannot be any point in the interior of set $\mathcal{C}$. If $\mathcal{C}$ is additionally a *convex* set, then the set of all non-inferior points on the edge of $\mathcal{C}$ is the set Γ of all points x_0 through which the hyperplanes can pass, and so the sets $\mathcal{C}$ and $\Sigma_>(x_0)$ are separated by these hyperplanes. (In Fig. 4.9, the set Γ is the heavy line on the edge of $\mathcal{C}$.)

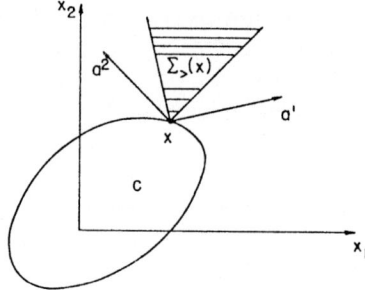

Fig. 4.8. Illustration of the definitions of C and $\Sigma_>(x)$

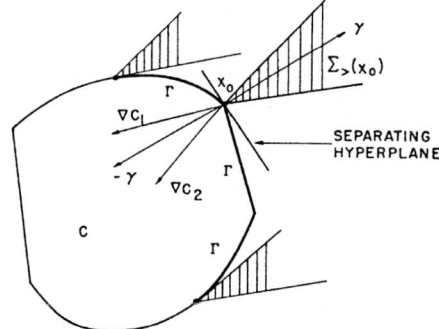

Fig. 4.9. The set of non-inferior points at the edge of C

If x_0 is a point such as this and γ is the normal (pointing away from the interior of the set) on the hyperplane in this point x_0, then γ belongs to the polar cone of set $\Sigma_>(x_0)$, since γ does not form an obtuse angle with the vectors in $\Sigma_>(x_0)$.

An important detail should be mentioned here with regard to the historical development of Fuzzy Set Theory, a detail that is hidden behind the above discussion of a convex set C. Zadeh had not sought the convexity of the set C without a very particular reason, since the following separation theorem exists for convex sets: If C_1 and C_2 are disjoint convex sets, then a hyperplane exists that separates them. Zadeh placed this information in a footnote with an additional reference to literature about mathematical programming.[68] Linear separability was a thoroughly interesting problem. Chap. 3 described how Minsky and Papert had proven in 1969 that Rosenblatt's perceptron failed at precisely this problem of linear separability. As Zadeh showed with his note about mathematical programming, restricting the scope to convex sets was very significant.[69]

---

[68] Zadeh mentioned [154] Egglestone, Convexivity (1958) as a source
[69] See the much more detailed explanation in Chap. 5.3

4.3 "Optimal Fetishism", Expanded Linearity Concepts and Adaptive Systems

We shall dispense with any further details from this text. Once the set Γ of the non-inferior systems in $\mathcal{C}$ is identified by the techniques outlined here or by other means, a system designer is now faced with the question of how he should select a single system from this class Γ if all of its elements are either incomparable or equivalent to one another. This can be done either by introducing a scalar-valued criterion that defines Γ but is not necessarily $\mathcal{C}$ or alternatively by rolling the dice. Zadeh advocates a two-phase process rather than – as in the conventional approach – choosing the scalar-valued criterion on Σ at the beginning, should the set Γ of the non-inferior systems be determined in a first step and a single system be selected in a second step.

In a footnote, Zadeh explains his concept of *non-inferiority*: There is a very similar concept of "admissibility" in statistics, which was borrowed from Abraham Wald[70], but then he writes: "We use the term "noninferior" because the terms "admissible" and "efficient" have become too fuzzy[71] through varying usage in different contexts. (An equivalent mathematical term is "maximal".)"[72]

If it is still hard to notice the similarity to mathematical statistics and decision theory in this text, possibly even because of that last footnote, this influence will become very clear in the following step.

4.3.4 Adaptivity

Toward the end of 1962 in his next correspondence contribution for the *Proceedings of the IEEE*, as the journal was now called, Zadeh was interested in adaptivity: *On the Definition of Adaptivity*.[73] This concept is also interesting because it has roots in biology[74], whereas the other concepts, like state and stability, had been carried over from the engineering-related system theory to biological, sociological and economic systems. The concept of the adaptive system had recently been playing an important role in theories of communication and control, Zadeh wrote in introducing his text, yet no satisfactory explanation or even a precise definition could be found in the literature. He referred to Bellman and Kalaba's paper *Dynamic Programming and Adaptive Control Processes: Mathematical Fundation*.[75] In this paper, the treatment of decision processes under conditions of uncertainty with respect to their underlying physical relationships was supposed to include a mathematical basis:

---

[70] [547] Wald, Decision (1950)
[71] The word "fuzzy" as Zadeh uses it here does not refer directly to the Fuzzy Set Theory which he introduced two years later. The use of this word, which was highly unusual at the time, in the sense of "frayed" or "ambiguous" is nonetheless notable. "You remind me something I forgot ... But I used that word in a non-technical sense," Zadeh said when asked about it in an interview. [706] R. S. interview with L. A. Zadeh (2001)
[72] [639] Zadeh, Optimality (1963), footnote 2
[73] [590] Zadeh, Adaptivity (1963)
[74] [190] Gaines, Past (1983)
[75] Zadeh refers to [45] Bellman et al., Adaptive (1960)

> We wish to study multi-stage decision processes, and processes which can be construed to be of this nature, for which we do not possess complete information. This lack of information takes various forms of which the following are typical. We may not be in possession of the entire set of admissible decisions; we may not know the effects of these decisions; we may not be aware of the duration of the processes and we may not even know the over-all purpose of the process. In any number of processes occurring in the real world, these are some of the difficulties we face.[76]

Bellman and Kalaba underscored that such processes, which they called "adaptive", occurred in practically all aspects of statistical studies, in operations research and also in many areas of communication theory and that they were encountered theoretically in game theory and in the field of learning processes. For their part, Bellman and Kalaba alluded to Zadeh's already two-year-old editorial *What Is Optimal?* "for a discussion of the dangers and difficulties inherent in any mathematical treatment"![77] In a correspondence text about the definition of "adaptivity", Zadeh then referred to the paper *Dynamic Programming* by Bellman and Kalaba and to "stimulating conversations" with them. Finally, he summed up: "In conclusion, the definition of adaptivity suggested in this note is merely a concretization of the vague statement: 'An adaptive system is insensitive to changes in its environment.' "[78]

As a first indicator of an adaptive system, he cited its ability to provide an acceptably good performance in a changing and/or incompletely known environment. In his model, a priori information about the incompletely known and/or changing functional conditions are represented by a relation. If A is a system whose performance is measured by the function P and S_γ is the input source of the system, then the performance is measured with a real-valued vector $P(\gamma)$. The "acceptability criterion", which Zadeh defined with the relation $P(\gamma) \in W$, then indicates that the performance of system A is acceptable when $P(\gamma)$ falls within the set W. His definition of adaptivity was thus was follows:

> A system A is adaptive with respect to $\{S_\gamma\}$ and W if it performs acceptably well (i. e., $P(\gamma) \in W$)), with every source in the family $\{S_\gamma\}, \gamma \in \Gamma$. More compactly, A is adaptive with respect to a W if it maps Γ into W.[79]

Zadeh naturally commented then and there that, based on this definition, every system was adaptive with respect to Γ and W, and the question should thus not be whether the system is adaptive but rather with respect to which sets Γ and W the adaptivity of the system exists.

---

[76] [45] Bellman et al., Adaptive (1960), p. 5
[77] [45] Bellman et al., Adaptive (1960), p. 7
[78] [590] Zadeh, Adaptivity (1963), p. 470
[79] [590] Zadeh, Adaptivity (1963), p. 470

4.3 "Optimal Fetishism", Expanded Linearity Concepts and Adaptive Systems

When a conference on the identification problem in communication and control engineering took place at Princeton during the same year of 1963, Zadeh tied the lecture he delivered there into his earlier deliberations on this subject.[80] New solution procedures had been proposed in the intervening years by Richard Bellman and Robert Kalaba[81], Herbert E. Robbins[82] and others. Zadeh was therefore writing with regard to his aforementioned earlier text and after he had quoted his definition of the problem at that time:

> Viewed from our present state of knowledge of the identification problem, the above formulation appears rather narrow and imprecise. Today we have a much better understanding of various facets of the identification problem and are in possession of a fairly large number of concrete results bearing on its solution in specific cases.[83]

By his definition of the matter he now called the "general identification problem" he was also making use of the system theory terminology and he began his paper with a reformulation of the problem from a system theory perspective:

> The so called general identification problem or – as it is more commonly referred to – the identification problem, is merely one of a broad class of problems in which an experimenter does not have complete a priori knowledge about a system under test and wishes to design an experiment which would provide the needed information about the systems.[84]

Zadeh then clarified the problem of reaction (the output) of a system based on a stimulus (the input) as follows:

> An experimenter is given a black box A (or possibly a set of n copies of A) in an initial state which may or may not be specified. Starting at time t_0 he applies A an input u drawn from the input function space of A and observes the output A through a window W over the interval $[t_0, t_1]$. (For convenience, $[t_0, t_1]$ will be referred to as the observation interval.) The result of such an experiment is an input-output pair $u_{[t_0,t_1]}, y_{t_0,t_1}$ or for short, (u, y). If the experimenter is given more than one copy of A, e. g., if he had at his disposal a set of n copies $A^1, ..., A^N$, then he may apply different inputs, say, $u^1, ..., u^N$ to these copies and observe the corresponding outputs $y^1, ..., y^N$. The result of such a multiple experiment is a set of N input-output pairs $(u^1, y^1), ..., (u^N, y^N)$. Based on a priori information about A the experimenter knows (or thinks that he knows) that A belongs to a

---

[80] See also the description in Sect. 2 of this chapter of Zadeh's ideas on the "identification problem" in 1956, which he abandoned at the time
[81] [590] Zadeh, Adaptivity (1963), p. 470
[82] [464] Robbins, Extension (1959)
[83] [619] Zadeh, Identification (1963), p. 1f
[84] [619] Zadeh, Identification (1963), p. 1

specified class of systems $\{A\}$. (e.g., A may be known to be a linear time-invariant differential system of order less than m; or, A may be known to be a finite state system having no more than k states; etc.) Having (u, y) or, more generally $(u^1, y^1), ..., (u^N, y^N)$ in hand, the experimenter's problem is to determine within $\{A\}$ a system A^* such that WA is equivalent to WA^*. Then he can treat A^* as if it were A, for under the assumption that the output of A is observed through W, A is indistinguishable from A^* if $WA = WA^*$.[85]

Without going any further into the mathematical details here, what should be retained from this explanation is that the distinctive feature of this reformulation of the identification problem lies in the introduction of the "window" W, which as an "observation window" essentially represents the limited accuracy of measurement for the system output. A "window" such as this can be considered a singular system, but it can also be interpreted as a "filter" or can lead to the "acceptability criterion". With a view toward future developments, Zadeh stated in the conclusion to his paper that he judged the theories by Bellman and Kalaba as well as the "stochastic approximation" and "empirical Bayes method" by Robbins[86] to be very promising and influential approaches to the problem of system identification and control theories.[87]

4.4 Zadeh's System Theory

The book project Zadeh had planned in Princeton came to fruition later in Berkeley when he completed it with his colleague Charles A. Desoer. The result was the 1963 work *Linear System Theory: The State Space Approach*.[88] Zadeh wrote the first four chapters, in which he laid the basis for a new approach to system theory.[89] Here, too, he began with the definition from Webster's Dictionary already quoted in Chap. 1: A system is an assemblage of objects united by some form of interaction or interdependence. To these

---

[85] [619] Zadeh, Identification (1963), p. 6
[86] [463] Robbins, Bayes (1955)
[87] See also the report Zadeh produced for the *IRE Transactions on Information Theory* about progress in the fields of prediction and filtering in the years 1957–1960. He begins by mentioning the various generalizations of Wiener's theory in view of general processes and filters. In the second section, he cites Bellman's Dynamic Programming as a new and very promising direction in determining optimal and adaptive filters and predictions. Even if these subject areas were still unrelated, he writes, Rudolf Kalman showed in 1960 that there was a "mathematical duality" between the problems of filtering on one hand and control on the other. [642] Zadeh, Prediction (1961), p. 139. Kalman had received his Ph.D. shortly before under John R. Ragazzini in New York, who had also been Zadeh's dissertation supervisor. See Chap. 1
[88] [639] Zadeh et al., System (1963)
[89] In addition, Zadeh wrote the appendix about Fourier and Laplace transformations

objects he attributed a finite number of measurable properties or attributes that characterize them. The interactions among the objects as well as the mutual dependence of the attributes was supposed to be expressed by well-defined mathematical values.

Fig. 4.10. C. Desoer

Giving his textbook the title *Abstract Model and Physical Realization*, Zadeh examined the similarity between mechanical systems and electrical networks: Identical systems of equations can serve to evaluate systems from completely different scientific fields. An electrical network will obey the same differential equations as those describing a mechanical system. Zadeh showed this analogy between mechanical and electrical systems using the following examples: An object is provided with two connection variables v_1 and v_2, which fulfill the following differential equation:

$$\frac{dv_2}{dt^2} = \frac{d^2}{v_1}dt^2 + v_1.$$

This object can be realized in different ways.

- The first implementation he mentions is an electrical network (see Fig. 4.11, left). Here the variable v_1 stands for the input voltage and the variable v_2 represents the current flowing into the network.
- The second form of implementation is a mechanical system (see Fig. 4.11, right). In this example, v_2 is identified with the force acting upon the particle M, while v_1 represents the velocity of M.

The principle of identical structures in the form of a differential equation that can be applied in two different areas of physics – this satisfied the basic thesis of Bertalanffy's idea of a General Systems Theory as described in Chap. 2.[90]

---

[90] It is really not right to argue that Zadeh – as McNeill and Freiberger have written – considered the people in Bertalanffy's camp to be "crackpots" ([370] McNeill et al., Fuzzy (1993), p. 22). In interviews with the author, Zadeh has never expressed any such opinion.

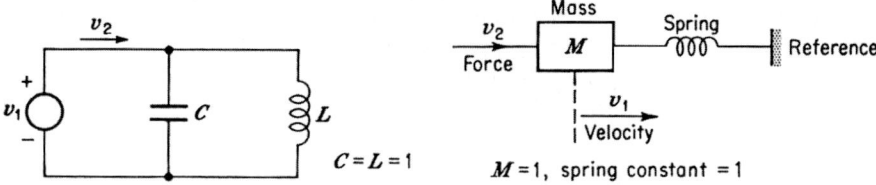

Fig. 4.11. Electrical and mechanical realization of a system

How much Zadeh actually was referring to the *General Systems Theory* cannot be discerned from this textbook, but this is apparent in the article *From Circuit Theory to System Theory*[91], which was written at the same time for the anniversary edition of the *Proceedings of the IRE* in May 1962 and which was quoted at the end of Chap. 1. This article shall be consulted in the section that follows.[92]

4.4.1 The State Space Approach

The invention of the transistor and the trends of the previous few years toward microminiaturization and integrated electronics, toward large communication networks and artificial neuronal networks, toward automata, the advent of information theory, game theory and dynamic programming and finally Pontryagin's Maximum Principle[93] all led to the subordination of the classical theory of circuits to a specific area of a much broader scientific discipline: "system theory – which, as the name implies, is concerned with all types of systems and not just electrical networks".[94]

Zadeh opened his article by giving the reader a summary of developments within electrical engineering and communication engineering over the previous two decades. System theory, he stated, was about the mathematical properties of systems – that was its main feature – and not about their physical forms.

---

[91] [662] Zadeh, System (1962)

[92] See Chap. 1

[93] On the subject of Pontryagin's Maximum Principle, I refer to the already cited work by Michael Plail: [446] Plail, Entwicklung (1998), pp. 173–189, footnote 117. Lev Semeonovich Pontryagin (1908–1988), Russian mathematician, studied at the University of Moscow from 1925 to 1929 despite the blindness he suffered in an accident when he was 14. Following his studies he was employed by the department of mechanics and mathematics. In 1934, he became a professor there and, one year later, head of the department of topology and functional analysis. Beginning in 1952, Pontryagin was occupied with differential equations, control theory and dynamic systems. He was elected to the Academy of Sciences in Moscow in 1939, and in 1959 became a full member. In 1970, he became the vice-president of the International Mathematical Union

[94] [662] Zadeh, System (1962), p. 856

4.4 Zadeh's System Theory

Fig. 4.12. L. Pontryagin

Thus, whether a system is electrical, mechanical or chemical in nature does not matter to a system theorist. What matters are the mathematical relations between the variables in terms of which the behavior of the system is described.[95]

In a footnote, Zadeh indicates after the sentence quoted above that the four following paragraphs in the article had appeared word-for-word, with very minor changes, in his article from 1954.[96] What was new, however, was the next paragraph, in which he made reference to the General Systems Theory of Ludwig von Bertalanffy, "who long ago perceived the essential unity of systems concepts and techniques in various fields of science". He followed this paragraph, in turn, with the one quoted at the end of Chap. 1, in which Zadeh wrote that a radically different form of mathematics, "the mathematics of cloudy quantities," was necessary to address living systems effectively.[97] Somewhat surprisingly, Zadeh's first paragraph on *State and State-Space Techniques* begins with a description of the Turing machine:

> Roughly speaking, a Turing machine is a discrete time ($t = 0, 1, 2, ...$) system with a finite number of states or internal configurations, which is subjected to an input having the form of a sequence of symbols (drawn from a finite alphabet) printed on a tape which can move in both directions along its length. The output of the machine at time t is an instruction to print a particular symbol in the square scanned by the machine at time t and to move in one or the other direction by one square. A key feature of the machine is that the output at time $t + 1$ and the state at time $t + 1$ are determined by the state and the input at time t.[98]

---

[95] [662] Zadeh, System (1962), p. 856
[96] [661] Zadeh, System (1954), cf. Chap. 1
[97] [662] Zadeh, System (1962), p. 857, cf. also Chap. 1
[98] [662] Zadeh, System (1962), p. 858

Using the designations s_t for the state, u_t for the input and y_t for the output, all at time t, the operations of a Turing machine can now be written as follows:
$$s_{t+1} = f(s_t, u_t), \quad t = 0, 1, 2, ...,$$
$$y_t = g(s_t, u_t),$$
where f and g are functions of pairs of the variables s_t and u_t.

Zadeh first notes the fact that these "Turing representations" and "state equations" had already been used by Shannon in his "epoch-making paper on the mathematical theory of communication" when he described channels with noise. In that paper, however, s_t and u_t had been used not to determine s_{t+1} and u_{t+1} but their probability distributions. Zadeh then cites John von Neumann's papers on automata theory. Finally, Zadeh introduces this concept of a system into the field of differential equation systems, and here the state equations take on the following form:
$$\dot{s}(t) = f(s(t), u(t)),$$
$$y(t) = g(s(t), u(t)).$$

In this equation, $\dot{s}(t) = d/dt\, s(t)$ and $s(t), y(t), u(t)$ are vectors representing the state of a system and the input and output of the system at time t. In the Soviet Union, these concepts had been used in the field of automatic control since the 1940s[99], but in the US, they had only been known since the publication of Richard Bellman's *Dynamic Programming*.[100]

Zadeh next addresses three principal problems of system theory which were the subject of on-going research and discussion at that time: the *characterization of systems*, the *classification* and the *identification* of systems. I will discuss the first problem in somewhat more detail.

System Characterization

If a system is a black box B at initial time t_0, then three types of time functions can be associated with it[101]:

- A controllable variable u, which is a time function whose values for all $t \geq t_0$ can be freely selected from a defined set. This set is called the *input space*.
- An initially controllable variable s, which is a time function whose values for all $t = t_0$ can be freely selected from a defined set, but not for later times. This set is called the *state space*.
- An observable variable y, which is a time function whose values for all $t \geq t_0$ can be observed, but whose values for whose values for $t \geq t_0$ cannot be controlled in any way.

---

[99] In this context, some of the names Zadeh mentioned were the following mathematicians and engineers: Anatolii Isakovich Lurie (1901–1980), Mark Aronovich Aizerman (born 1913), Alexandr M. Letov, Nikolai Nikolaevich Krasovskii (born 1924), Ioel Gilevich Malkin (1907–1958), Lev S. Pontryagin
[100] [48] Bellman, Dynamic (1957)
[101] The following assumptions apply for all t_0

Provided the above assumptions are fulfilled, a system B is considered *completely characterized* if for all $t \geq t_0$ the value $y(t)$ of the output at time t is clearly established by the value of the state s at time t_0 and the values of u are established in the closed interval $[t_0, t]$. Zadeh expressed this relationship symbolically like this:

$$y(t) = B(s(t_0); u_{[t_0, t]}).$$

Here he used the same designations as in his paper on *The General Identification Problem* in Princeton the same year[102]:

- $u_{[t_0, t]}$ is the segment of the time function u over the closed interval $[t_0, t]$;
- $s(t_0)$ is the assumed value of s at time t_0;
- $B(\cdot\,;\cdot)$ is a univalent function of their arguments. [103]

The above equation should be true for all $s(t_0)$ and $u_{[t_0, t]}$, and for every possible input-output pair $(u_{[t_0, t]}, y_{[t_0, t]})$ there is a state $s(t_0)$ in the state space of B. In this case, $u(t), y(t)$ and $s(t)$ stand for the *input, output* and *system state* of the system at time t, and to

$$s(t) = f(s(t_0); u_{[t_0, t]}), \quad t \geq t_0,$$
$$y(t) = g(s(t_0); u(t)),$$

These designations are, in fact, continuous analogues to the Turing representation

$$s_{t+k} = f(s_t u_t, ..., u_{t+k-1}),$$
$$y_t = g(s_t u_t).$$

System Classification
In the event that there is a family of systems $\mathcal{C}_1, \mathcal{C}_2, \mathcal{C}_3, ...$ and a system belongs to one of these classes, such as to class $\mathcal{C}_\lambda$, as a black box B, then the question arises as to how $\mathcal{C}$ can be identified via various inputs by observing the response behavior of B.

System Identification
In system identification, about which Zadeh had already written in several earlier texts, the goal was that of identifying system characteristics based on an observation of the system response behavior to test inputs. Given a system B and a family of systems $\mathcal{C}_1, \mathcal{C}_2, \mathcal{C}_3, ...$, which of these systems is equivalent to B?[104]

In the previous ten years there had been numerous articles written on all three problems of system theory by scientists from different areas who had different methods, not only of solving them but also of formulating them. Zadeh highlighted in particular the proposals by representatives of automata

---

[102] [639] Zadeh, Identification (1963)
[103] Note: $B(\cdot\,;\cdot)$ is a functional of $u_{[t_0, t]}$ and a usual function of $s(t_0)$; $s(t_0)$ is a usually finite-dimensional vector
[104] Zadeh noted here that the identification problem could be seen as a special case of system characterization but that this point of view was not very useful

theory here. As the theory developed, he expected an expansion of this program of research into the neighboring fields, as well, very much in the spirit of Bertalanffy's General Systems Theory:

> Nevertheless, it is certain that problems centering on the characterization, classification and, specially, identification of systems as well as *signals and patterns* will play an increasingly important role in system theory in the years to come.[105]

4.4.2 A Renaissance of General Systems Theory?

When the second Systems Symposium took place at the Case Institute of Technology in Cleveland (Ohio) in the spring of 1963, the 17 presenters and over 200 participants included not only scientists representing General Systems Theory and cybernetics but also engineering scientists and philosophers. Soon thereafter Mihajlo D. Mesarovic[106] published the conference proceedings under the title *Views on General Systems Theory*[107] and this book does indeed contain some very diverse approaches. In the foreword, however, Mesarovic pointed out that all of the contributors had been in agreement on one point: "necessity for the development of a general systems theory." There were many different opinions, however, when it came to the "how"!

> First of all, some of the participants took a definite stand, venturing to define a system and then discussing the consequences of such a definition. A second group of participants argued that the general systems theory should not be formalized since this very act will limit its generating power and make it more or less specific. A third group proposed to consider systems theory as a view point taken when one approaches the solution of a given (practical) problem. Finally, it was expressed that a broad-enough collection of powerful methods for the synthesis (design) of systems of diverse kinds should be considered as constituting the sought-for theory and any further integration was unnecessary. There were also participants that shared the viewpoints of more than one of the above groups.[108]

Kenneth E. Boulding[109], one of the founders of *General Systems Theory*, found himself inspired during some of the presentations to compose little poems, which were printed in the proceedings as introductions to the papers. Boulding treated Zadeh's contribution *The Concept of State in System Theory* thusly:

---

[105] [662] Zadeh, System (1962), p. 863
[106] Mihajlo D. Mesarovic, Yugoslavian-American engineer and scientist, has been the Cady-Staley Professor of Systems Engineering and Mathematics at Case Western since 1978. Mesarovic is a member of the Club of Rome
[107] [399] Mesarovic, Views (1964)
[108] [399] Mesarovic, Views (1964), p. xiv
[109] The economist Kenneth E. Boulding was one of the founding members of the Society for General Systems Research in 1954. See Chap. 3

4.4 Zadeh's System Theory

> A system is a big black box
> Of which we can't unlock he locks,
> And all we can find out about
> Is what goes in and what goes out.
> Perceiving input-output pairs,
> Related by parameters,
> Permits us, sometimes, to relate
> An input, output, and a state.
> If this relation's good and stable
> Then to predict we may be able,
> But if this fails us – heaven forbid!
> We'll be compelled to force the lid!   K. B.[110]

In Zadeh's presentation, he proposed pursuing a system theory that was very general and completely detached from the meanings of the individual components.

> I said that in system theory you are not interested in what is the physical interpretation of these variables, whether x is a voltage, or force, or mass, you are not interested in that. You are working with these variables. So in that sense system theory is abstract. So, in economy people talk about systems but x is dollars and this is income and this is inflation, whatever. But system theory then is an abstraction. That's the point that I made about system theory. System theory is abstract. It is not concerned with the physical identity of these elements. In circuit theory you are concerned with this – this is resistance, this is inductance, this is capacitor, this is voltage, this is current, all these have certain meaning. ... The only person at that time who did not think that position was Wiener. Wiener talked about x, y, z, but he didn't say x is voltage, and y is current.[111]

As is evident even from Boulding's little poem, Zadeh based his abstract remarks on the concepts of the input, output and state of a system. While input and output were not expected to offer any difficulties, the concept of state appeared problematic. Zadeh noted that the idea of states had played an important role in the physical disciplines for a long time. It referred to a set of numbers that contain all of the information about a system's past and that determine its future behavior. The names Poincaré[112], Birkhoff[113], Markov[114], Nemytskii[115] and Pontryagin stood for developments of more

---
[110] [399] Mesarovic, Views (1964), S. 39
[111] [705] R. S. interview with L. A. Zadeh (1999)
[112] For information on Jules Henri Poincaré, see footnote 49 in Chap. 2
[113] George David Birkhoff (1884–1944), American mathematician
[114] Andrei Andreyevich Markov (1856–1922), Russian mathematician
[115] Victor Nemytskii, Russian mathematician

precise definitions of the concept of state in the fields where it was applied, such as dynamic systems and optimal control.

However, the growing complexity and variety of the systems studied in the sciences had led to the necessity of framing the concept of state more broadly so that it could also be utilized for those systems that were not described by differential equations. Coming up with a *general* state concept was surely difficult, maybe even impossible, and so Zadeh limited himself "to sketch[ing] an approach that seems to be more natural as well as more general than those employed heretofore, but still falls short of complete generality".[116]

As Boulding's poem predicted, Zadeh presented a proposal that was certainly difficult for the non-mathematical system scientists to understand. It is mentioned here because it serves as the bridge between Zadeh's earlier scientific works and the theory of fuzzy sets he was soon to devise. First, Zadeh recapitulated the state concepts of dynamic system theory and automata theory.

Dynamic Systems[117]

A "dynamic system" can be defined as an abstract mathematical structure which satisfies the following axioms:

1. Γ is a *state space* and θ is a set of *time values* in which the behavior of the system is defined; Γ is a topological space and θ is an ordered topological space, a subset of the real numbers.
2. There is a topological space of functions in time Ω, which is defined on θ; this is the set of admissible *inputs* of the system.
3. For every initial time t_0 in θ, every initial state x_0 in Σ and every input u in Ω, defined for $t \leq t_0$, the future states of the system are identified by a *transition function*:

$$\Phi : \Omega \times \theta \times \theta \times \Sigma \times \Sigma; \qquad \Phi_u(t; t_0; x_0) = x_t.$$

For all $t_0 \leq t_1 \leq t_2$ from θ, every x_0 in Σ and every solid u in Ω, defined via $[t_0, t_1] \cap \theta$ the following relationship applies:

$$\Phi_u(t_0; t_0, x_0) = x_0.$$

$$\Phi_u(t_2; t_0; x_0) = \Phi_u(t_2; t_1, \Phi_u(t_1; t_0; x_0)).$$

Furthermore, the system must be non-anticipatory, i. e. the following applies for $u, v \in \Omega$ and $u \equiv v$ on $[t_0, t_1] \cap \theta$:

$$\Phi_u(t; t_0; x_0) = \Phi_v(t; t_0; x_0).$$

---

[116] [658] Zadeh, State (1964), p. 40
[117] The following definition follows [658] Zadeh, State (1964), p. 41

4. Every output of the system is a function $\Psi : \theta \times \Sigma \to \mathbf{R}$.
5. The functions Φ and Ψ are continuous with respect to the topologies in Σ, θ and Ω.

Automata theory[118]
A "finite state system" or an *automaton* A is a triple of sets Σ, U and Y with a pair of transition functions:

$$f : \Sigma \times U \times T \to \Sigma,$$
$$g : \Sigma \times U \times T \to Y.$$

Here let T be the set of integers $\{..., -1, 0, 1, ...\}$.
The elements of Σ are the states of A, the elements of U are its inputs and the elements of Y are its outputs. The time t runs over T and state, input and output of A at time t are indicated as follows: s_t, u_t, y_t. The transition functions f and g relate the state at time $t+1$ and output at time t to the state and input at time t:

$$s_t = f(s_t, u_t, t),$$
$$y_t = g(s_t, u_t, t).$$

A comparison of the above definitions of dynamic systems and "finite state systems" allowed Zadeh to subordinate the "finite state systems" to the dynamic systems, provided that Ψ is a transition function of $\theta \times \Sigma \times \Omega$ in $\mathbf{R}$, and not only of $\theta \times \Sigma$ in $\mathbf{R}$. Both system definitions appeared inadequate to him, though, and so he suggested his own approach.

4.4.3 System States as Input-Output State Relation Pairs

Zadeh used the terms that were introduced in the earlier section regarding his critique of the concepts of linearity and adaptivity: In a physical experiment with a real object, during a particular observation interval $[t_0, t_1]$, an input-output function will act upon this object and an output function is measured thereupon. Zadeh therefore dubbed the result of such an experiment an *input-output pair* $((u_{[t_0,t_1]}, y_{[t_0,t_1]}))$.

However, if the same input function is applied to other copies of this physical object in further experiments, then the respective output functions can be different, in other words, several outputs $y_{[t_0,t_1]}$ are possible for a given input $u_{[t_0,t_1]}$. The set of all ordered pairs of time functions over the observation interval $[t_0, t_1]$ he characterized with:

$$R_{[t_0,t_1]} = \{(u_{[t_0,t_1]}, y_{[t_0,t_1]})\}.$$

---
[118] The following definition follows [658] Zadeh, State (1964), p. 42

An *oriented abstract object* A Zadeh defined as a family $R_{[t_0, t_1]}, t_0, t_1 \in (-\infty, \infty)$ of sets of ordered pairs as well as a generating pair (u, y) whose first component $u = u_{[t_0, t_1]}$ is an input segment and whose second component $y = y_{[t_0, t_1]}$ is an output segment[119] : "Thus, an oriented abstract object A can be identified with the totality of input-output pairs which belong to A".[120]

Every section of an input-output pair of A should also be an input-output pair of A again. This demanded a requirement of consistency from the elements of $R_{[t_0, t_1]}$: If $(u_{[t_0, t_1]}, y_{[t_0, t_1]})$ is an input-output pair of A, then every pair also has the form $(u_{[\tau_0, \tau_1]}, y_{[\tau_0, \tau_1]})$, where $t_0 \leq \tau_0 \leq t_1$, $\tau_0 \leq \tau_1 \leq t_1$, and for the interval $[\tau_0, \tau_1]$, $u_{[\tau_0, \tau_1]} = u_{[t_0, t_1]}$ and $y_{[\tau_0, \tau_1]} = y_{[t_0, t_1]}$.

- He described the set $R[u]$ of all segments u on $[t_0, t_1]$ with $(u, y) \in A$ as the *input-output space* of A and
- he described the set $R[y]$ of all segments y on $[t_0, t_1]$ with $(u, y) \in A$ as the *output-output space* of A.
- The set $R_{[t_0, t_1]}$ of all ordered pairs of input segments and output segments is thus a subset of the product space $R[u] \times R[y]$.

These definitions led to a concept of state that is associated with a relation, since for each input there were many outputs, and vice versa. Zadeh solved this problem by parameterizing:

Given an input-output pair (u, y) to A: $u = u_{[t_0, t_1]}$ and $y = y_{[t_0, t_1]}$. Additionally, α is a parameter over $\Sigma = \mathbf{R}^n$. Parameter α is said to *parameterize* A when a function $\bar{A}$ exists on the product space $\Sigma \times R[u]$, and so an α can be found in Σ for all (u, y) belonging to A and for all t_0 and t, and therefore: $y = \bar{A}(\alpha; u)$. Every pair (u, y) for which such an α exists in Σ should then be an input-output pair of A and, conversely, every input-output pair of A with an α from Σ should satisfy the condition $y = \bar{A}(\alpha; u)$. For all α in Σ and for all u in $R[u]$, $(u, \bar{A}(\alpha, u))$ is thus an input-output pair of A.

The fact that α parameterizes A is not enough, however, to justify calling α a state of A. For that, $\bar{A}$ must satisfy one more property, which Zadeh named the *response separation property*. In order to make sense of this property, Zadeh introduced the following convention:

- A segment $u = u_{(t_0, t_1]}$, to which a segment $v = v_{[t, t_1]}$ is joined, is referred to as uv. In particular: if $u = u_{(t_0, t]}$ and $u' = u_{(t, t_1]}$, then $uu' = u_{(t_0, t_1]}$.[121]
- A function $\bar{A}(\alpha; u)$ has the *response separation property* if an element α^* that is clearly defined by α and u exists in Σ for all α in Σ and all uu' in $R[uu']$ such that the following is true:

$$\bar{A}(\alpha; uu') = \bar{A}(\alpha; u)\bar{A}(\alpha^*; u').$$

Using this tool, Zadeh now prepared the definition of the state of a system A: If α parameterizes system A and if $\bar{A}(\alpha; u)$ possesses the *response separation*

---
[119] Hereafter, the short forms input and output will be used
[120] [658] Zadeh, State (1964), p. 44
[121] Note the half-open intervals here

property, then the elements of Σ constitute the states of A. Σ is the state space of A and $y = \bar{A}(\alpha; u)$ is an *input-output state relation* for A. For $u = u_{(t_0, t]}$, α in $\bar{A}(\alpha; u)$ is called the *initial state* of A at time t_0. It is denoted as $s(t_0)$. With $u_{(t_0, t]}$ as the input segment, $s(t_0)$ as the initial state and $y_{(t_0, t]}$ as the corresponding response segment, the *input-output state relation* thus appears in this way:

$$y_{(t_0, t]} = \bar{A}(s(t_0); u_{(t_0, t]}).$$

This means that the initial state of A at time t_0 and the input segment $u_{(t_0, t]}$ definitively identify the response segment $y_{(t_0, t]}$. The *response separation property* is now utilized to define the state of a system A at time t:

Definition 3. *If the system A begins in state $s(t_0) = \alpha$, and if an input $u_{(t_0, t_1]}$ acts upon A, then the response separation property implies that an α^* exists in Σ such that the following is true for every $u = u_{(t, t_1]}$: $\bar{A}(uu') = \bar{A}(\alpha; u)\bar{A}(\alpha^*; u')$.*

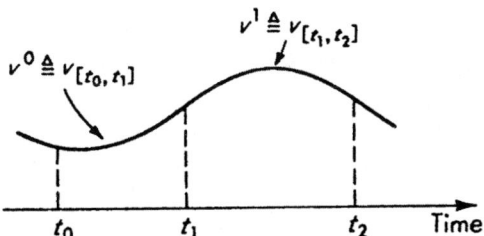

Fig. 4.13. Two consecutive segments

This element α^*, which is identified definitively by $s(t_0)$ and $u_{(t_0, t]}$, is called the state of A at time t and it is designated $s(t)$.

The state of system A at time t is therefore definitively identified by state A at time t_0 and by those values that the input takes on between t_0 and t. The following *state equation*, as it is termed, expresses this circumstance:

$$s(t) = s(s(t_0); u_{(t_0, t]}).$$

4.5 Zadeh's Decision

It became clear that the initially very promising attempts to integrate the technically oriented system theory into a multidisciplinary General Systems Theory and to establish it permanently had failed. Joint conferences attended by representatives of technical system theory and General Systems Theory, like the ones Mesarovic had organized at the Case Institute of Technology in

Cleveland, became more infrequent. The scientists were socialized differently, after all; they talked differently and each had different ideas about science and technology. Only a few managed to get along well in both camps:

> Wiener was also a sort of philosopher a little bit, so that he was a mathematician but he was also a philosopher. He wrote Cybernetics the way philosophers would write, and so there was a group of people: Norbert Wiener, Bertalanffy, Rapoport, von Foerster ... these people were doing certain kind of philosophical work. ... They were not engineering types and so to them systems science was a little bit like some sort of – difficult to characterize – but it was a little bit like a religion. There are things like that in science, for example, general semantics. It's a little bit like a religion.[122]

To this was added the great success enjoyed by the computer sciences starting in the 1960s, a field which quickly claimed a leading role in the scientific and university systems.

> System Theory came grown up but then computers and computers then took over. In other words: the center of attention shifted. ... So, before that, there were some universities that started departments of system sciences, departments of system engineering, something like that, but then they all went down. They all went down because computer science took over.

Once Zadeh had become the chairman of the Department of Electrical Engineering at Berkeley in 1963, he experienced these shifts very intensively, for it was during his five-year tenure in this position that his department was renamed the Department of Electrical Engineering and Computer Science.[123] There were other changes looming, though. As demonstrated by the short discussion contributions referenced in the earlier sections of this chapter, Zadeh had extensively criticized the relationship between mathematics and his own scientific-technical engineering discipline. The tool offered by mathematics was not appropriate to the problems that needed to be handled in the engineering sciences.

Information and communications technology had led to the construction and design of systems that were so complex that it took much more effort to measure and analyze these systems than had been the case just a few years before.

Much more exact methods were now required to identify, classify or characterize such systems or to evaluate and compare them in terms of their performance or adaptivity.

---

[122] [706] R. S. interview with L. A. Zadeh (2000).

[123] "When I was chair of the department in '63, '64, the question was what kind of computer should we get and at that time people wanted to get analog computers – at that time! ... Huge installations, you know. ... They wanted to spend billions of dollars getting an analog computer. I said no, let's go digital! " [706] interview with L. A. Zadeh (2000).

4.5 Zadeh's Decision

In order to provide a mathematically exact expression of experimental research with real systems, it was necessary to employ meticulous case differentiations, differentiated terminology and definitions that were adapted to the actual circumstances, things for which the language normally used in mathematics could not account. The circumstances observed in reality could no longer simply be described using the available mathematical means.

> In science you always have simple definitions. One-line definitions, two-line definitions, something like that. I say it can not be done. You can not take a complicated concept and define it at one line, two lines. You can not do that![124]

This discrepancy between theory and practice had occurred to Zadeh more than ten years earlier during his discussions with Guillemin.[125] He was clearly and manifestly conscious of them when he wrote down the basics of his theory of linear systems for the planned textbook. The suggestions on the specification and expansion of commonly used definitions that he made during this phase are a testament to the fact that Zadeh was working constructively to bridge the gap between theory and practice. They are attempts at remedying the inadequacies in the mathematical theory by being more precise in order to adapt to the actual system conditions, which had become more complex. Zadeh was forced to recognize, however, that these attempts would not be successful:

> As a mathematically oriented system theorist, I had been conditioned to believe that the analytical tools based on set theory and two-valued logic are all that is needed to build a framework for a precise, rigorous and effective body of techniques for the analysis of almost any kind of man-made or natural system. Then, in 1961–1963, in the course of writing a book on system theory (with C. A. Desoer), I began to feel that complex systems cannot be dealt with effectively by the use of conventional approaches largely because the description languages based on classical mathematics are not sufficiently expressive to serve as a means of characterization of input-output relations in an environment of imprecision, uncertainty and incompleteness of information.[126]

This was a bitter realization. Zadeh had always considered himself a mathematically oriented electrical engineer.

> I was always interested in mathematics, even when I was in Iran, in Teheran, but I was not sufficiently interested to become a pure

---

[124] [705] R. S. interview with L. A. Zadeh (1999)
[125] See Chap. 1
[126] Undated two-page type-written manuscript by Lotfi Zadeh, written after 1978; hereinafter cited as "Autobiographical Note I"

> mathematician. In other words, I never felt that I should pursue pure mathematics or even applied mathematics. So this mixture of an engineer was perfectly suited for me. So, essentially, I'm sort of a mathematical engineer, that's the way I would characterize myself. But I'm not a mathematician. I was somewhat critical of the fact that mathematics has gone away from the real world. ... I criticized the fact that mathematics has gone too far away from the real world.[127]

There were two ways to overcome this situation. In order to describe the actual systems appropriately, he could try to increase the mathematical precision even further, but Zadeh failed with this course of action.

> When I tried to define adaptivity and things of this kind, because as I said at some places my objective was to come up with precise definitions, consistent with my belief that mathematics is good as answer on questions. But gradually I began to realize that it cannot be done! Adaptivity, linearity, precise definitions.[128]

The other way presented itself to Zadeh in the year 1964, when he discovered how he could describe real systems as they appeared to people. "I'm always sort of gravitated toward something that would be closer to the real world".[129] What was closer to the real world was the Theory of Fuzzy Sets that Zadeh created in this year. The story of its genesis is told in the next chapter.

---

[127] [706] R. S. interview with L. A. Zadeh (2000)
[128] [706] R. S. interview with L. A. Zadeh (2000)
[129] [706] R. S. interview with L. A. Zadeh (2000)

5
Fuzzy Sets and Fuzzy Systems

Richard Bellman and Lotfi Zadeh had become acquainted in New York City in 1954 when Bellman had traveled to Columbia to sound out his prospects of obtaining a position at the university. He was very happy working at RAND, but some of his wife's relatives had had contact with communists. In this era of McCarthyism, since RAND was a research organization with ties to the military, Bellman was expecting to get fired. This situation never arose, however, and Bellman was able to stay at RAND. When he and Zadeh first met in New York in 1954, Bellman talked about his work on *Dynamic Programming* and Zadeh encouraged him to write an article about it, which he then wanted to submit to the IRE. Even though the manuscript that Bellman prepared was eventually rejected by the IRE, an acquaintance began here which became a close friendship when Zadeh transferred to the University of California at Berkeley in 1958; this friendship would last until Bellman's death in 1984.[1] Bellman remembered their encounter as follows:

Fig. 5.1. R. Bellman

---

[1] "Richard Bellman was my best friend and I always considered him to be a really great man." [711] R. S. interview with L. A. Zadeh (2002). Zadeh also wrote the obituary for Bellman that appeared in the *IEEE Transactions on Automatic Control*: [593] Zadeh, Bellman (1984), p. 961

On one of my trips to New York, I visited Columbia at the invitation of Merrill Flood. I met Lotfi Zadeh and Rudy Kalman. Rudy was working in differential equations. I convinced him that control theory was the natural extension of differential equations and he started to work in control theory. I now had a very busy time in New York. I saw several publishers. Peter Lax at the Courant Institute, John Jacques at SKI, Lotfi Zadeh at Columbia and Norman and Fay in the evening. Some years later, Lotfi moved to Berkeley and started his work on fuzzy systems. We had many discussions and I saw the importance of his work immediately. With the aid of fuzzy systems, it was now possible to apply mathematics to social systems and to handle types of uncertainty which escaped classical probability completely.[2]

Just how did the "creation" of fuzzy sets come to pass? What were the reasons that inspired Lotfi Zadeh to come up with a theory for these "unsharp sets"? In the summer of 1964, Bellman and Zadeh decided to do some scientific work together at RAND in Santa Monica. Zadeh had taken July and August off for this purpose; his only remaining obligation was to deliver a paper on pattern recognition at a conference being held at Wright-Patterson Air Force Base in Dayton, Ohio.[3]

Fig. 5.2. Lotfi A. Zadeh with his parents

---

[2] [49] Bellman et al., History (1984), p. 223. Fay and Norman are Zadeh's wife and their son

[3] During the interview, Zadeh was unfortunately no longer certain about the location and other facts regarding this conference

> I was concerned with I was supposed to give a talk in pattern recognition. So there were two things that were in my mind: it was the talk that I was going to give and then after that I was supposed to go to Santa Monica.[4]

The flight to Dayton made a stopover in New York, where Zadeh wanted to meet his parents for dinner. They were unable to get together, however, and so Zadeh had free time.

> I was by myself and so I started thinking about some of these issues. And it was during that evening that the thought occurred to me that when you talk about pattern and things of this kind, ... that the thing to do is to use grade of membership. I remember that distinctly and once I got the idea, it became grow to be easy to develop it. So going back I'm thinking about it, it's quite possible that if that dinner took place, I wouldn't have got these ideas. Because I was free, you know, I could think about it. It's one of those things, you know, that came up with you take controlled.[5]

5.1 The Genesis of Fuzzy Set Theory

Recognizing patterns means deciding whether an object or its points match or at least closely resemble a pattern. Such a decision is often not a simple one to make – it is much easier not to commit with 100% certainty but instead just to give an approximate answer. This, in turn, means gradually evaluating the membership of points to a pattern and replacing the yes/no decision with a softer, vague, unsharp statement. From this small idea, Zadeh developed a mathematical theory whose basic modules are sets with elements that are capable of belonging to the sets to only a certain degree and functions that give these elements their degrees of membership to the sets.

The history of fuzzy sets begins with a problem of pattern recognition! As was described in Chap. 3, it was extremely questionable at this time whether machines could be capable of recognizing patterns. Rosenblatt had introduced his perceptron in 1958, and the book by Minsky and Papert would not be published until 1969. During this period, when machines were euphorically celebrated as *Global Problem Solvers*[6], some proponents of artificial

---

[4] [709] R. S. interview with L. A. Zadeh (2001)
[5] [709] R. S. interview with L. A. Zadeh (2001)
[6] At the Dartmouth conference in 1956, Allen Newell, John Clifford Shaw and Herbert Alexander Simon had introduced the Logic Theorist program, which used a heuristic search technique operating by the principle of "trial and error" to prove 38 of 52 theorems of the Principia Mathematica. (See Chap. 3.) In 1959, they presented the program *Global Problem Solver (GPS)*

intelligence research considered it just a matter of time before machines would make brilliant contributions in the field of pattern recognition.[7]

When a 26-page description of the *State of the Art in Pattern Recognition* appeared in the *Proceedings of the IEEE* in 1968, the author George Nagy[8] introduced his overview by stating that pattern recognition could be considered a reservoir for very diverse problems, and that every approach to the solutions to these problems should be seen as a contribution to pattern recognition. Vocabulary, concepts, useful ideas had been provided by many different disciplines, and he enumerated them: statistical decision theory, circuit theory, automata theory, set theory, control theory, linguistic analysis, information theory, Mathematical Programming and lastly Rosenblatt's "nerve net studies".[9] With respect to Mathematical Programming, he referred to the 1965 article *Pattern Separation by Convex Programming* by Judah Ben Rosen.[10] A technical report by the same author and published under the same title by the Applied Mathematics and Statistics Laboratories at Stanford University had already appeared in 1963, though, and in this first scientific text that Lotfi Zadeh wrote on the subject of fuzzy sets, he cited this report by Rosen, which begins with the following sentence[11]:

> The separation of sets of points in an n-dimensional Euclidean space is the basic mathematical problem associated with pattern recognition.[12]

As Chap. 3 described, Minsky and Papert showed six years later that it was precisely this mathematical problem that, in principle, Rosenblatt's perceptron could not solve. This set of problems should be considered more closely using Rosen's observations.

Each pattern can be represented mathematically by a single point in the n-dimensional real vector space $\mathbf{R}_n$.[13] If this space $\mathbf{R}_n$ can be separated into regions such that each region contains only one point from a particular pattern set and none from any other pattern set, then the problem of pattern

---

[7] "Within a generation, the problem of creating 'artificial intelligence' will be substantially solved," said Marvin Minsky in 1967. [404] Minsky, Computation (1967), p. 2. See also the statements by Frank Rosenblatt, quoted in Chap. 3

[8] George Nagy (born 1937), Hungarian-American electrical engineer. Nagy worked with Rosenblatt on an audio perceptron in 1962 and 1963 and he was later involved with the pattern recognition of Chinese characters and the use of pattern recognition on neurophysiological data. See the author information in: *Proceedings of the IEEE*, Vol. 56, No. 5, May 1968, p. 863

[9] [411] Nagy, Pattern (1968)

[10] [470] Rosen, Pattern (1965). Judah Ben Rosen, American electrical engineer. Rosen was a professor at the Universities of Wisconsin and Minnesota. Previously (1952–1955) he had been at Princeton University and after that (1955–1962) headed the Applied Mathematics Department at Shell. From 1944 to 1947, he was involved with the Manhattan Project

[11] Interestingly, Zadeh cites Rosen's text only here and not in any of his later articles

[12] [469] Rosen, Pattern (1963), p. 1

[13] Rosen's notation shall be retained here, i.e. $\mathbf{R}_n$ instead of $\mathbf{R}^n$, etc

recognition has been solved. A pattern is then classified by determining in which region of $\mathbf{R}_n$ each of its points lies.

Chapter 3 also showed how a Euclidean space can be separated by hyperplanes when it is two-dimensional. Generally, Rosen proceeded as follows: A number l of point sets in an n-dimensional Euclidean space are to be separated by an appropriate number of hyperplanes. The m_i points in the ith set (where $i = 1, \ldots, l$) are denoted by n-dimensional vectors $p_{ij}, j = 1, \ldots, m_i$. Then the following matrix describes the points in the ith set:

$$P_i = p_{i1}, p_{i2}, \ldots, p_{im_i}.$$

In the simplest case at which the Rosenblatt perceptron failed, two points P_1 and P_2 are to be separated. Rosen provides this definition:

Definition 4. *The point sets P_1 and P_2 are linearly separable if their convex hulls do not intersect.*[14]

An equivalent statement is the following:

The point sets P_1 and P_2 are linearly separable if and only if a hyperplane $H = H(z, \alpha) = \{p | p'z = \alpha\}$ exists such that P_1 und P_2 lie on the opposite sides of H.[15]

The orientation of the hyperplane H is thus specified by the n-dimensional unit vector z and its distance from the origin is determined by a scalar α. The linear separation of P_1 and P_2 was therefore equivalent to demonstrating the existence of a solution to the following system of strict inequalities. (Here $\| \|$ denotes the Euclidean norm and e_i is the m_i-dimensional unit vector.):

$$p_{1j}z > \alpha \quad j = 1, \ldots, m_1$$
$$p_{2j}z < \alpha \quad j = 1, \ldots, m_2 \qquad \|z\| = 1,$$

$$p_{1j}z > \alpha e_1$$
$$p_{2j}z < \alpha e_2 \qquad \|z\| = 1.$$

Rosen came to the conclusion "that the pattern separation problem can be formulated and solved as a convex programming problem, i.e., the minimization of a convex subject to linear constraints".[16]

He considered the two linearly separable sets P_1 and P_2. The Euclidean distance δ between these two sets is then indicated by the maximum value of a γ, for which z and α exist such that:

$$p'_1 z \geq (\alpha + \tfrac{1}{2}\gamma)e_1$$
$$p'_2 z \leq (\alpha + \tfrac{1}{2}\gamma)e_2 \qquad \|z\| = 1.$$

---

[14] In mathematics, the convex hull of a set is defined as the set of all convex combinations of its points. In other words, given the points p_i from P and given λ_i from $\mathbf{R}$, then the following set is the convex hull of P: $conv(P) = \lambda_1 \cdot p_1 + \lambda_2 \cdot p_2 + \ldots + \lambda_n \cdot p_n$
[15] p' refers to the transpose of p
[16] [469] Rosen, Pattern (1963), p. 1

The task is therefore to determine the value of the distance δ between the sets P_1 and P_2 formulated as the nonlinear programming problem that can find a maximum γ for which the above inequalities are true. Rosen was able to reformulate it into a convex quadratic programming problem that has exactly one solution when the points P_1 and P_2 are linearly separable. To do so, he introduced a vector x and a scalar β for which the following applies: $\gamma = 2/\sqrt{\|x\|}$, $\alpha = \beta/\|x\|$ and $z = x/\|x\|$. Maximizing γ is thus equivalent to minimizing the convex function $\|x\|^2$:

$$\sigma = \min_{x,\beta} \left\{ \tfrac{1}{4}\|x\|^2 \,\middle|\, \begin{array}{l} p_1' x \geq (\beta+1)e_1 \\ p_2' x \leq (\beta-1)e_2 \end{array} \right\}$$

After introducing the $(n+1)$-dimensional vectors $y = \begin{pmatrix} x \\ \beta \end{pmatrix}$, $q_{ij} = \begin{pmatrix} p_{ij} \\ -1 \end{pmatrix}$ and the $(n+1) \times m_i$-Matrices $Q_i = [q_{i1}, q_{i2}, ... q_{im_i}]$, Rosen could use the standard form of convex quadratic programming and formulate the following theorem of linear separability:

Theorem 1. *The point sets P_1 and P_2 are linearly separable if, and only if, the convex quadratic programming problem*

$$\sigma = \min_y \left\{ \tfrac{1}{4} \sum_{i=1}^n y_i^2 \,\middle|\, \begin{array}{l} Q_1' y \geq e_1 \\ -Q_2' y \leq e_2 \end{array} \right\}$$

has a solution. If P_1 and P_2 are linearly separable then the distance δ between them is given by $\delta = \tfrac{1}{\sqrt{\sigma}}$, and a unique vector $y_0 = \begin{pmatrix} x_0 \\ \beta_0 \end{pmatrix}$ achieves the minimum σ. The separating hyperplane is given by $H(x_0, \beta_0 = \{p \mid p' x_0 \geq \beta_0\}$.[17]

Zadeh had not been working in the field of pattern recognition – "Not really, but I was interested."[18] He had been interested two years earlier, as is evident from Chap. 4, for when he wrote the brief correspondence contribution on *Optimality and Non-Scalar-Valued Performance Criteria*[19], he had used the separation theorem for convex sets, which states that for two convex disjoint sets there exists a hyperplane that separates them. He now returned to this mathematical problem of pattern recognition, "which plays an important role in communication and control theories".[20] The problems of pattern recognition were naturally multifaceted. Even the process of *transforming* (or *representing*) the object patterns into a set of real variables which represent these patterns correctly and which would also be accepted by a computer proved to be problematic. However, possibilities were found for most practical

---

[17] See [469] Rosen, Pattern (1963), p. 4ff
[18] [711] R. S. interview with L. A. Zadeh (2002)
[19] [639] Zadeh, Optimality (1963)
[20] [44] Bellman et al., Abstraction (1966), p. iii

problems, such as representing sound patterns by their spectrum or representing character patterns using n-dimensional vectors of ones and zeros. Rosen had also proceeded from this type of representation and Zadeh followed him in this respect in order to deal with two other problems of pattern recognition; these were problems that were well-known from the field of statistics:

> There are two basic operations: abstraction and generalization, which appear under various guises in most of the schemes employed for classifying patterns into a finite number of categories.[21]

Abstraction was to be understood as the problem of identifying a decision function on the basis of a randomly sampled set, and *generalization* referred to the use of the decision function identified during the abstraction process in order to classify the pattern correctly.

Although these two operations could be defined mathematically on sets of patterns, Zadeh proposed another way: "a more natural as well as more general framework for dealing with these concepts can be constructed around the notion of a "fuzzy" set – a notion which extends the concept of membership in a set to situations in which there are many, possibly a continuum of, grades of membership".[22]

Fig. 5.3. R. Bellman and his wife Nina

Whether Zadeh spoke about this and the ideas described below in his lecture in Dayton, Ohio in 1964 can unfortunately no longer be determined.[23]

---

[21] [675] Bellman et al., Abstraction (1964), p. 1
[22] [675] Bellman et al., Abstraction (1964), p. 1
[23] Neither a manuscript nor any other sources exist. Lotfi Zadeh no longer remembers this lecture exactly and so he did not want to either confirm or rule out this detail. [705] R. S. interview with L. A. Zadeh (1999), [709] R. S. interview with L. A. Zadeh (2001), [712] R. S. interview with L. A. Zadeh (2003)

At any rate, within a short period of time he had further developed his little theory of "gradual membership" into an appropriately modified set theory: "Essentially the whole thing, let's walk this way, it didn't take me more than two, three, four weeks, it was not long."[24] When Zadeh finally met with Bellman in Santa Monica, he had already worked out the entire theoretical basis for his theory of fuzzy sets:

> A month later, while I was visiting RAND in Santa Monica, I described a preliminary version of the theory of fuzzy sets to Richard Bellman. His immediate reaction was highly encouraging and he has been my strong supporter and a source of inspiration ever since.[25]

Fig. 5.4. R. Kalaba

Once he had thoroughly discussed all of the points with Bellman, Zadeh submitted the text to the journal *Information and Control* in November 1964. Since he himself was on the editorial board, there was only a brief reviewing process and as a result, it was possible for the article entitled *Fuzzy Sets* to make it into the June edition in 1965. This article is widely regarded as the first text about the theory of fuzzy sets, but even during the period between when it was written and when it was published, Zadeh was producing other texts about his new theory. These were not journal publications, however, and so they went largely unnoticed at the time and for the most part remain unknown today. Zadeh's text *Fuzzy Sets* had been pre-published in November 1964 as part of the report series by the Electronics Research Laboratory of his own university in Berkeley.[26] In addition, he had also jotted down the results of his discussions with Bellman in Santa Monica on aspects of pattern recognition in conjunction with his theory of fuzzy sets and had sent it to him. On September 9, 1964, Bellman answered on the stationery of the *Journals of Mathematical Analysis and Applications (JMMA)* and, writing in his role as

---

[24] [709] R. S. interview with L. A. Zadeh (2001)
[25] Autobiographical note I
[26] [678] Zadeh, ERL-Fuzzy (1964)

Fig. 5.5. Letter from R. E. Bellman to L. A. Zadeh dated September 9, 1964

the editor, told Zadeh that he would be glad to publish this text as an article in his journal (Fig.5.5).[27]

Zadeh's article first appeared in the *Journal of Mathematical Analysis and Applications* in 1966 under the title *Abstraction and Pattern Classification*, and its by-line included the names of Bellman and Robert Kalaba in addition to Lotfi Zadeh (who was listed third!).[28] However, this text was identical in content to a "preliminary paper"[29] that was published with the same authors' names as a memorandum at the RAND Corporation[30] back in October of 1964 (Fig. 5.6) and thus before the journal article *Fuzzy Sets*. Therefore, it is here that Zadeh defines a fuzzy set in a scientific text for the first time:

> To be more specific, a fuzzy set A in a space $\Omega = \{x\}$ is represented by a characteristic function f which is defined on Ω and takes values in the interval $[0,1]$, with the value of f at x, $f(x)$, representing the "grade of membership" of x in A. Thus, if A is a set in the usual sense,

---

[27] [696] Letter from Richard Bellman to Lotfi Zadeh dated 9 September 1964. Private archive of Lotfi Zadeh

[28] [44] Bellman et al., Abstraction (1966)

[29] [675] Bellman et al., Abstraction (1964). For the journal version, a number of details not pertinent to the content (commas, index numbers) were changed

[30] *United States Air Force Project RAND*. Without Bellman as a co-author, it would surely have been impossible to publish the text as a RAND memorandum

MEMORANDUM
RM-4307-PR
OCTOBER 1964

ABSTRACTION
AND PATTERN CLASSIFICATION

R. Bellman, R. Kalaba and L. A. Zadeh

PREPARED FOR:
UNITED STATES AIR FORCE PROJECT RAND

Fig. 5.6. Title page of the RAND memorandum by Bellman, Kalaba and Zadeh

$f(x)$ is 1 or 0 according as x belong or does not belong to A. When A is a fuzzy set, then the nearer the value of $f(x)$ to 0, the more tenuous is the membership of x in A with the "degree of belonging" increasing with increase in $f(x)$.[31]

It is notable here that Zadeh had not yet made use of the term "membership function". It should likewise be noted that the following argumentation, which led Zadeh's Fuzzy Set Theory closer to three-valued logic, was later not used by him.

Since it is sometimes helpful to illustrate the membership of a point in a fuzzy set more concretely by using a choice between two levels ε_1 and ε_2 (where ε_1 and $\varepsilon_2 \in [0,1]$), Zadeh declared later:

(a) A point x belongs to A if $f(x) = 1 - \varepsilon_1$,
(b) x does not belong to A if $f(x) \leq \varepsilon_2$, and
(c) x is undefined with respect to A $\varepsilon_2 < f(x) < 1 - \varepsilon_1$.

This basically amounted to a three-valued characteristic function with:

- $f(x) = 1$ if $x \in A$,
- $f(x) = \frac{1}{2}$ if x is undefined with respect to A,
- $f(x) = 0$ if $x \notin A$.

Zadeh next defined the operation for forming unions and intersections in these fuzzy sets (see Fig. 5.7):

$C = A \cap B$, $f_C(x) = \max[f_A(x), f_B(x)]$, or abbreviated: $f_C = f_A \vee f_B$.
$C = A \cup B$, $f_C(x) = \min[f_A(x), f_B(x)]$, or abbreviated: $f_C = f_A \wedge f_B$

---

[31] [675] Bellman et al., Abstraction (1964), p. 1

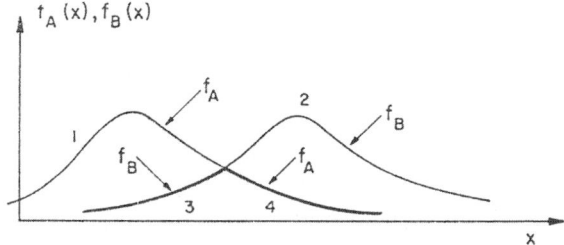

Fig. 5.7. Illustrations of the union (1, 2) and intersection (3, 4) of two fuzzy sets A and B

He observed that these definitions would be reduced to the definitions of the logical "or" and "and" in the case of ordinary sets and that, if the characteristic functions are three-valued, they lead to the three-valued logic of Steven Kleene.[32]

Throughout the rest of the work, no more mention was made of reducing the range of the characteristic function to just three values. Instead, Zadeh emphasized that he was dealing with n values: $n \in |N$.

With the fuzzy sets, Zadeh now considered himself in a position to substantiate in "a more natural as well as more general" way the meaning of *abstraction*, a term he had heretofore informally defined as the identification of those properties of a set's elements $x^1, \ldots, x^n$ which they have in common and which, in aggregate, define the set.

- All of the pairs (x^i, f^i) that characterize a fuzzy set A, formed from each object and its degree of membership in A are referred to by Zadeh as a "collection" of *samples* or *observations* from A.
- The *abstraction* of this "collection" $\{(x^i, f^i)\}^n$ is the estimation of the characteristic function of A from the samples $(x^1, f^1), \ldots \ldots$
  $\ldots, (x^n, f^n)$.
- Once an estimate of f has been constructed, a *generalization* is performed on the collection $\{(x^i, f^i)\}^n$ when the estimate is used to compute the values of f at points other than $x^1, \ldots, x^n$.

The problem lies in constructing one "good" estimate of f, i.e. using a priori information about the class of functions to which f belongs, such that this information in combination with the samples from A would be sufficient to enable one to construct a "good" estimate of f. In this context, "good" means finding the f in the class of functions which "best" (in a specified sense of "best") fits the given selection $(x^1, f^{i_1}), \ldots, (x^n, f^{i_n})$. As a special case of this statistical problem, "which applies to ordinary rather than fuzzy sets", Zadeh

---

[32] [675] Bellman et al., Abstraction (1964), p. 3. Zadeh had apparently become acquainted with this three-valued logic during his stay in Princeton; Zadeh was not yet familiar with the multi-valued logic calculi established by Jan Łukasiewicz and popularized by the logicians of the Lemberg-Warsaw school

considered "the widely used technique for distinguishing between two sets of patterns via a separating hyperplane."[33]

If we have a set A of patterns which we want to classify, then a hyperplane should be found, if one exists, that passes through the origin of the e-dimensional Euclidean space $\mathbf{R}^e$ such that the given points $x^1, \ldots, x^n$, which belong to set A, all lie on the same side of the hyperplane.[34] So the following applies to the estimate f of the membership function:

$$\mathbf{f}(x; \lambda) = 1 \quad for \quad \langle x, \lambda \rangle \geq 0,$$
$$\mathbf{f}(x; \lambda) = 0 \quad for \quad \langle x, \lambda \rangle < 0,$$

with $\langle x, \lambda \rangle$ denoting the scalar product of x and λ. The problem is now to find a λ in R^1 such that:

$$\langle x, \lambda \rangle \geq 0, \quad for \quad i = 1, \ldots, n.$$

Any $\mathbf{f}(x^i; \lambda)$ whose λ satisfies this last equation qualifies as an "abstracting function". The corresponding generalization on $(x^1, 1), \ldots, (x^n, 1)$ takes on the form of the statement:

> *Any x satisfying $\langle x, \lambda \rangle \geq 0$*
> *belongs to the same set as the samples $x^1, \ldots, x^n$.*

Zadeh then mentioned the program developed by Rosen, which can be used as needed to maximize the distance between the hyperplane and the set of all points $x^1, \ldots, x^n$.

Since in most instances the a priori information about the characteristic function of a fuzzy set is not sufficient to construct an estimate of $f(x)$ which is "optimal" in a meaningful sense, one is forced to resort to a heuristic rule for *estimating $f(x)$*, and the only means of judging the "goodness" of the estimate yielded by a heuristic rule such as this lies in experimentation.[35]

5.2 Fuzzy Sets and Fuzzy Systems

On April 20, 21 and 22, 1965, a Symposium on System Theory took place at the Polytechnic Institute in Brooklyn and Zadeh presented the attendees with *A New View on System Theory*.[36] For this new view, Zadeh introduced the

---

[33] [675] Bellman, et al., Abstraction (1964), p. 4

[34] For ordinary sets, of course, the following applies: $f^{i_1} = f^{i_2} = \ldots = f^{i_n} = 1$. [675] Bellman et al., Abstraction (1964), p. 5

[35] In the third part of this text, Zadeh described a heuristic rule of pattern classification such as this and he showed that it is equivalent to a special case of the minimal distance principle, "which is frequently employed in signal discrimination and pattern recognition." [675] Bellman, et al., Abstraction (1964), p. 6

[36] A shortened version of the paper delivered at this symposium appeared in the proceedings under the title *Fuzzy Sets and Systems* (Fig. 5.8): [664] Zadeh, Systems (1965)

FUZZY SETS AND SYSTEMS*

L. A. Zadeh

*Department of Electrical Engineering, University of California,
Berkeley, California*

> The notion of fuzziness as defined in this paper relates to situations in which the source of imprecision is not a random variable or a stochastic process, but rather a class or classes which do not possess sharply defined boundaries, e.g., the "class of bald men," or the "class of numbers which are much greater than 10," or the "class of adaptive systems," etc.
>
> A basic concept which makes it possible to treat fuzziness in a quantitative manner is that of a fuzzy set, that is, a class in which there may be grades of membership intermediate between full membership and non-membership. Thus, a fuzzy set is characterized by a membership function which assigns to each object its grade of membership (a number lying between 0 and 1) in the fuzzy set.
>
> After a review of some of the relevant properties of fuzzy sets, the notions of a fuzzy system and a fuzzy class of systems are introduced and briefly analyzed. The paper closes with a section dealing with optimization under fuzzy constraints in which an approach to problems of this type is briefly sketched.

Fig. 5.8. First page (excerpt) of Zadeh's contribution to the Symposium on System Theory in Brooklyn in April 1965

concepts of his Fuzzy Set Theory (Fig. 5.8), "which provide a way of treating fuzziness in a quantitative manner".[37]

He underscored the fact that he was talking about situations in which instances of inexactness were not ascribed to random variables or stochastic processes but rather to a class or classes that do not have sharply defined boundaries. "An example of such a "class" is the "class" of adaptive systems."[38] A simpler example was the "class" of real numbers that were much greater than 10, and other was the "class" of bald men. The word *class* appeared in quote marks each time because these were not classes or sets in the usual mathematical sense of the word, "since they do not dichotomize all objects into those that belong to the class and those that do not."[39]

5.2.1 Fuzzy Sets

Zadeh briefly outlined the fundamental properties of fuzzy sets and also mentioned the comfortable way of using them to handle the abstraction process in pattern recognition, which he indicated with an example:

> For example, suppose that we are concerned with devising a test for differentiating between handwritten letters $\mathcal{O}$ and $\mathcal{D}$. One approach to this problem would be to give a set of handwritten letters and indicate their grades of membership in the fuzzy sets $\mathcal{O}$ and $\mathcal{D}$. On performing abstraction on these samples, one obtains the estimates $\tilde{\mu}_\mathcal{O}$ and $\tilde{\mu}_\mathcal{D}$ of $\mu_\mathcal{O}$ and $\mu_\mathcal{D}$ respectively. Then given a letter x which is not one of the given samples, one can calculate its grades of membership in $\mathcal{O}$ and $\mathcal{D}$, and, if $\mathcal{O}$ and $\mathcal{D}$ have no overlap, classify x in $\mathcal{O}$ or $\mathcal{D}$.[40]

---

[37] [664] Zadeh, Systems (1965), p. 29
[38] [664] Zadeh, Systems (1965), p. 29
[39] [664] Zadeh, Systems (1965), p. 29
[40] [664] Zadeh, Systems (1965), p. 30

One confusing aspect of the abstraction problem, he wrote, was that human intellect was capable of very effective abstractions precisely when the corresponding problems were not mathematically well defined. Because we did not understand this fact, we were also not capable of designing devices which could perform as well as human intellect in the areas of heuristic programming, pattern recognition and related problem fields. Zadeh then listed some of the tools of his theory of fuzzy sets:

- Two fuzzy sets A and B in a space X are *equal*, $A = B$, if and only if $\mu_A(x) = \mu_B(x)$ for all $x \in X$.
- A fuzzy set A is a *subset* of a fuzzy set B, $A \subseteq B$, if and only if $\mu_A(x) \leq \mu_B(x)$ for all $x \in X$.
- The *complement* A' of a fuzzy set A is defined by: $\mu'_A(x) = 1 - \mu_A(x)$ for all $x \in X$.
- The *union* of two fuzzy sets A and B is denoted $A \cup B$. It is defined as the smallest fuzzy set that contains both A and B. An immediate consequence arises from the fact that the membership function of $A \cup B$ has the following membership function: $\mu_{A \cup B}(x) = \max[\mu_A(x), \mu_B(x)]$.
- The *intersection* of two fuzzy sets A and B is denoted $A \cap B$. It is defined as the largest fuzzy set contained in both A and B. The membership function of $A \cap B$ is: $\mu_{A \cap B}(x) = \min[\mu_A(x), \mu_B(x)]$.

Zadeh then defined the concept of the "shadow" of a fuzzy set and the property of convexity of fuzzy sets:

- A fuzzy set A is characterized in the Euclidean n-dimensional space $\mathbf{E}^n$ by the membership function $\mu_A(x_1, x_2, \ldots, x_n)$, $x = (x_1, x_2, ..., x_n)$.
- Given a hyperplane H in E^n. The orthogonal *shadow* [41] of A on H is then a fuzzy set $S_H(A)$ with the following property:

Let L be a line orthogonal to H and h its point of intersection with the hyperplane H, then the following is true:

$$\mu_{S_H(A)}(h) = \sup_{x \in L} \mu_A(x), \quad \mu_{S_H(A)}(x) = 0 \quad \text{for} \quad x \notin H.$$

If $H_1 = x|x_1 = 0$ is a coordinate hyperplane, then the membership function of A is given on this hyperplane H_1 as follows:

$$\mu_{S_{H_1}}(A) = (x_2, \ldots, x_n) = \sup_{x_1}(x_1, x_2, ..., x_n), \quad \text{for} \quad x_1 = 0$$

$$\mu_{S_{H_1}}(A) = 0, \quad \text{for} \quad x_1 = 0.$$

---

[41] The term "shadow" is suggestive if one thinks about the shows cast, for example, by a cloud or an airplane. "Projections" will also be mentioned later

$S_H^*(A)$ denotes a so-called cylindrical fuzzy set which has the following membership function:

$$\mu_{S_H^*(A)}(x) = \mu_{S_{H_1}(A)}(x_2, ..., x_n), \qquad x \in \mathbf{E}^n.$$

Then A itself is also naturally its subset: $A \subset S_H^*(A)$ and with the definition of the hyperplanes $H_i = \{x|x_i = 0\}, i = 1, ..., n$, A is limited from above by the intersection of all of these cylindrical fuzzy sets:

$$A \subset \bigcap_{i=1}^n S_H^*(A).$$

– A fuzzy set is *convex* if and only if the sets $\Gamma_\alpha = \{x|\mu_A(x) \geq \alpha\}$ are convex for all α in the interval $[0, 1]$. This is equivalent to the following statement:

$$\mu_A(\lambda x_1 + (1-\lambda)x_2) \geq \min[\mu_A(x_1), \mu_A(x_2)].$$

For example, the fuzzy set of real numbers that are "approximately equal to 1" is a convex set in E^l. In his article, Zadeh provided an illustration of the membership function of this convex fuzzy set as well as that of a non-convex fuzzy set (see Fig. 5.9).[42]

Zadeh described his "new view" on system theory over three more sections, and they will be discussed in the following paragraphs.

5.2.2 Fuzzy Systems

Given a system S with input $u(t)$, output $y(t)$ and state $x(t)$, it is a *fuzzy system* if $u(t)$ or $y(t)$ or $x(t)$ or any combination of them includes fuzzy sets. For instance, if an *input* of S is specified at time t as "significantly more than 5", then the output is a fuzzy set and a system that can operate on the basis of such imprecise inputs is a fuzzy system. Similar examples occur when the states of a system S are described by fuzzy adjectives like *light, heavy, not very heavy, very light* etc.; if these are fuzzy sets then S is a fuzzy system.

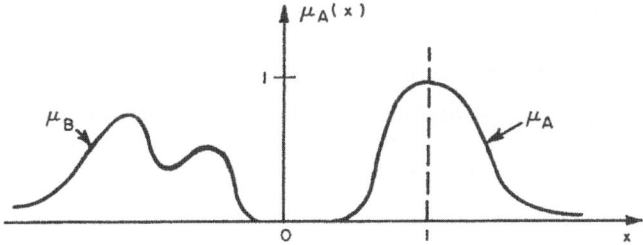

Fig. 5.9. Membership functions of a convex and a non-convex fuzzy set

---

[42] For the sake of completeness, it is mentioned that a fuzzy set A is *concave* if and only if its complement A' is convex

If S is a system with a discrete time and if t runs through all of the integers, then S can be characterized by the usual state equations[43]:

$$x_{t+1} = f(x_t, u_t)$$
$$y_t = g(x_t, u_t).$$

Fuzzy systems are systems whose inputs, outputs and/or states are fuzzy sets: "The difference between fuzzy and non-fuzzy systems, then, lies in the nature of the ranges of the variables $u(t)$, $y(t)$ and $x(t)$."[44]

The two equations above, which are known by their principal structure, were interpreted by Zadeh to be state equations of fuzzy systems. Of course, their structure is also identical to the state equations of stochastic systems; yet their interpretation showed $u(t), y(t)$ and $x(t)$ to be *probability distributions* of the *input, output* and *state* of the system:

> The difference between stochastic and fuzzy systems is that in the latter the source of imprecision is nonstatistical in nature and has to do with the lack of sharp boundaries of the classes entering into the descriptions of the input, output or state . Mathematically, however, these two types of systems are substantially similar and present comparable and, unfortunately, great difficulties in their analyses.[45]

5.2.3 Fuzzy Classes of Systems

At the beginning of the article *Fuzzy Sets and Systems*[46], Zadeh had mentioned "adaptive systems" as a typical example of a fuzzy system. As Chap. 4 revealed, adaptivity was one of those concepts in the realm of system classification of which Zadeh had been critical in the past. Other "unsharp" concepts having to do with the characterization of system properties had been *linearity* and *optimality*. With the help of fuzzy sets, this fuzziness could now be conceptualized:

> In fact, one may argue that most of the adjectives used in system theory to describe various types of systems, such as: linear, nonlinear, adaptive, time-invariant, stable, etc. are in reality names for fuzzy classes of systems. If one accepts this point of view, then one is freed from the necessity of defining these terms in a way that dichotomizes the class of all systems into two classes, e. g. systems that are linear and systems that are nonlinear. ... More realistic, then, one would regard, say, the "class" of adaptive systems as a fuzzy class, with each system having a grade of membership in it which may range from zero to one.[47]

---

[43] See Chap. 4
[44] [664] Zadeh, Systems (1965), p. 33
[45] [664] Zadeh, Systems (1965), p. 33f
[46] [664] Zadeh, Systems (1965), p. 29
[47] [664] Zadeh, Systems (1965), p. 34

Zadeh introduced fuzzy classes of systems. In his state space approach to system theory, he had defined systems as sets of input-output pairs. Every system is therefore a subset of the product space $U \times Y$, where U represents the set of all inputs and Y is the set of all outputs. Accordingly, a fuzzy class of systems is a fuzzy set in this product space $U \times Y$.

5.2.4 Optimization under Fuzzy Conditions

"x should not be significantly greater than 5" or "x should be nearly 10" or "x should lie approximately between 5 and 10" – Conditions like this had been placed on variables in some of the optimization problems – primarily in human-machine systems. Until this point, restrictions defined in such a "soft" way had to be replaced with "hard" restrictions, but this approach was not satisfactory when the *fuzzy* restrictions were by their nature not well suited for approximating *hard (crisp)* restrictions. Zadeh now recommended the "natural and possibly also more efficiently calculable" use of fuzzy restrictions, e. g. on the standard problem of maximizing a non-negative objective function $f(x_1, ..., x_n)$ over a normally defined *(crisp)* set A in E_n.[48] A denotes the set of restrictions and this results in the condition $x \in A$. The usual characteristic function $\mu_A(x)$ of the set A has the properties

$$\mu_A(x) = 1 \quad \text{for} \quad x \in A,$$
$$\mu_A(x) = 0 \quad \text{for} \quad x \notin A.$$

The maximization problem is therefore equivalent to the following formulation:

$$f^*(x) = f(x)\mu_A(x),$$

without any other constraints needing to be fulfilled.

If A is a fuzzy set, then the maximization of f by means of the restricting fuzzy set A can be interpreted as the average value of the maximization of the modified function f^*, over E_n. Then the maximization of $f(x)$, which is subject to a fuzzy restriction represented by a fuzzy set A, is reduced to an unrestricted maximization of the following function:

$$f^*(x_1, ..., x_n) = f(x_1, ..., x_n)\mu_A(x_1, ..., x_n).$$

Problems of optimal control often involve restrictions of the type $x_1 \in A_1, x_2 \in A_2, ..., x_n \in A_n$, wherein $A_i (i = 1, ..., n)$ are normal (crisp) sets in E_n. If A is then the direct product of the sets A_i, in other words $A = A_1 \times A_2 \times ... \times A_n$, then the individual sets A_i can be considered to be projections (shadows) of set A on its individual coordinate axes.

If A is a fuzzy set, then it is not unequivocally determined by its shadow; it can, however, be restricted from above by the intersection of all cylindrical

---

[48] Hereafter, the word "crisp" will be used to mean the opposite of "fuzzy"

fuzzy sets $A_1^*, A_2^*, ..., A_n^*$, the membership function of $A_i^*(i = 1, ..., n)$ being given as follows:

$$\mu_{A_i}^*(x) = \sup_{x_2} \sup_{x_3} ... \sup_{x_n} \mu_A(x_1, ..., x_n) = \mu_i(x_i).$$

Therefore: $A \subset \bigcap_{i=1,...,n}^{n} \mu_i(x_i)$ or equivalently:

$$\mu_A(x) \leq \min_{i=1,...,n} [\mu_1(x_1), ... \mu_n(x_n)].$$

With this approximate identification of $\mu_A(x)$, the modified objective function can now be found with the correct $\mu_i(xi)$ as follows:

$$f^*(x_1, ..., x_n) = f(x_1, ..., x_n) \min_{i=1,...,n} [\mu_1(x_1), ..., \mu_n(x_n)].$$

5.2.5 First Papers on Fuzzy Sets

In May of 1965, the short text *Shadows of Fuzzy Sets* appeared as Zadeh's second ERS Report on his Fuzzy Set Theory. In it he listed the properties of fuzzy sets and examined them in depth.[49] Zadeh prepared another paper also entitled *Shadows of Fuzzy Sets*, albeit in Russian this time, the content of which was somewhat more comprehensive and in which Zadeh had further developed his thoughts on the use of fuzzy sets in pattern recognition. This paper he sent to Moscow in August 1965, where it was published in 1966 in the journal *Problemy Peredachi Informatsii*.[50] That year, Zadeh had received two invitations to the Soviet Union. This was – particularly at that time and above all for a former citizen of a Soviet republic who had become a US citizen in 1958 – highly unusual. Invitations to speak had come from both the Popov Society and the Cybernetics Congress and Zadeh did not wish to turn either of them down. So he traveled with his wife to the Soviet Union twice in 1965. In June, at the invitation of the Popov Society, he delivered a lecture for the Academy of Sciences in Moscow; the publication of the aforementioned Russian language article is probably a result of this lecture.[51] The following September, he spoke about his new theory at the Congress on Cybernetics.

Cybernetics had been ignored in the Soviet Union before Stalin's death, but afterward the situation changed incredibly fast and cybernetics became very popular. The Cybernetics Congress in 1965 offered Lotfi Zadeh an outstanding opportunity to make his first official presentation on fuzzy sets, and the location and ambiance had been selected to be extraordinarily original:

---

[49] [684] Zadeh, ERL-Shadows (1965)
[50] [653] Zadeh, Shadows (1966)
[51] This lecture appears to have been delivered before a very small audience. An image in [574] Zadeh F., Life (1998) shows 11 people in front of the Academy building, including guests Claude E. Shannon, the Russian professor of electrical engineering Vladimir Ivanovich Siforov (1904–1993) and Lotfi A. Zadeh

The congress took place aboard the ship *S. S. Admiral Nachimov*, which carried more than 1121 passengers (1040 Russian scientists and 60 guests, including 21 Americans, six of whom were wives of the participants) on a six day journey across the Black Sea.[52] This nautical excursion consisted of the congress sessions themselves and a program of cultural events. The journey began and ended in Odessa, and Zadeh's lecture was a great success, not least of all because he had delivered it in Russian, a language he has spoken since childhood.[53] The many attendees from the Soviet Union appeared exceedingly interested in fuzzy sets. Meanwhile, the famous first article *Fuzzy Sets* had also been published in June of 1965.

5.3 The Article *Fuzzy Sets*

"More often than not, the classes of objects encountered in the real physical world do not have precisely defined criteria of membership."[54] – The very first sentence of this introduction encapsulates the conclusion to which Zadeh had come after decades of working on real physical objects with mathematically precise methods.

Here, though, he offered examples which do not make clear how this theory developed: The class of animals, for example, certainly included dogs, horses, birds etc., and of course mountains, liquids and plants did not belong to this class. Yet there were objects whose status was unclear in this regard: starfish, bacteria etc. The same kind of unclarity arose in the case of a number like 10 with respect to the "class" of all real numbers which are much greater than 1. Of course, the "class of all real numbers which are much greater than 1" or the "class of all beautiful women" or the "class of all tall men" were not classes or sets in the usual mathematical sense of these terms. Nevertheless, Zadeh wrote, it was a fact that such imprecisely defined "classes" played an important role in human thinking, especially in the fields of pattern recognition, communication of information and abstraction.

The definition of fuzzy sets and that of their properties resemble the corresponding passages from the earlier texts, except that Zadeh was now explicitly using the term "membership function"!

> A *fuzzy set (class)* A in X is characterized by a *membership function (characteristic function)* $f_A(x)$ which associates with each point in X a real number in the interval [0, 1], with the value of $f_A(x)$ at x representing the 'grade of membership' of x in A.[55]

---

[52] [574] Zadeh F., Life (1998), p. 76
[53] See [574] Zadeh F., Life (1998), p. 76ff
[54] [612] Zadeh, Fuzzy (1965), p. 338
[55] [612] Zadeh, Fuzzy (1965), p. 339. In this text, Zadeh denoted the membership function with the letter f. In the following, I will make consistent use – even in verbatim quotes – of the Greek letter μ, which he used later and which has been

The article goes on to define an empty *fuzzy set*[56], the *equality* of two *fuzzy sets*, the *complement* of a fuzzy set, *fuzzy subsets* and the formation of *unions* and *intersections* for fuzzy sets (see Fig. 5.10).[57] As to the assertion that all fuzzy sets of a universal set X constitute a distributive lattice with a 0 and 1, he referred to the lattice theory by Garrett Birkhoff[58] and merely mentioned that de Morgan's laws[59] and the distributive laws are valid for the fuzzy sets in X with the defined operations.

Algebraic Operations
After the union and intersection of fuzzy sets, Zadeh also defined other combinations of fuzzy sets, which will be described only briefly here:

INFORMATION AND CONTROL 8, 338–353 (1965)

Fuzzy Sets*

L. A. ZADEH

Department of Electrical Engineering and Electronics Research Laboratory, University of California, Berkeley, California

A fuzzy set is a class of objects with a continuum of grades of membership. Such a set is characterized by a membership (characteristic) function which assigns to each object a grade of membership ranging between zero and one. The notions of inclusion, union, intersection, complement, relation, convexity, etc., are extended to such sets, and various properties of these notions in the context of fuzzy sets are established. In particular, a separation theorem for convex fuzzy sets is proved without requiring that the fuzzy sets be disjoint.

Fig. 5.10. First page (excerpt) of Zadeh's first article about fuzzy sets

---

standard for decades now, to denote the membership function in order to avoid confusion

[56] Two fuzzy sets A and B are equal, $A = B \Leftrightarrow \mu_A(x) = \mu_B(x)$ for all $x \in X$

[57] With regard to the union and intersection of fuzzy sets, it is notable that Zadeh now transposed definitions and subsequent observations that had appeared before ([675] Bellman et al., Abstraction (1964)). Here he defines the union of two fuzzy sets A and B as a fuzzy set $C = A \cup B$ with the membership function $f_C(x) = \max[\mu_A(x), \mu_B(x)]$. Only afterward does he note and prove that this is the smallest fuzzy set containing both A and B. Accordingly, he defines the intersection of two fuzzy sets A and B as fuzzy set $C = A \cap B$ with the membership function $\mu_C(x) = \min[\mu_A(x), \mu_B(x)]$. That this is the largest fuzzy set that is contained in both A and B was also easy to demonstrate

[58] [80] Birkhoff, Lattice (1948)

[59] Augustus de Morgan (1806–1871), British mathematician and logician, founder of the theory of relations who also formulated important laws of propositions and the logic of classes, including the rules named after him: $\neg (p \wedge q) \Longrightarrow \neg p \vee \neg q$ und $\neg (p \vee q) \Longrightarrow \neg p \wedge \neg q$. In 1826, De Morgan obtained a professorship at the newly founded University College in London

- The *algebraic product* $A \cdot B$ of two fuzzy sets A and B is defined by the membership function[60] $f_{A \cdot B} = f_A \cdot f_B$
- The *algebraic sum* $A + B$ of two fuzzy sets A and B is defined by the membership function $f_{A+B} = f_A + f_B$
- The *absolute difference* $|A - B|$ of two fuzzy sets A and B is defined by the membership function $f_{|A-B|} = |f_A - f_B|$.

Finally, Zadeh also generalizes the concept of the *convex combination*[61] of two vectors f, g to the *convex combination* of fuzzy sets A, B and for all Λ by the following definition:

$$(A, B; \Lambda) = \Lambda A + \Lambda' B, \qquad \text{where } \Lambda', \text{ is the complement of } \Lambda.$$

The corresponding membership function reads:

$$f_{(A,B;\Lambda)}(x) = f_\Lambda(x) f_A(x) + [1 - f_\Lambda(x)] f_B(x).$$

As a consequence of the easily verified inequalities

$$\min[f_A(x), f_B(x)] \leq \Lambda f_A(x) + (1 - \Lambda) f_B(x) \leq \max[f_A(x), f_B(x)], x \in X,$$

the following basic property of the *convex combination* of fuzzy sets follows:

$$A \cap B \subset (A, B; \Lambda) \subset A \cup B, \qquad \text{for all } \Lambda.$$

Fuzzy Relations

In this article, Zadeh also extended the concept of relation to include the fuzzy relation. He borrowed from Paul Richard Halmos[62] the definition of a relation as a set of ordered pairs (x, y) such that $x, y \in X$. A fuzzy relation he defined as a fuzzy set in the product space $X \times X$, or just as generally: an n-ary fuzzy relation is a fuzzy set A in the product space $X \times X \times \ldots X$, with the membership function $f_A(x_1, ..., x_n)$, where $x_i \in X$, $i = 1, \ldots, n$. In the case of binary fuzzy relations, Zadeh also explained the composition of two fuzzy relations, as the fuzzy relation whose membership function is related to those of A and B by

$$f_{A \circ B}(x, y) = \sup_\nu \min[f_A(x, \nu), f_B(\nu, y)].$$

Convex Fuzzy Sets

He wrote a new section about convexity, for he required the properties of convex fuzzy sets in order to address the problems of pattern recognition later on.

---

[60] Zadeh omitted the argument of the membership function if he did not think there was any chance of misunderstanding
[61] If C is a convex subspace in the vector space V, then the following applies for all: $f, g \in V : \{(1 - \lambda)f + \lambda g, 0 \leq \lambda \leq 1\} \subset C$, for all $\lambda \in \mathbf{R}$
[62] [226] Halmos, Naive (1960); Paul Richard Halmos (born 1916), Hungarian-American mathematician. In 1940, Halmos became John von Neumann's assistant at the IAS in Princeton and later a professor at Indiana University

As will be seen in the sequel, the convexity can readily be extended to fuzzy sets in such a way as to preserve many of the properties which it has in the context of ordinary sets. This notion appears to be particularly useful in applications involving pattern classification, optimization and related problems.[63]

He first defined *convex* fuzzy sets[64] (Fig. 5.11):

A fuzzy set A is convex $\Leftrightarrow$
$\Gamma_\alpha = \{x | f_A(x) \geq \alpha\}$ is convex for all $\alpha \in (0, 1]$.

He then defined *bounded* fuzzy sets:

A fuzzy set A is *bounded* $\Leftrightarrow \Gamma_\alpha = \{x | f_A(x) \geq \alpha\}$ is bounded $\forall \alpha > 0$
that is, $\forall \alpha > \exists$ a finite $R(\alpha)$ such that $\|x\| \leq R(\alpha) \forall x \in R(\alpha)$

Therefore, if A is a bounded fuzzy set, then for each $\varepsilon > 0$ there exists a hyperplane H such that for all x on the side of H which does not contain the origin: $f_A(x) \geq \varepsilon$.

Pattern Recognition

Once he had handled the convexity of fuzzy sets in this article, Zadeh also turned his attention to separating sets from one another, a problem associated with the field of pattern recognition. Ordinary convex sets A and B can be separated easily if they are disjoint since in this case, according to the separation theorem, there exists a separating hyperplane H with the set A lying on one side and the set B on the other side. Zadeh now expanded the validity of this theorem for *bounded* fuzzy sets A and B, although, of course, he had to do so without disjointedness.

He began with two bounded fuzzy sets A and B and the hyperplane H in the Euclidean space E^n defined by the equation $h(x) = 0$ such that all

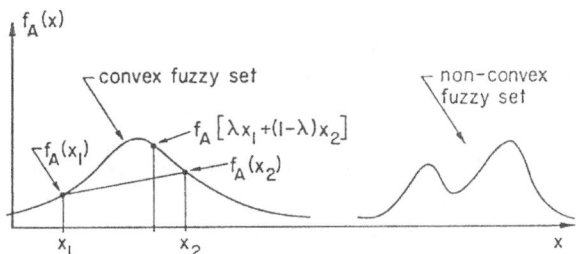

Fig. 5.11. Convex and non-convex fuzzy sets in $\mathbf{E}^1$

---

[63] [612] Zadeh, Fuzzy (1965), p. 346

[64] The following equivalent definition comes from Zadeh's colleague Elwyn Berlekamp (born 1940), electrical engineer and mathematics professor at Berkeley: A is convex $\Leftrightarrow f_A[\lambda x_1 + (1 - \lambda)x_2] \geq min[f_A(x_1), f_A(x_2)], \forall x_1, x_2 \in X, \forall \lambda \in [0, 1]$

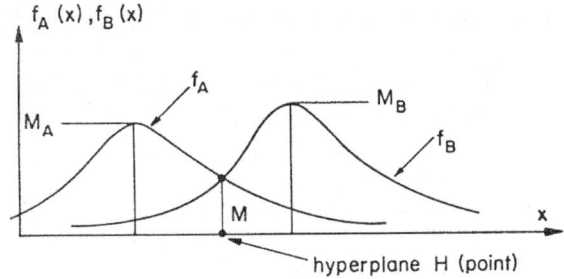

Fig. 5.12. Illustration of the separation theorem for fuzzy sets in $\mathbf{E}^1$

x for which $h(x) \geq 0$ lie on one side of H and all x for which $h(x) \leq 0$ lie on the other side. For each hyperplane H there exists a number K_H such that $f_A(x) \leq K_H$ on the one side of H and $f_B(x) \geq K_H$ on the other side of H.

Zadeh denoted the infimum of these two numbers with M_H and its complement $D_H = 1 - M_H$ as the *degree of separation* of A and B through H.

Since in general an entire family of hyperplanes is given, Zadeh formulated the separation problem thusly: to find the particular hyperplane from a family of hyperplanes $\{H_\lambda\}$ which realizes the highest possible degree of separation. For the special case of this problem where the hyperplanes $\{H_\lambda\}$ lie within the Euclidean space E^n and λ is a ranging variable in E^n, Zadeh defines the degree of separation of A and B by the following relation, with the index λ being omitted for simplicity:

$$D = 1 - M,$$

where $M = \inf_H M_H$ is the infimum of M_H of all hyperplanes H.

At the end of the article, Zadeh proved that the highest degree of separation of two convex and bounded fuzzy sets A and B that can be achieved by means of a hyperplane in the Euclidean space E^n is the complement of the maximum degree in the intersection $A \cap B$. (This is illustrated in the above Fig. 5.12 for $n = 1$.)

Theorem 2. *Let A and B be bounded convex fuzzy sets in E^n, with maximal grades M_A and M_B, respectively $[M_A = \sup_x f_A(x)$ and $M_B = \sup_x f_B(x)]$. Let M be the maximal grade for the intersection $A \cap B (M = \sup_x \min[fA(x), fB(x)])$. Then $D = 1 - M$.*[65]

---

[65] [612] Zadeh, Fuzzy (1965), p. 352

5.4 An Interpretation for Unions and Intersections

In a small subsection of the article *Fuzzy Sets*, Zadeh provided an interpretation of the union and intersection operations of fuzzy sets he had introduced which permits an assumption to be made about the origins of Fuzzy Set Theory. Zadeh's interpretation goes back to Shannon's discovery in the use of electrical circuits to model logical statements, which was discussed in Chap. 1.[66]

In the case of conventional sets, as Shannon had showed in 1938 and Zadeh had highlighted in his early text from 1950[67], every set C from sets $A_1, \ldots, A_i, \ldots, A_n$ can be combined with one another using the conjunctions $\cup$ und $\cap$ and such that it represents a network of circuits $\alpha_1, \ldots, \alpha_n$ ($i, j = 1, \ldots, n$). By this logic, $A_i \cup A_j$ and $A_i \cap A_j$ are, respectively, series and parallel combinations of the circuits α_i and α_j.

For the analogous interpretation in the case of fuzzy sets, Zadeh employed the concept of the sieve. He provided the membership function $\mu_i(x)$ of A_i at x with a "sieve" $S_i(x)$ with mesh size $\mu_i(x)$. This interpretation results in immediate and clearly evident correlations of the parallel combinations of sieves $S_i(x)$ and $S_j(x)$ with $\mu_i(x) \vee \mu_j(x)$ and of their series combinations with $\mu_i(x) \wedge \mu_j(x)$ (see Fig. 5.13):

Zadeh shed light on this analogy in a conversation nearly 35 years later:

> Well, if you have a sieve and you have things of different size, so it will pass all those that are smaller than the opening. And it will not pass those that are not. So that it has that property, it is based at max-min. If we have two sieves, you see, then the two sieves will be the equivalent of parallel, it will be the larger of the two. If we have things and we put it into two sieves, one of which has larger things than the other one, then is still the one of with the larger things that will be effective. Now, if we put them in series it will be the one with the smaller thing effective. So parallel is max and series is min. So this is what I meant.[68]

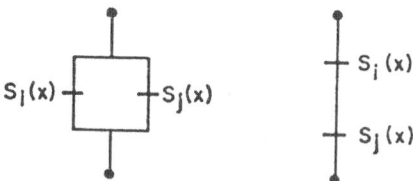

Fig. 5.13. Illustration of the parallel and serial connection of sieves

---

[66] See the statements in Chap. 1
[67] [667] Zadeh, Thinking (1950), p. 30
[68] [699] R. S. interview with L. A. Zadeh (2000)

5.4 An Interpretation for Unions and Intersections

The point of this analogy is not what residue is left behind or which objects are allowed to pass through; the intent is to discover which of the sieves determines the effectiveness of the entire system of sieves. Zadeh defined the union of fuzzy sets as the smallest fuzzy set that contains all original fuzzy sets and the intersection as the largest fuzzy set that is still contained in all original fuzzy sets. In the case of conventional set theory, the membership functions are identical to the indicator functions of sets that take only 0 and 1 as values. An object not belonging to a set receives the value 0, since it is not an element of the set, and an object that does belong to this set is given the value 1, since it is an element of the set. The union of sets is the set of objects which belong to at least one of these sets. This is surely the smallest set containing all of the original sets, and its indicator function provides each object that is an element of one of the original sets with the membership value 1. This is the maximal membership function.

The intersection of sets is the set of objects which belong to all of these sets without exception. This is surely the largest set that is a subset of all of the original sets, and its membership function provides only those objects that are elements of all of the original sets with the membership value 1. This is the minimal membership function.

Zadeh sought a way to extend these structures to include fuzzy sets:

> The question was in that paper: How to generalize various concepts, union, intersection and so forth? And so then I pointed out that conventional definitions of them don't work. But if you define in terms of "largest" set that's intersection: the largest set that contain in both, then you came up with proper generalizations![69]

In order to identify the largest fuzzy set contained in all original fuzzy sets and the smallest fuzzy set containing all original fuzzy sets, it is first necessary to clarify the meaning of "to be contained" for fuzzy sets. According to Zadeh's definition, a fuzzy set A is contained in a fuzzy set B, $A \subseteq B$ if for every object the membership function of A is less than (or equal to) the membership function of B:

$$\mu_A(x) \leq \mu_B(x) \text{ for all } x \in X.\text{[70]}$$

- The smallest fuzzy set that contains all of the original fuzzy sets is therefore the fuzzy set whose membership function gives each possible object the highest value it has received from the membership functions of the individual fuzzy sets.
- The largest fuzzy set that is contained in all of the original fuzzy sets is thus the fuzzy set whose membership function gives each possible object the lowest value it has received from the membership functions of the individual fuzzy sets.

---

[69] [699] R. S. interview with L. A. Zadeh (2000).
[70] In this function, X is the universal set of all possible objects.

The union and intersection of fuzzy sets thus emerge as those fuzzy sets whose membership functions are the maximum and the minimum of the membership functions of the original fuzzy sets.

As to how these definitions of the union and intersection of fuzzy sets came about, I would like to propose an interpretation at this point that establishes a connection in a science history context to Zadeh's works on electrical circuits, networks and particularly filters: In his work on electrical filters in the early 1950s, Zadeh had initially transferred the algebraic properties of projection operators in vector spaces and the resulting functional calculus to ideal filters. He considered the parallel and serial connection of ideal filters to be analogous to the algebraic operations of vector space projectors. Since in reality there were no ideal filters, however, Zadeh next looked at real filters and optimal filters, as well, to try to create an optimal approximation of their conditions in an ideal filter.[71]

Electrical filters have a "stopband" and a "passband" for the frequencies to be filtered, and ideally there is a single "threshold frequency" at which these two bands are separated.[72] The filter obstructs all frequencies above this threshold frequency, while all frequencies up to said threshold frequency are supposed to pass through the filter. As was explained in Chap. 1, Zadeh was very much aware of the fact that the performance of real filters did not correspond to that of ideal filters and the graphic representations of the filtration performance of both showed a more or less large environment around the threshold frequency; the frequencies lying within this "fuzzy" range were allowed to pass or were blocked – though with reduced intensity. Thus the graphic representations do not show any leaps in performance at the threshold frequency in either filter but instead display a smooth *(fuzzy)* transition from the passband to the stopband.

If one interprets a filter's performance curve as a membership function of possible frequencies to the effectiveness of the filter, then almost all of the frequencies in the passband will receive a membership value of 1 and almost all of the frequencies in the stopband will receive the membership value 0. In the environment around the threshold frequency, however, there are membership values other than 0 and 1 for those frequencies that are allowed to pass even though they should be blocked and those that are blocked although they should pass. With a membership function such as this, the set of frequencies to be filtered can be interpreted as a fuzzy set.

After Zadeh had established a calculus for ideal filters which he could use to deal with combinations of them in analogy to the algebraic operations of vector space projectors, he wanted to generalize the work he had done for

---

[71] For more on Zadeh's ideas with regard to optimality criteria, see Chap. 4

[72] See Chap. 1. So that the argumentation does not become unnecessarily complicated, I have limited myself here to low-pass filters, which are supposed to allow only frequencies below a threshold frequency to pass. The argumentation is similar for other filter types

5.4 An Interpretation for Unions and Intersections

optimal filters. To do so he also needed more general vector space operators, and his deliberations quickly led him to statistical observations, although they did not seem particularly promising to him.[73]

As was discussed in Chap. 4, Zadeh was working very intensively in the 1950s with abstract system theory and primarily with the differences between ideal and real systems. Chapter 4 showed how Zadeh eventually founded the theory of fuzzy sets and fuzzy systems on the basis of this discrepancy. Yet Fuzzy Set Theory now offered a calculus that also permitted an interpretation for the combinations of real filters, and Zadeh provided it himself in his article *Fuzzy Sets* – he just called them *sieves* rather than filters. If real filters are combined in series or in parallel, then the following intensity functions result:

– In a parallel connection, the resulting intensity function assigns each possible frequency the highest value that it has received from one of the intensity functions of the individual filters.
– In a serial connection, the resulting intensity function assigns each possible frequency the lowest value that it has received from one of the intensity functions of the individual filters.

The similarity of the performance functions of real filters to the membership functions of fuzzy sets leads to my thesis that the curves of real filtration performances have played an important role in the evolutionary history of fuzzy sets. One indication of this in the article can be seen in the example that follows Zadeh's explanation that simple parallel and serial connections of sieves correspond to the union and intersection of two fuzzy sets. Here he is making explicit reference to electrical circuits – even though he does not expressly mention sieve circuits, i. e., filter circuits – but he is discussing "sieve networks":

> More generally, a well-formed expression involving $A_1, \ldots, A_n, \cup$, and $\cap$ corresponds to a network of sieves $S_1(x), \ldots, S_n(x)$ which can be found by the conventional synthesis techniques for switching circuits. As a very simple example,
>
> $$C = [(A_1 \cup A_2) \cap A_3] \cap A_4$$
>
> corresponds to the network shown in Fig. 5.14.
> Note that the mesh sizes of the sieves in the network depend on x and that the network as a whole is equivalent to a single sieve whose meshes are of size $f_C(x)$.[74]

With this interpretation, the developments described in the previous chapters leading up to the founding of Fuzzy Set Theory are folded into in the history of communication technology and system theory, fields that had been molded

---

[73] See Chap. 1
[74] [612] Zadeh, Fuzzy (1965), p. 343. The reference numbers in the illustration were adapted to the order of the illustrations in this book

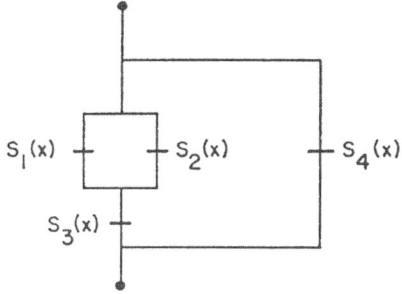

Fig. 5.14. Combination of filter circuits

by electrical engineering. The next section reveals that Zadeh definitely recognized in his theory of fuzzy sets and fuzzy systems the potential for a general system theory.

5.5 System Theory and Fuzzy Systems

Five years after the appearance of the textbook *Linear System Theory: The State Space Approach*[75], Zadeh and Elijah Polak co-published the anthology *System Theory* as volume 8 of the *Inter-University Electronics Series*.[76]

Fig. 5.15. E. Polak

This series was an attempt to portray the state of the art in electrical engineering. Zadeh's contribution *The Concepts of System, Aggregate and State*

---

[75] [586] Zadeh et al., System (1963)
[76] [587] Zadeh, System (1969). The co-author Elijah Polak has been teaching at Berkeley since 1961. His specializations were optimal control, mathematic programming, system theory and computer aided design. He had first studied at the University of Melbourne in Australia and then went to Berkeley, where he also earned his doctorate

5.5 System Theory and Fuzzy Systems

in System Theory was the first article in this book and it thus also introduced the first section, which was called *General System Theory*. This was then followed by parts II to V: *Linear Systems, Nonlinear Systems, Stochastic and Learning Systems and Optimal Systems*. Zadeh was apparently now interpreting his state-space approach as an approach to a general system theory.

Only one section of the text will be discussed here, but it shows how intensely Zadeh's deliberations about system theory correlated to his thoughts on fuzzy sets. Zadeh considered abstract objects $\mathcal{A}_1, \mathcal{A}_2, \ldots, \mathcal{A}_N$, where some of the inputs or outputs of one object are simultaneously also outputs or inputs, respectively, of some of the other objects. He called an aggregation of objects such as this a system.[77] He provided the following figure to illustrate a simple combination of abstract objects that form a system:

The following conditions apply here between the inputs and outputs of objects $\mathcal{A}_1$ und $\mathcal{A}_2$:

$$y^1 = u^2,$$
$$u = u^1 \quad \text{and} \quad y = y^2.$$

If one worked with combinations of two or more systems like this, the question would arise of how the respective input-output pairs and the combinations of $\mathcal{A} = \mathcal{A}_1, \mathcal{A}_2, \ldots, \mathcal{A}_N$ could result from the input-output pair that constitutes the overall system $\mathcal{A}$. This was one of the central questions of system theory and Zadeh called it the problem of input-output analysis.

There were two basic ways of formulating this problem, he wrote, and they differed from each other by how the sets of input-output pairs are defined in each instance. In the *explicit form*, these sets are defined by their characteristic function. Each input-output pair is given the value 1 if it belongs to a component system; it receives the value 0 if this is not the case. In concrete terms, this means: If an input-output pair that is provided with a component system $\mathcal{A}_i (i = 1, \ldots, n)$ is denoted as (u^i, y^i), then $\mathcal{A}_i$ is defined explicitly by the characteristic function $\mu(u^i, y^i)$:

$$\mu(u^i, y^i) = 1 \quad \text{for} \quad (u^i, y^i) \in \mathcal{A}_i$$
$$\mu(u^i, y^i) = 0 \quad \text{for} \quad \text{all ordered pairs of time functions,}$$
$$\text{not belonging to } \mathcal{A}_i.$$

Knowing the characteristic function of the object $\mathcal{A}_i$ is therefore equivalent to having a list or an explicit characterization of all input-output pairs that belong to $\mathcal{A}_i$.

In the *implicit form*, the component systems are defined by their respective input-output relationships. To determine the input-output pairs belonging to

---

[77] Zadeh noted here that with this definition each abstract object is a system and vice versa, which is why he henceforth referred both to abstract objects and to their aggregations as systems

$\mathcal{A}_i$ in this case, it is necessary to "solve" the input-output relationship that defines $\mathcal{A}_i$. What exactly does that mean?

1. The *problem of input-output analysis in explicit form* is expressed as follows: The system $\mathcal{A}$ is given as a combination of the component systems $\mathcal{A}_1, \mathcal{A}_2, \ldots, \mathcal{A}_N$, and each $\mathcal{A}_i (i = 1, \ldots, n)$ is defined by its characteristic function (u^i, y^i). The characteristic function of $\mathcal{A}$ should then be determined from knowledge 1) about the characteristic functions of $\mathcal{A}_1, \mathcal{A}_2, \ldots, \mathcal{A}_N$ and 2) about the conditions placed on each of the inputs and outputs of $\mathcal{A}_1, \mathcal{A}_2, \ldots, \mathcal{A}_N$, since they are linked together.
2. The *problem of input-output analysis in implicit form*, on the other hand, is expressed in this way: The system $\mathcal{A}$ is given as a combination of the component systems $\mathcal{A}_1, \mathcal{A}_2, \ldots, \mathcal{A}_N$, and each $\mathcal{A}_i (i = 1, \ldots, n)$ is defined by an input-output relationship, e.g. a differential equation. The input-output relationship associated with $\mathcal{A}$ should then be determined from knowledge 1) about the input-output relationships of $\mathcal{A}_1, \mathcal{A}_2, \ldots, \mathcal{A}_N$ and 2) about the conditions placed on each of the inputs and outputs of $\mathcal{A}_1, \mathcal{A}_2, \ldots, \mathcal{A}_N$, since they are linked together.

The first problem is much easier to solve than the second. The latter problem can be solved analytically only for particular system types, such as when the systems are defined by differential equations with constant coefficients.

Zadeh illustrated the conditions using an example that will also serve as clarification here: Given a system $\mathcal{A}$ consisting of the combination of two systems $\mathcal{A}_1, \mathcal{A}_2$, as shown in Fig. 5.16, the systems $\mathcal{A}_1, \mathcal{A}_2$ have the following characteristic functions:

$$\begin{array}{llll}
\mu_1(u^1, y^1) = 1 & \text{for} & (u^1, y^1) \in \mathcal{A}_1, \\
\mu_1(u^1, y^1) = 0 & \text{for} & (u^1, y^1) \notin \mathcal{A}_1, \\
\mu_2(u^2, y^2) = 1 & \text{for} & (u^2, y^2) \in \mathcal{A}_2, \\
\mu_2(u^2, y^2) = 0 & \text{for} & (u^2, y^2) \notin \mathcal{A}_2.
\end{array}$$

The condition binding the systems together is expressed by the equation $u^2 = y^1$. It necessarily follows then that (u^1, y^2) is an input-output pair for the system $\mathcal{A}$ if and only if a time function y^1 exists such that (u^1, y^1) and (y^1, y^2) are the input-output pairs for $\mathcal{A}_1$ and $\mathcal{A}_2$, respectively. This statement

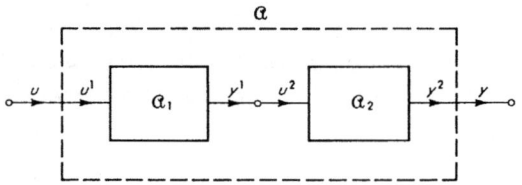

Fig. 5.16. Tandem connection of two systems to an overall system

"$(u^1, y^1) \in \mathcal{A}_1$ and $(y^1, y^2) \in \mathcal{A}_2$" Zadeh now expressed by the following equation:

$$\min[\mu_1(u^1, y^1), \mu_2(u^1, y^2)] = 1.$$

In this equation, $\min[a, b]$ denotes the smaller of the two numbers a, b. He then expressed the statement "There exists a y^1 such that $(u^1, y^1) \in \mathcal{A}_1$ and $(y^1, y^2) \in \mathcal{A}_2$" as follows:

$$\max_{y^1} \min[\mu_1(u^1, y^1), \mu_2(u^1, y^2)] = 1.$$

This implies in turn that the characteristic function of the system $\mathcal{A}$ can be described using terms from the characteristic functions of systems $\mathcal{A}_1$ and $\mathcal{A}_2$, specifically:

$$\mu_1(u^1, y^2) = \max_{y^1} \min[\mu_1(u^1, y^1), \mu_2(u^1, y^2)] = 1.$$

In so doing, Zadeh had defined the set of input-output pairs of the "tandem combination" of $\mathcal{A}_1$ and $\mathcal{A}_2$ using terms of the sets of input-output pairs of $\mathcal{A}_1$ and $\mathcal{A}_2$.

What is interesting here with regard to Fuzzy Set Theory is the introduction of the min function. Zadeh provided the paragraph in question with a footnote in which he explained:

> Alternatively and more simply, we could write $\mu_1(u^1, y^1)\mu_2(u^1, y^2) = 1$. We use min rather than the product because the former has wider generality (e. g., is applicable when, as in the case of fuzzy systems, the characteristic functions take values in the interval $[0, 1]$). Note that for binary variables $\min[a, b] = ab$ and $\max[a, b] = a + b - ab$, where $+$ denotes sum. More generally, one can employ the notation $\min[a, b] = a \wedge b$ and $\max[a, b] = a \vee b$ to simplify the writing of equations [...].[78]

Using the explicit form of the input-output problem of systems, Zadeh again found a calculus with which he could analyze combined systems on the basis of their subsets (Fig. 5.17).

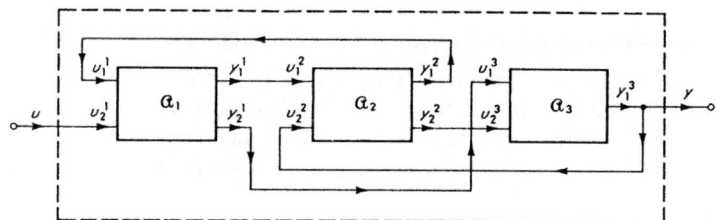

Fig. 5.17. Example of a system as a combination of three subsystems

---

[78] [600] Zadeh, Concepts (1969), p. 20, footnote 1

The function μ used here, which evaluates each input-output pair, he referred to as the "characteristic function" of systems. In normal set theory, sets are "characterized" in a similar fashion by the *indicator function*. For his Fuzzy Set Theory, Zadeh had to expand this function because the fuzzy memberships that are expressed as values between 0 and 1 also needed to be accounted for. Later – as has already been mentioned – Zadeh gave the name "membership functions" to these functions which characterize the fuzzy sets.

As the quote above makes clear, the "min function" introduced here corresponds to the formation of intersections when applied to membership functions of fuzzy sets, and the "max function", as a membership function, corresponds to the union of fuzzy sets.

Finally, one further point in Zadeh's line of argumentation should be highlighted here. As has just been discussed, Zadeh expressed the statement "$(u^1, y^1) \in \mathcal{A}_1$ and $(y^1, y^2) \in \mathcal{A}_2$" by the following minimum:

$$\min[\mu_1(u^1, y^1), \mu_2(u^1, y^2)] = 1.$$

The statement "There exists a y^1 such that $(u^1, y^1) \in \mathcal{A}_1$ and $(y^1, y^2) \in \mathcal{A}_2$" he expressed by placing a maximum in front:

$$\max_{y_1} \min[\mu_1(u^1, y^1), \mu_2(y^1, y^2)] = 1.$$

This is justified by the following thought: If the statement "There exists a y^1 such that $(u^1, y^1) \in \mathcal{A}_1$" is asserted, this does not mean that it is also possible to specify such a y^1. It is merely clear that such a y^1 exists. Put differently: Of all existing y^1, one of them fulfills the condition $(u^1, y^1) \in \mathcal{A}_1$. Assuming the set of all existing y^1 is $\{y_1^1, y_2^1, \ldots, y_n^1\}$ then the above statement means:

$$(u^1, y_1^1) \in \mathcal{A}_1 \quad \text{or} \quad (u^1, y_2^1) \in \mathcal{A}_1, \ldots, \quad \text{or} \quad (u^1, y_n^1) \in \mathcal{A}_1.$$

Zadeh had used the maximum in his Fuzzy Set Theory in place of this logical "or", which in set theory corresponds to a union.

The "max-min equation" that appears here would play an important role in the further development of Fuzzy Set Theory, particularly in its applications. This is true for the areas of fuzzy control and fuzzy relations that will be treated in the next two chapters. Zadeh had no idea that his Fuzzy Set Theory would find application in technological fields as quickly as turned out to be the case a short time later. In 1973, the first real system was controlled by means of an algorithm that included fuzzy sets. In the 1980s, fuzzy control principles were implemented in numerous industrial settings, at first in Japan and then later in other countries. In the late '60s, however, Zadeh was still expecting the fuzzy sets to be seized upon primarily in the non-exact sciences.

> What we still lack, and lack rather acutely, are methods for dealing with systems which are too complex or too ill-defined to admit of

precise analysis. Such systems pervade life sciences, social sciences, philosophy, economics, psychology and many other "soft" fields.[79]

His appraisal of the situation would turn out to be right at least as far as the life sciences were concerned, for applications of Fuzzy Set Theory were already being considered, first theoretically and then practically as well, in the years immediately following this. A description of this development will be reserved for the final chapter. In a 1994 interview with the newspaper *Azerbaijan International*, Zadeh was asked, "How did you think Fuzzy Logic would be used at first?" He answered:

> In many, many fields. I expected people in the social sciences – economics, psychology, philosophy, linguistics, politics, sociology, religion and numerous other areas to pick up on it. It's been somewhat of a mystery to me why even to this day, so few social scientists have discovered how useful it could be. Instead, Fuzzy Logic was first embraced by engineers and used in industrial process controls and in "smart" consumer products such as hand-held camcorders that cancel out jittering and microwaves that cook your food perfectly at the touch of a single button. I didn't expect it to play out this way back in 1965.[80]

The following chapter should clarify the transitions between the theoretical development of Fuzzy Set Theory that has heretofore been discussed and the practical implementations that began shortly thereafter. It deals with the fuzzifications of theoretical and real systems.

---

[79] [687] Zadeh, ERL-Toward (1969), p. 1 and [668] Zadeh, Toward (1971), p. 469
[80] Interview with Lotfi Zadeh, Creator of Fuzzy Logic, by Betty Blair, *Azerbaijan International*, Winter 1994 (2.4): $http://www.azer.com/aiweb/categories/magazine/24\_folder/24\_articles/24\_fuzzylogic.html$ (accessed on May 2, 2003, 3:51 P.M.)

6
Fuzzifications

In the first three years following the appearance of *Fuzzy Sets*, Lotfi Zadeh did not publish any further developments of his new theory. He had become chairman of the department of electrical engineering at Berkeley in 1963 and was busy with the restructuring that was going on there.

6.1 Reactions

What was probably the first publicized reaction by a scientist to Zadeh's *Fuzzy Sets* came from the USSR: In its March/April 1966 edition, the Russian journal *Engineering Cybernetics* published a two-page contribution by Vasiliy I. Loginov, who interpreted Zadeh's membership function as a "likelihood function".[1] This view of memberships as "conditional probabilities" is relatively common today for the construction of membership functions using experimental methods. However, Loginov's text is also an early attestation to the immediate objection that probability theoreticians and statisticians raised against Fuzzy Set Theory: This theory was superfluous, they said, since it did not offer any greater potential for problem solving than probability theory and statistics.[2]

Zadeh traveled to Moscow once again in 1966 to deliver a lecture about Fuzzy Set Theory before the approximately 4,000 attendees at the International Congress of Mathematicians[3], and it was predominantly Soviet scientists who

---

[1] [324] Loginov, Probability (1966); Loginov had already submitted his manuscript to the editors of *Engineering Cybernetics* on Christmas Eve of 1965

[2] "Fuzzy sets" are not mentioned anywhere in the official English translation of Loginov's article. Loginov ignored the concept of Zadeh's new theory entirely. He only used Zadeh's term "membership function" and this he attributed to probability theory. It does not appear to be a clerical or typographical error that Zadeh's article is cited in Loginov's bibliography as "Hussy Sets"

[3] The invitation came about at the suggestion of Stephen Kleene. [709] R. S. interview with L. A. Zadeh (2001)

proved to be open to fuzzy sets, as they recognized the contribution the concept could make to cybernetics:

> You see, in the Soviet Union they were against cybernetics until the death of Stalin. The word cybernetics was a bad word. But after the death of Stalin things changed and cybernetics became public. That's why cybernetics was used much more frequently. Cybernetics became popular in the Soviet Union but not in the United States ... And so, what I want to say is that my first paper was presented in the Soviet Union in 1965. That was my first presentation of Fuzzy Sets. My second presentation was in 1966 in Moscow at a congress of mathematicians. They were very interested – very – they learned quickly.[4]

The situation was different in the United States, though. By far the most reactions here were very negative, and some of the criticisms stooped to personal attacks. On September 13, 1965, Zadeh, in his capacity as chairman, received a letter from within his own department of electrical engineering at the University of California-Berkeley. It read as follows:

> We disagree on a very fundamental question. There is no use in kidding ourselves in saying that it is very minor. Since the disagreement lies at the very root of the academic policies of our Department, we should try to clarify our respective positions the best we can. I would say that you have a tendency of confusing our academic objectives with those appropriate for a Couturier of World Renown. There are fashions in engineering and applied science and this is in several ways very healthy. However, academic objectives are badly distorted when they can be described as "be interested only in that which is in the highest fashion today or even better in that which will be in fashion tomorrow." Such objectives are appropriate for Ives Saint Laurent but not for a chairman of an engineering department in a university. It is after many hours of reflection that I have reached the conclusion that I would be derelict in my duty if I would not oppose the appointment of a Computer Sciences candidate to any of the Assistant Professor positions that will open to us.[5]

At the *Interdisciplinary Colloquium on Mathematics in the Behavioral Sciences* no. 3 (1966–67) in Los Angeles[6], Zadeh spoke on November 18, 1966 on the subject of *Fuzzy Sets and Concepts*. Here, too, there was criticism:

---

[4] [706] R. S. interview with L. A. Zadeh (2000). See also Chap. 5
[5] Zadeh, received letter (1965), no further details available
[6] The Ukrainian-American economist Jacob Marshak (1898–1977) organized this series of meetings at the University of California, Los Angeles (UCLA) from 1960 until his death

Arthur M. Geoffrion[7], who at that time was a mathematician at the Western Management Science Institute (WMSI), said that this theory could only be used when the membership function of a fuzzy set was known exactly:

> It is applicable when there is a standardized and perfectly accurate method of measuring degree of class membership, but it seems to be considerably less applicable otherwise, as when opinion is involved.[8]

In contrast, the linguist H. P. Edmundson of the University of California, Los Angeles[9] supported Zadeh's view that the fuzzy sets were needed in many disciplines, as commentaries by specialists in the fields of psychology, economics and logic had confirmed. He himself underscored

> that fuzzy sets also arise in linguistics. In particular, in the study of semantics, attempts to formulate satisfactorily the notion of a semantic space using crisp sets have essentially failed. As a consequence, the modeling of meaning as a set of senses or the modeling of synonymy in terms of equivalence classes has proved difficult to justify either theoretically or empirically. It seems likely that the concept of fuzzy set will provide a way to account for what has been called a "semantic space" and lead to a suitable metric or pseudometric. Similarly, it also may lead to a satisfactory way to replace the strict dichotomy of sentences as grammatical or ungrammatical, by a more natural concept involving grade of membership.[10]

In addition to requests for offprints, Zadeh also received encouraging feedback; the cognitive psychologist Jerome S. Bruner[11] wrote him: "But you are certainly right that we are in need of models to represent the case in which uncertainty derives from ambiguity in criterion". Philosopher Max Black[12] also responded. He had published a paper entitled *Vagueness – An Exercise in Logical Analysis*[13] back in 1934 in which he anticipated a vague idea from Zadeh's theory, for he wrote:

---

[7] Arthur Minot Geoffrion (born 1937) earned his Ph.D. at Stanford University in 1965 and is now a professor of management at UCLA's Anderson School

[8] [698] Zadeh, Discussion (1966)

[9] H. P. Edmundson has been a professor of computer sciences and mathematics since 1953. He works in the field of computational linguistics

[10] [698] Zadeh, Discussion (1966)

[11] Letter from Bruner to Zadeh dated 21 June 1967; Zadeh private archive. Jerome Bruner (born 1915), professor of psychology, Harvard (1952–1972) and Oxford (1972–1980)

[12] Max Black (1909–1988), American mathematician and philosopher. Black was born – like Zadeh – in Baku, the capital of Azerbaijan, but his family emigrated soon after his birth, first to Paris and then in 1912 to London. Black became an American citizen in 1948. He was employed at the University of Illinois at Urbana and became a professor of philosophy at Cornell University in New York in 1946

[13] [83] Black, Vagueness (1937). See also [81] Black, Concepts (1963)

> The vagueness of the word chair is typical of all terms whose application involves the use of the senses. In all such cases, "borderline cases" and "doubtful objects" are easily found to which we are unable to say either that the class name does or does not apply.[14]

He now told Zadeh:

> You were good enough to send me, some time ago, some of your recent papers on topics connected with "Fuzzy Sets". If I have not written before, the reason has not been lack of interests, but an inescapable press of other duties. Now that I have had a chance, at least, to study your work, I want to express my admiration and interest. I believe that your ingenious construction promises to provide intellectual tools of great value. In case you have not come across it, I might draw your attention to an early article of mine, entitled "Vagueness" (*Philosophy of Science*, Vol. 4, 427–455; reprinted in my book, *Language and Philosophy*, Cornell University Press, 1949). A more recent article on similar topics is "Reasoning with Loose Concepts" (*Dialogue*, Vol. 2, June 1963, 1–12). I would be happy to see offprints of any of your further publications.[15]

Zadeh also presented his Fuzzy Set Theory at the First International Conference on Man and Computer, which took place in Bordeaux from June 22 to 26, 1970.[16] The title of his paper was *Fuzzy Languages and their Relation to Human and Machine Intelligence*, and his thesis was that the difference between human and mechanical intelligence lay in the ability of the human brain – "an ability which present-day digital computers do not possess – to think and reason in imprecise, non-quantitative terms". Humans could

Fig. 6.1. R. Kalman

---

[14] [83] Black, Vagueness (1937), p. 434
[15] Black, Letter (1967)
[16] The First International Conference on Man and Computer was organized by the Institut de la Vie at the University of Bordeaux. The conference proceedings were not published until 1972

understand inexact instructions, whereas inputs for a computer had to be defined with precision. He suggested devising *fuzzy languages* which functioned such that commands formulated in a language like this could also be processed and carried out by future computers. In the discussion which followed, Rudolf Kalman[17], who had studied in New York when Zadeh was an assistant professor, rose to speak.[18]

> Kalman was a student of mine at Columbia University so I knew him quite well. But he never changed his position ever since. He stayed with the continuum. He never wrote anything on digital discrete, never. All of his themes were connected with continuous, except for his PhD thesis where he considered Markov processes, except for that it was continuous. And so he was a strong believer in the power of mathematics and so when I started writing something was fuzzy he was very antagonistic to all that and he is to this date.[19]

Kalman's comments were blunt; it was his opinion that fuzzy sets were lacking in any scientific importance:

> We do *talk* about fuzzy things but they are not scientific concepts. Instead, let us view the development of science as something like the following. You look at a vast mass of facts – fuzzy or not – and you would like to make some sense out of it. This is usually done through rising to a higher conceptual level, by working harder than the average person. Some people in the past have discovered certain interesting things, formulated their findings in a *non-fuzzy way*, and therefore we have progressed in science.[20]

Kalman's statement, quoted here from the conference proceedings, is practically sugarcoated compared to Zadeh's recollection of the incident.[21] Almost 30 years later, Zadeh published a different account of the course of this discussion, of which I can reprint only excerpts here[22]:

> Kalman: I would like to comment briefly on Professor Zadeh's presentation. His proposals could be severely, ferociously, even brutally

---

[17] Rudolf E. Kalman (born 1930), Hungarian-American electrical engineer. Professor at Stanford University (1964–1971), graduate research professor and director at the Center for Mathematical System Theory, University of Florida, Gainesville (1971–1992)

[18] For more on Rudolf Kalman, see also Chap. 1

[19] [706] R. S. interview with L. A. Zadeh (2000)

[20] [262] Kalman, Discussion (1972), p. 169

[21] [706] R. S. interview with L. A. Zadeh (2000)

[22] Zadeh used quotes from this argument to introduce the speech he delivered when he was awarded the Honda Prize in 1989. It was reprinted as part 1 of *The Birth and Evolution of Fuzzy Logic – A Personal Perspective* in the *Journal of Japan Society for Fuzzy Theory and Systems*. [705] R. S. interview with L. A. Zadeh (1999)

criticized from a technical point of view. This would be out of place here. But a blunt question remains: Is Professor Zadeh presenting important ideas or is he indulging in wishful thinking? The most serious objection of "fuzzification" of system analysis is that lack of methods of system analysis is not the principal scientific problem in the "system" field ... No doubt Professor Zadeh's enthusiasm for fuzziness has been reinforced by the prevailing political climate in the U.S. – one of unprecedented permissiveness. "Fuzzification" is a kind of scientific permissiveness; it tends to result in socially appealing slogans unaccompanied by the discipline of hard scientific work and patient observation. I must confess that I cannot conceive of "fuzzification" as a viable alternative for the scientific method; I even believe that it is healthier to adhere to Hilbert's naïve optimism, "Wir wollen wissen, wir werden wissen"...

Zadeh: ... I realize, of course, that I am challenging scientific dogma ... Now, when one attacks dogma, one must be prepared to become the object of counterattack on the part of those who believe in the status quo. Thus, I am not surprised when in reaction to my views I encounter not only enthusiasm and approbation, but also criticism and derision. Nevertheless, I believe that, in time, the concepts that I have presented will be accepted and employed in a wide variety of areas ...

Kalman: ... The question, then, is whether Professor Zadeh can do better by throwing away precise reasoning and relying on fuzzy concepts and algorithms. There is no evidence that he can solve any nontrivial problem ... Professor Zadeh's fears of unjust criticism can be mitigated by recalling that the alchemists were not prosecuted for their beliefs but because they failed to produce gold ...

Zadeh: I will rest on what I have said, since only time can tell whether or not my ideas may develop into an effective tool for the analysis of systems which are too complex or too ill-defined to be susceptible of analysis by conventional techniques. Conceding that I am not an unbiased arbiter, my belief is that, eventually, the answer will turn out to be in the affirmative.[23]

Since then, this argument has been settled definitively in Zadeh's favor, although it smoldered until the 1990s. When Kalman spoke of "fuzzifications", he was very likely using this word disparagingly, and Zadeh's decision to call his unsharp groups "fuzzy sets" amounted to an invitation to do so. On numerous occasions, Zadeh has explained his thinking on this issue:

I realized from the outset that the adjective "fuzzy" will create some problems because it is both pejorative and unscientific. Nevertheless,

---

[23] [606] Zadeh, Evolution (1999), p. 892f

I settled on it because it described more accurately than any other term I could think of the type of imprecision which is associated with unsharp class boundaries.[24]

Today, "fuzzification" is a commonly used term both in mathematics and in technology. This chapter deals with the first "fuzzifications" of mathematical concepts and technical systems.

6.2 Fuzzy Automata

The first dissertation about fuzzy sets was an advancement of Zadeh's thoughts on the separation problem in pattern recognition. In *Fuzzy Sets and Pattern Recognition*[25], completed in December 1967, the Chinese student Chin-Liang Chang worked with a convergence proof for perceptrons which Albert B. I. Novikoff[26] of the Stanford Research Institute had presented in April of 1962.[27] Chang examined the Novikoff algorithm and the version generalized by Mark Aizerman, Emmanuel Braverman and Lev Rozonoér[28] and he expanded them to include the separation of sets by a hyperplane for the case of fuzzy sets![29]

In August of the same year, William Go Wee[30] at Purdue University in Indiana had also submitted his dissertation *On Generalizations of Adaptive Algorithms and Application of the Fuzzy Sets Concept to Pattern Classification*. Wee had written this work under King Sun Fu[31], one of the pioneers in the field

---

[24] Zadeh, Lotfi A., Autobiographical note. See interviews with Zadeh: *IEEE Spectrum* 6, 1995, p. 34; in *KI* 3, 2001; see also the website: $http://www.kuenstliche-intelligenz.de/Artikel/InterviewwithProfLotfiAZadeh.htm$ or at the website of *Azerbaijan International*, winter 1994 (2.4): $http://www.azer.com/aiweb/categories/magazine/24\_folder/24\_articles/24\_fuzzylogic.html$

[25] [121] Chang, Fuzzy (1967)

[26] Albert Boris J. Novikoff, American mathematician, earned his doctorate at Stanford and today is a professor at the Courant Institute of Mathematical Sciences at New York University

[27] Novikoff spoke at the Symposium on Mathematical Theory of Automata, which took place April 24 to 26, 1962 at the Polytechnic Institute of Brooklyn. [427] Novikoff, Convergence (1962)

[28] [15] Aizerman et al., Problem (1964), [13] Aizerman et al., Foundations (1964), [14] Aizerman et al., Method (1964), [96] Braverman, Method (1965), [85] Blaydon, Pattern (1966)

[29] [427] Novikoff, Convergence (1962)

[30] William Go Wee (born 1937), Filipino electrical engineer. Wee arrived at Purdue University in Indiana in 1963

[31] King Sun Fu (1930–1985) was the founding president of the International Association for Pattern Recognition and was a professor at the Purdue School of Electrical Engineering, West Lafayette, Indiana (1960–1985). See Fig. 6.8

Fig. 6.2. M. Aizerman

of pattern recognition.[32] Wee had applied the fuzzy sets to iterative learning procedures for pattern classification and had defined a finite *automaton* based on Zadeh's concept of the fuzzy relation as a model for learning systems:

> The proposed model represents a nonsupervised learning system if a proper performance evaluator can be selected. The decision maker operates deterministically. The learning section is a fuzzy automaton. The performance evaluator serves as an unreliable "teacher" who tries to teach the "student" to make right decisions.[33]

The fuzzy automaton representing the learning section implemented a "nonsupervised" learning fuzzy algorithm and converged monotonously. Wee showed that this fuzzy algorithm could not only be used in the area of pattern classification but could also be translated to control and regulation problems. He also demonstrated that the fuzzy automaton he had defined contained the concepts of deterministic and non-deterministic automata as special cases:

> Based on the concept of fuzzy relation defined by Zadeh, a class of fuzzy automata is formulated similar to that of Mealy's definition. A fuzzy automaton behaves in a deterministic fashion. However, it has many properties similar to that of a probabilistic automaton.[34]

Working with his doctoral advisor, Wee presented his findings in the article *A Formulation of Fuzzy Automata and its Applications as a Model of Learning Systems*.[35] In it he defined the fuzzy automaton – in conformity with automata theory – as a quintuple[36]:

---

[32] Chin-Liang Chang, who wrote a dissertation on pattern recognition under Zadeh, had also had contact with Professor Fu. Chang expresses gratitude to Fu for the conversations they shared
[33] [551] Wee, Generalization (1967), p. 101
[34] [551] Wee, Generalization (1967), p. 88
[35] [550] Wee et al., Automata (1969)
[36] See the definitions given in Chap. 3

A (finite) *fuzzy automaton* A is a quintuple (U, V, S, f, g) with

U as a finite, non-empty input set,
V as a finite, non-empty output set,
S as a finite, non-empty state set,
f as a membership function of a fuzzy set in $S \times U \times S$ to $[0,1]$,
g as a membership function of a fuzzy set in $V \times U \times S$ to $[0,1]$.

They named the functions f and g the *fuzzy transition function* and the *fuzzy output function*, respectively. In particular, there is a state transition relation here for every fuzzy input, and Wee expressed them as transition matrices. The matrix for a sequence of n inputs was thus an n-ary fuzzy relation in the product space of the n transition matrices. At this point, Wee borrowed the concept of the fuzzy relation from Zadeh[37]:

An n-valued fuzzy relation is a fuzzy set A in the product space $X \times X \times \ldots \times X$, with the membership function $f_A(x_1, \ldots, x_n)$, where $x_i \in X$, $i = 1, \ldots, n$.

In the case of binary fuzzy relations, Zadeh had defined the composition of two fuzzy relations, as those fuzzy relations whose membership function arises from those of A and B as follows:

$$f_{A \circ B}(x, y) = \sup_{\nu} \min[f_A(x, \nu), f_B(\nu, y)].$$

Since Wee only considered finite automata, he substituted the supremum in Zadeh's *sup-min composition rule* with the maximum, and by the iterative application of this *max-min composition rule* he obtained the membership function for the n transition of the fuzzy automaton.

It was while he was still in his dissertation phase that Wee co-authored with Eugene S. Santos[38] the article *General Formation of Sequential Machines*, which was published in *Information and Control* in 1968, in order to classify the different automata. Santos generalized this approach at the same time as the so-called "maximin automata".[39] Some of the definitions of such automata as systems are shown here[40]:

---

[37] See Chap. 5
[38] Eugene S. Santos studied at the Mapua Institute of Technology in Manila until 1963 and afterward at Ohio State University in Columbus, where he earned the Ph.D. in 1965. In 1974, he became a professor in the department of computer sciences and information systems at Youngstown State University in Youngstown, Ohio
[39] The two papers [486] Santos, Formulation (1968) and [487] Santos, Maximin (1968) were submitted at almost the same time (25 and 26 April 1968, respectively) and were published back-to-back in the same issue
[40] In contrast to the definitions in Chaps. 3 and 4, Santos considered only automata with no outputs

A *pseudo automaton* A is a system (U, S, f, F, h) for which the following is true:

U is a finite, non-empty input set,
S is a finite, non-empty state set,
f is a function from $S \times U \times S \times T$ to [0,1], where T is a subset of the real straight line, i.e. $f(s, u, s', t) \in [0, 1]$, for all $s, s' \in S, u \in U$ und $t \in T$.
  f denotes the *state transition function* of A at time t.
F is a subset of S and as such is the set of *end states*.
h is a function from $S \times T$ to [0,1] and as such is the starting *distribution* of A.

If $T = \mathbf{N}$ is the set of all natural numbers, then A is called *discrete*; if f and h are independent of T, then A is called *stationary*. Elements of U are referred to as input symbols and their finite series are called *input tapes*. The "empty tape" e has the property $xe = x = ex$ for all tapes x. The denomination U^* for the set of input tapes and $lg(x)$ for the length of a tape[41] brought Santos and Wee to their next definition:

An automaton A^* is a system (U, S, f^*, F, h), with U, S, F and h being defined as above. f^* is defined like f, except that U is replaced by U^*.

For every automaton A^*, a pseudo automaton A can be found in a natural way; in other words, f is the restriction of f^* on $S \times U \times S \times T$. Santos classified the pseudo automata by looking at various restrictions f and h: A pseudo automaton is "of the class $C(C)$" if f and h satisfy a set C of conditions. If the automaton can be obtained from a pseudo automaton by an expansion rule R that is consistent with C and that clearly expands f according to f^*, then the automaton A^* is said to be "of the class $C(C, R)$". For instance, the definition of probabilistic automata as automata of the class $C(C_p, R_p)$ appeared as follows.

Probabilistic Automata $C(C_P, R_P)$
The restrictions in C_P: For all $s \in S, u \in U$ and $n \in N$, it should be true that

$$\sum_{s' \in S} f(s, u, s', n) = 1 \text{ and } \sum_{s' \in S} h(s', n) = 1.$$

The expansion rule R_P: f^* is defined by induction, $lg(u^*), u^* \in U^*$

$$f(s, e, s', n) = \begin{cases} 1 & \text{if } s = s' \\ 0 & \text{if } s \neq s' \end{cases}$$

$$f^*(s, u^*u, s', n) = \sum_{s'' \in S} f^*(s, u^*, s'', n) f(s'', u, s', n + lg(u^*)).[42]$$

The definition of deterministic automata turned out to be very simple.

---

[41] The tape y is not a k suffix of tape x if $x = zy$ for a tape z and $lg(y) = k$
[42] [487] Santos, Maximin (1968), p. 364f

Deterministic Automata $\mathbf{C}(C_D, R_D)$
C_D contains the same conditions as C_P, but also includes the restriction that the functions f and h can assume only the two values 0 and 1. The expansion rule R_D is equal to the expansion rule R_A. With the *max-min composition rule*, Santos defined the *maximin automata* as the most general of all of the automata he had studied, and he noted that "various other interesting classes of automata are given which have never appeared in any other paper before".[43]

Maximin Automata $\mathbf{C}(C_A, R_A)$
C_A is the empty set, and so there are no restrictions on the functions f and h. The expansion rule R_A for maximin automata was defined thusly:

$$f^*(s, e, s', n) = \begin{cases} 1 & \text{if} \quad s = s' \\ 0 & \text{if} \quad s \neq s' \end{cases}$$

$$f^*(s, u^*u, s', n) = \max_{s'' \in S} \min f^*(s, u^*, s'', n) f(s'', u, s', n + lg(u^*)).^{44}$$

In his very abstract text, Santos noted here only that, with this definition, the maximin automata interestingly contained both the deterministic and the non-deterministic automata as special cases. Furthermore, the state transition function f and the starting distribution h could both be interpreted as grades of membership in fuzzy sets.

6.3 Fuzzy Algorithms

Zadeh viewed his theory of fuzzy systems as a general system theory. In 1966 he had spoken in Trieste on the topic of *The Concept of State in System Theory*[45]; three years later he wrote *Towards a Theory of Fuzzy Systems*.[46] His goal was a theory for all systems – including those that were too complex or poorly defined to be accessible to a precise analysis. Alongside the systems of the "soft" fields, the "non-soft" fields were replete with systems that were only "unsharply" defined, namely "when the complexity of a system rules out the possibility of analyzing it by conventional mathematical means, whether with or without the computers".[47] As he would also do a year later in Bordeaux, Zadeh was already pointing out here the usefulness of fuzzy sets in the computer sciences: In describing their fields of application, he enumerated

---
[43] [487] Santos, Maximin (1968), p. 363
[44] [487] Santos, Maximin (1968), p. 364f
[45] [659] Zadeh spoke at a NATO Advanced Study Conference about network and switching theory, which took place in Trieste from August 28 to September 12, 1966; the conference proceedings were not published until 1968. [79] Biorci, Network (1968)
[46] [687] Zadeh, ERL-Toward (1969) and [668] Zadeh, Toward (1971). Hereafter, the latter publication will be cited, although the text is identical to the earlier report
[47] [668] Zadeh, Toward (1971), p. 469f

the problems that would be solved by future computers. Alongside pattern recognition, these included traffic control systems, machine translation, information processing, neuronal networks and games like chess and checkers. We had lost sight of the fact that the class of non-trivial problems for which one could find a precise solution algorithm was very limited, he wrote. Most real problems were much too complex and thus either completely unsolvable algorithmically or – if they could be solved in principle – not arithmetically feasible. In chess, for instance, there was in principle an optimal playing strategy for each stage of the game; in reality, however, no computer was capable of sifting through the entire tree of decisions for all of the possible moves with forward and backward repetitions in order to then decide what move would be the best in each phase of the game. The set of good strategies for playing chess had fuzzy limits similar to the set of tall men – these were fuzzy sets. By far the most systems that remained to be solved were fuzzy systems, and in a footnote Zadeh remarks that the maximin automata proposed by Wee and Santos were also considered examples of fuzzy systems.[48]

"How can fuzziness be made a part of system theory?" – To answer this question, Zadeh had an ace up his sleeve: In 1969 he presented "fuzzy algorithms". With that, he had fuzzified *the* central concept of computer sciences.

> The concept in question will be called a fuzzy algorithm because it may be viewed as a generalization, through the process of fuzzification, of the conventional (nonfuzzy) conception of an algorithm.[49]

Algorithms depend upon precision. An algorithm must be completely unambiguous and error-free in order to result in a solution. The path to a solution amounts to a series of commands which must be executed in succession. Algorithms formulated mathematically or in a programming language are based on set theory. Each constant and variable is precisely defined, every function and procedure has a definition set and a value set. Each command builds upon them. Successfully running a series of commands requires that each result (*output*) of the execution of a command lies in the definition range of the following command, that is, in other words, an element of the *input set* for the series. Not even the smallest inaccuracies may occur when defining these coordinated definition and value ranges.

Once Zadeh had fuzzified input, output and state in system theory and had thus founded a theory of fuzzy systems[50], it was obvious to him how to go about fuzzifying algorithms. The commands needed to be fuzzified and so, of course, did their relations!

> I began to see that in real life situations people think certain things. They thought like algorithms but not precisely defined algorithms.[51]

---
[48] [668] Zadeh, Toward (1971), p. 471, footnote 1
[49] [591] Zadeh, Algorithms (1968), p. 94
[50] See Chap. 5
[51] [709] R. S. interview with L. A. Zadeh (2001)

Inspired by this idea, he wrote an article for *Information and Control* in 1968 which uncharacteristically contained neither theorems nor proofs[52]:

> Essentially, its purpose is to introduce a basic concept which, though fuzzy rather than precise in nature, may eventually prove to be of use in a wide variety of problems relating to information processing, control, pattern recognition, system identification, artificial intelligence and, more generally, decision processes involving incomplete or uncertain data. The concept in question will be called *fuzzy algorithm* because it may be viewed as a generalization, through the process of fuzzification, of the conventional (nonfuzzy) conception of an algorithm.[53] ...
>
> To illustrate, fuzzy algorithms may contain fuzzy instructions such as:
> (a) "Set *y approximately equal to 10* if *x* is *approximately equal to 5*," or
> (b) "If *x* is *large*, increase *y* by *several* units," or
> (c) "If *x* is *large*, increase *y* by *several* units; if *x* is *small*, decrease *y* by *several* units; otherwise keep *y* unchanged."
>
> The sources of fuzziness in these instructions are fuzzy sets which are identified by their underlined names.[54]

All people function according to fuzzy algorithms in their daily life, Zadeh wrote – they use recipes for cooking, consult the instruction manual to fix a TV, follow prescriptions to treat illnesses or heed the appropriate guidance to park a car. Even though activities like this are not normally called algorithms: "For our point of view, however, they may be regarded as very crude forms of fuzzy algorithms".[55]

In January 1973, Zadeh published his very carefully devised and comprehensive *Outline of a New Approach to the Analysis of Complex Systems and Decision Processes*[56], in which he not only treated fuzzy algorithms but also integrated the other fuzzifications into a new approach that was supposed to bring about a completely new form of system analysis based on his Fuzzy Set

---

[52] "That paper appeared in *Information and Control* even though it is not really a mathematical paper. And the reason why it appeared there is because, again, I was on the editorial board. So it could be published quickly. And I do say it's not a mathematical paper but the idea. But then other people who were mathematicians have developed that and added more mathematical and so forth. So, my function was not that of coming up with very precise. It's just an idea. That's little bit like a composer who just hums something, a sort of orchestrazing ..." [709] R. S. interview with L. A. Zadeh (2001)

[53] [591] Zadeh, Algorithms (1968), p. 694

[54] [591] Zadeh, Algorithms (1968), p. 94f. In the original, the names were not underlined but printed in italics like they are here

[55] [591] Zadeh, Algorithms (1968), p. 95

[56] [641] Zadeh, Outline (1973)

Theory: "The approach described in this paper represents a substantial departure from the conventional quantitative techniques of system analysis."[57] This new way of going about system analysis differed from the conventional approach in three ways:

- "Linguistic variables" are used instead of or in addition to numerical variables.
- Simple relationships between variables are characterized as "fuzzy IF-THEN rules" (*fuzzy conditional statements*).
- Complex relationships are characterized as "fuzzy algorithms".

Linguistic Variables:

Zadeh defined *linguistic variables* as those variables whose values are words or terms from natural or artificial languages. For instance, "not very large", "very large" or "fat", "not fat" or "fast", "very slow" are terms of the linguistic variables *size*, *fatness* and *speed*. Zadeh represented linguistic variables as fuzzy sets whose membership functions map the linguistic terms onto a numerical scale of values. In Fig. 6.3, the linguistic variable "age" is displayed with the terms "very young," "young" and "old".

Fuzzy IF-THEN Rules:

Fuzzy IF-THEN rules are composite statements of the form IF A THEN B, where A and B are fuzzy expressions, "terms with a fuzzy meaning, e. g., 'IF John is nice to you THEN you should be kind to him,' are used routinely in everyday discourse. However, the meaning of such statements when used in communication between humans is poorly defined."[58]

In those cases when the relationships among linguistic variables are more complicated than can be represented by simple fuzzy IF-THEN rules, Zadeh proposed the fuzzy algorithms.

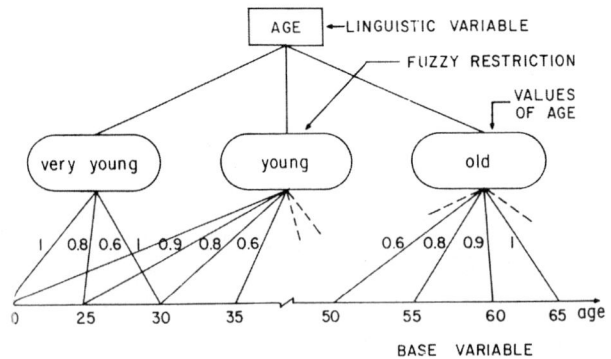

Fig. 6.3. Hierarchical structure of the linguistic variable "Age"

---

[57] [641] Zadeh, Outline (1973), p. 28
[58] [641] Zadeh, Outline (1973), p. 29

Fuzzy Algorithms
With reference to his earlier article about fuzzy algorithms, Zadeh provided the following definition:

> Essentially, a fuzzy algorithm is an ordered sequence of instructions (like a computer program) in which some of the instructions may contain labels of fuzzy sets, e. g.:
> Reduce x *slightly* if y is *large*
> Increase x *very slightly* if y is *not very large* and *not very small*
> If x is *small* then stop; otherwise increase x by 2.[59]

Zadeh divided the commands of a fuzzy algorithm into the following three classes, to which he added examples in each case:

1) *Assignment Statements*:
   $x \approx 5$
   $x = small$
   x is *small*
   x is *not large* and *not very small*.

2) *Fuzzy-IF-THEN-Rules*:
   IF x is *small* THEN y is *large* ELSE y is *not large*.
   IF x is *positive* THEN decrease y *slightly*.
   IF x is *much greater* than 5 THEN stop.
   IF x is *very small* THEN go to 7.

3) *Unconditional Commands*:
   multiply x by y
   decrease x *slightly*
   delete the first *few* occurrences of 1
   go to 7
   print x
   stop.

Of course, not all commands have to be fuzzy, but they can be fuzzy; we deal most often with combinations of commands and fuzzy commands. He gave examples:

– I am requested to "take several steps"; the word "several" can then be defined as the following fuzzy set[60]: $0.5/3 + 0.8/4 + 1/5 + 1/6 + 0.8/7 + 0.5/8$,

or

---

[59] [641] Zadeh, Outline (1973), p. 30
[60] Zadeh's notation at the time signifies in this case that the following overall numbers of steps are taken into consideration: 3, 4, 5, 6, 7 and 8, and that these numbers of steps have the following values of membership in the fuzzy set "several steps": $\mu_{\text{several steps}}(3) = 0.5$; $\mu_{\text{several steps}}(4) = 0.8$; $\mu_{\text{several steps}}(5) = 1.0$; $\mu_{\text{several steps}}(6) = 1.0$; $\mu_{\text{several steps}}(7) = 0.8$; $\mu_{\text{several steps}}(8) = 0.5$

– Someone is told IF x is *small* THEN stop ELSE go to 7, and "small" refers to the following fuzzy set: $1/1 + 0.8/2 + 0.6/3 + 0.4/4 + 0.2/5$.

Finally, Zadeh classified the fuzzy algorithms by how they are used; these categories should be sketched out briefly.

Fuzzy Definitional Algorithms
Fuzzy sets can be defined with the aid of fuzzy definitional algorithms.

> Since a fuzzy concept may be viewed as a label for a fuzzy set, a fuzzy definitional algorithm is, in effect, a finite set of possibly fuzzy instructions which define a fuzzy set in terms of other fuzzy sets (and possibly itself, i.e. recursively) or constitute a procedure for computing the grade of membership of any element of the universe of discourse in the set under definition.[61]

To illustrate, Zadeh presented the following (simplified) fuzzy algorithm *OVAL*, which is intended to verify whether or not an object T is an oval. The word *oval* itself stood for a fuzzy concept: "The fuzzy set *oval* is the intersection of the fuzzy and nonfuzzy sets whose labels appear on the right-hand side".[62] The expression CALL *CONVEX* represents the invocation of the sub-algorithm called *CONVEX*, which is likewise a fuzzy definitional algorithm and is intended to verify whether or not T is *convex*, and finally, IF A THEN B is interpreted as follows: IF A THEN B or go to next command.

Fuzzy Algorithm OVAL

1) IF T is not closed THEN T is not *oval*; stop.
2) IF T is self-intersecting THEN T is not *oval*; stop.
3) IF T is not CALL *CONVEX* THEN T is not *oval*; stop.
4) IF T does not have two *more or less* orthogonal axes of symmetry THEN T is not *oval*; stop.
5) IF the major axis of T is not much longer than the minor axis THEN T is not *oval*; stop.
6) T is oval; stop.

Fuzzy Sub-Algorithm CONVEX[63]

1) $x = a$ (some initial point on T).
2) Choose a direction of movement along T.
3) $t \approx$ direction of tangent to T at x.
4) $x' \approx x + 1$ (move from x to a neighboring point).
5) $t' \approx$ direction of tangent to T at x'.
6) $\alpha \approx$ angle between t' and t.
7) $x \approx x'$.

---

[61] [641] Zadeh, Outline (1973), p. 40
[62] [641] Zadeh, Outline (1973), p. 40
[63] The fuzzy sub-algorithm *CONVEX* verifies whether or not the curvature of T retains its algebraic sign if one moves along T in a randomly selected direction

8) $t \approx$ direction of tangent to T at x.
9) $x' \approx x + 1$
10) $t' \approx$ direction of tangent to T at x'.
11) $\beta \approx$ angle between t' and t.
12) IF β does not have the same sign as α THEN T is convex; return.
13) IF $x' \approx a$ THEN T is convex; return.
14) Go to 7).

Fuzzy Generational Algorithms
These fuzzy algorithms generate fuzzy sets; in Zadeh's example below, the letter "P" is generated with the height h and the base b and, for the sake of simplicity, as a dot pattern consisting of eight dots:

Fuzzy-Algorithm P(h, b)

1) $i = 1$.
2) $x(i) = b$ (first dot at base).
3) $x(i+1) \approx x(i) + \frac{h}{6}$ (put dot approximately $\frac{h}{6}$ units of distance above $x(i)$).
4) $i = i + 1$.
5) IF $i = 7$ THEN make right turn and go to 7).
6) Go to 3).
7) Move by $\frac{h}{6}$ units; put a dot.
8) Turn by 45°; move by $\frac{h}{6}$ units; put a dot.
10) Turn by 45°; move by $\frac{h}{6}$ units; put a dot.
11) Turn by 45°; move by $\frac{h}{6}$ units; put a dot; stop.

Zadeh borrowed a recipe program for chocolate sauce from a Fortran IV book by Robert S. Ledley[64] to use as an example of a fuzzy generational algorithm with feedback (see Fig. 6.4):

Fuzzy Relational and Behavioral Algorithms
Relationships between fuzzy variables can be described by fuzzy relational and behavioral algorithms to illustrate, for example, the behavior of complex systems approximatively. Using the given fuzzy sets *small* and *large*,

$$\text{small} = \{(1, \mu_{\text{small}}(1) = 1), (2, \mu_{\text{small}}(2) = 0.8),$$
$$(3, \mu_{\text{small}}(3) = 0.6), (4, \mu_{\text{small}}(4) = 0.4),$$
$$(5, \mu_{\text{small}}(5) = 0.2)\}$$

$$\text{large} = \{(1, \mu_{\text{small}}(1) = 0.2), (2, \mu_{\text{small}}(2) = 0.4),$$
$$(3, \mu_{\text{small}}(3) = 0.6), (4, \mu_{\text{small}}(4) = 0.8),$$
$$(5, \mu_{\text{small}}(5) = 1)\}$$

a relation can now be described by the

---
[64] [309] Ledley, Fortran (1966); see also Chap. 7 for information on Robert S. Ledley

252   6 Fuzzifications

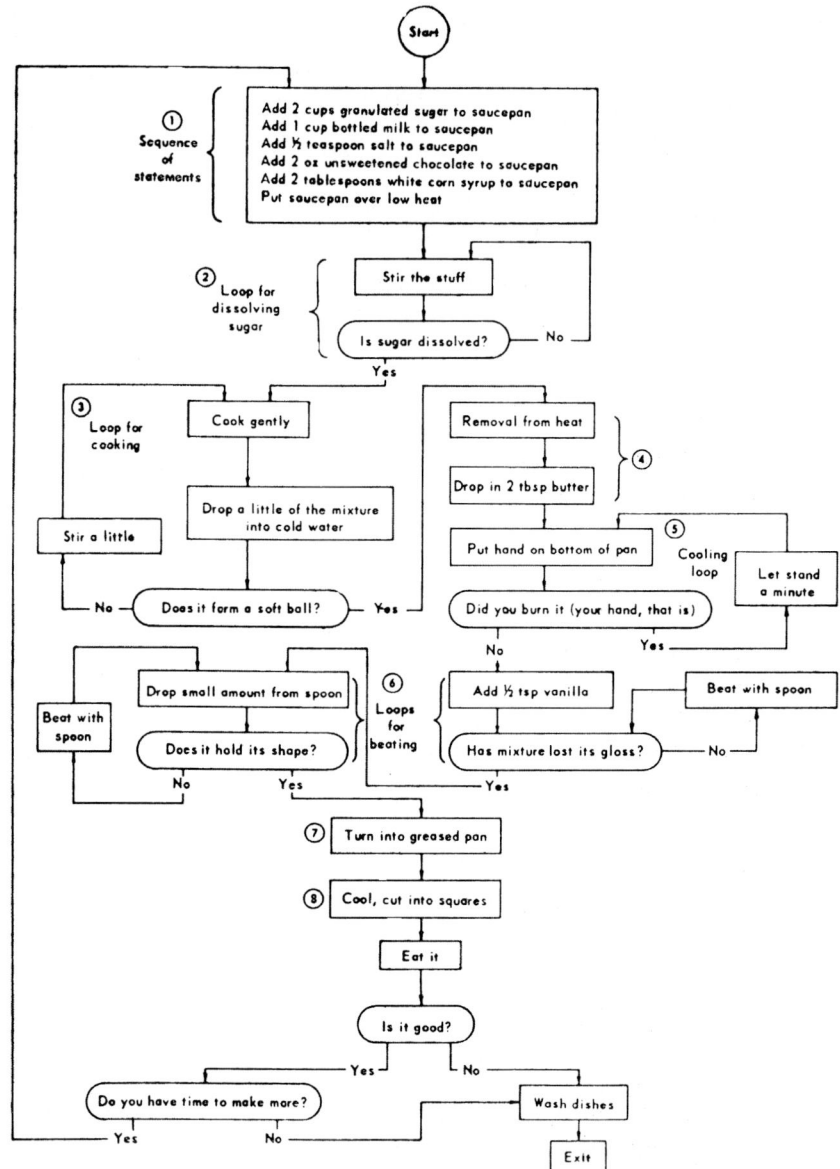

Fig. 6.4. Recipe for a chocolate sauce Zadeh adopted from R. S. Ledley

Fuzzy algorithm **R**(x, y, z):

1) IF x is *small* and y is *large* THEN z is *very small* ELSE z is *not small*.
2) IF x is *large* THEN (IF y is *small* THEN z is *very large* ELSE z is *small*) ELSE z and y are *very small*.

Using other fuzzy relational and behavioral algorithms, an attempt is made to describe system behavior approximatively by means of fuzzy relations. The relationships among inputs, states and outputs are often represented by state transition tables instead of IF-THEN rules, and this approach accordingly leads to fuzzy automata.

Fuzzy Decisional Algorithms

As examples of fuzzy decisional algorithms, Zadeh mentioned parking a car, passing through an intersection, transporting an object and buying a house. Such algorithms describe strategies or decision rules. In the example of passing through an intersection, a sub-algorithm was once again constructed for each intersection type. In his article, Zadeh considered only one of those sub-algorithms, called *SIGN*, which is responsible for the case that there is a stop signal at the intersection[65]:

Algorithm INTERSECTION
IF signal lights THEN *SIGNAL*

> ELSE if stop sign THEN call *SIGN*
>> ELSE if blinking light THEN call *BLINKING*
>>> ELSE call *UNCONTROLLED*.

Sub-algorithm SIGN
1) IF no stop sign on your side THEN
   IF no cars in the intersection THEN cross at *normal* speed
   ELSE wait for cars to leave the intersection and then cross.
2) IF not *close* to intersection THEN continue approaching at normal speed for a *few* seconds; go to 2).
3) *Slow down.*
4) IF in a great hurry and no police cars in sight and no cars in the intersection or its *vicinity* THEN cross the intersection at *slow* speed.
5) IF very *close* to intersection
   THEN stop; go to 7).
6) Continue *approaching* at very *slow* speed; go to 5).
7) IF no cars *approaching* or in the intersection THEN cross.
8) Wait a *few* seconds; go to 7).

---

[65] Zadeh often refers to the fact that he is working with simplifications here; realistic fuzzy algorithms would have to be substantially more complex in structure

6.4 Fuzzy Turing Machines

In his 1968 article *Fuzzy Algorithms for Information and Control*, Zadeh also fuzzified the Turing machine. *Fuzzy Turing machines* differ from standard Turing machines primarily by the type of dependencies the (fuzzy) state has at time $n+1$, both upon the (fuzzy) state at time n and upon the (fuzzy) input at time n. Zadeh considered a Turing machine with $Q = q_0, q_1, \ldots, q_r$ as the set of all of its states and $U = u_0, u_1, \ldots, u_m$ as the set of its tape symbols.

- In the case of a "nonfuzzy deterministic" Turing machine, the state at time $n+1$ is a function of the state at time n and of the tape symbol at time n, and therefore:

$$q^{n+1} = f(q^n, u^n),$$

  wherein f is a function from $Q \times U$ to Q and q_n are variables ranging over Q and U, respectively.
- In the case of a "nonfuzzy nondeterministic" Turing machine, f is a multi-valued rather than a single-valued function. The dependency can thus also more commonly be described by a relation:

$$R = \{(q^{n+1}, q^n, u^n)\}$$

  wherein R is a subset of the product space $Q \times Q \times U$.
- In the case of a fuzzy Turing machine, the above relation is a fuzzy set of the product space $Q \times Q \times U$ and is characterized by the membership function $\mu_R(q^{n+1}, q^n, u^n)$, which associates with each triplet (q^{n+1}, q^n, u^n) a grade of membership in the relation R.

So a nonfuzzy nondeterministic Turing machine can be viewed as a special case of a fuzzy Turing machine whose membership function $\mu_R(q^{n+1}, q^n, u^n)$ can take only the values 1 or 0: Either (q^{n+1}, q^n, u^n) belongs to R or not. Zadeh also explained the fuzzy Turing machine as a special case of a fuzzy system. He neatly summarized his idea of understanding "fuzziness as a part of system theory" and of recommending fuzzy systems as a field of application for the computer in this way:

> In short, an algorithm corresponds to a Turing machine, a nondeterministic algorithm corresponds to a nondeterministic Turing machine, and a fuzzy algorithm corresponds to a fuzzy Turning machine.[66]

6.5 Fuzzifications of Elements of Mathematical Theory

Richard Bellman left the RAND Corporation in 1965 and was appointed Professor of Mathematics, Electrical Engineering and Medicine at the University of California, Los Angeles (UCLA).[67] This unusual title alone illustrates Bellman's multifaceted gifts and interests.

---

[66] [591] Zadeh, Algorithms (1968), p. 99
[67] [709] R. S. interview with L. A. Zadeh (2001); for biographical information on Bellman, see also the *IEEE History Center* website:
$http://www.ieee.org/organizations/history\_center/legacies/bellman.html$

6.5 Fuzzifications of Elements of Mathematical Theory

For a time, Bellman's work was largely ignored by the fraternity of mathematicians, for he was working primarily with applied mathematics – "he was a sort of excommunicated, ... he was a mathematician like Poincaré"[68] – yet he later received much acclaim for his efforts.[69]

Fig. 6.5. R. Bellman

His close friendship with Zadeh resulted in a number of co-authored publications. Bellman had been invited to deliver a lecture at the International Symposium on Multiple-Valued Logic at Indiana University in Bloomington, Indiana in May of 1975. He spoke on the subject of "Local Logics". The symposium proceedings includes only a 27-page abstract and a note indicating that the print version would be published in the book *Modern Uses in Multiple-Valued Logic*.[70] This 1977 tome does include the 62-page-long paper *Local and Fuzzy Logics* by Bellman and Zadeh[71], in which the concept of fuzzy sets is carried over to fuzzy logic. Here the authors postulate the following properties of fuzzy logic:

– Truth values here are fuzzy sets of the unit interval that has denominations like "true", "very true", "not very true", "false", "more true" or "less true", etc.

---

[68] [711] R. S. interview with L. A. Zadeh (2002)
[69] Among other honors, Bellman was the first to receive the *Norbert Wiener Prize in Applied Mathematics* from the American Mathematical Society and the Society for Industrial and Applied Mathematics in 1970; he was awarded the first *Dickson Price* from Carnegie-Mellon University in 1970 and the *John von Neumann Theory Award* in 1976, which was endowed jointly by the Institute of Management Sciences and the Operations Research Society of America. In 1979, Bellman was awarded the IEEE Medal of Honor, "For contributions to decision processes and control system theory, particularly the creation and application of dynamic programming." See the IEEE History Center website: $http://www.ieee.org/organizations/history\_center/legacies/bellman.html$
[70] [52] Bellman, Local (1975)
[71] [51] Bellman et al., Local (1977)

- These truth values are generated by a grammar and they can be interpreted by means of semantic rules.
- Fuzzy logic is local, i. e. both the truth values and their conjunctions such as "AND", "OR" and "IF-THEN" have variable rather than fixed meanings.
- The interference rules of fuzzy logic are not exact but rather appoximative.

The concepts introduced in this article trace back to works Zadeh had published in the early '70s; they should be summarized here because they will be utilized in the next sections and the last chapter.

Fuzzy Relations

In 1971, Zadeh defined *similarity relations* and *fuzzy orderings*.[72] In doing so, he was proceeding from the concept of fuzzy relations as a fuzzification of the relation concept known in conventional set theory that he had already defined in his first text[73] on fuzzy sets[74]: If X and Y are conventional sets and if $X \times Y$ is their Cartesian product, then:

- $L(X)$ is the set of all fuzzy sets in X in X,
- $L(Y)$ is the set of all fuzzy sets in Y and
- $L(X \times Y)$ is the set of all fuzzy sets in $X \times Y$.

Relations between X and Y are subsets of its Cartesian product $X \times Y$, and the composition $t = q * r$ of the relation $q \subseteq X \times Y$ with the relation $r \subseteq Y \times Z$ into the new relation $t \subseteq X \times Z$ is given by the following definition:

$$t = q * r = \{(x,y) \exists y : (x,y) \in q \wedge (y,z) \in r\}.$$

Fuzzy relations between sets X and Y are subsets in $L(X \times Y)$. For three conventional sets X, Y and Z, the fuzzy relation Q between X and Y and the fuzzy relation R between Y and Z are defined: $Q \in L(X \times Y)$ and $R \in L(Y \times Z)$. The combination of these two fuzzy relations into a new fuzzy relation $T \in L(X \times Z)$ between X and Z can then be combined from the fuzzy relations Q and R into the new fuzzy relation $T \in L(X \times Z)$ when the logical conjunctions are replaced by the corresponding ones of the membership functions.

- The above definition of the composition of conventional relations includes a logical AND ($\wedge$), which, for the "fuzzification", is replaced by the minimum operator that is applied to the corresponding membership functions.[75]
- The above definition of the composition of conventional relations includes the expression "$\exists y$" ("there exists a y"). The existing $y \in Y$ is the first or the second or the third ... (and so on); written logically: ($\vee$) $\sup_{y \in Y}$. In the

---

[72] [654] Zadeh, Similarity (1971)
[73] [612] Zadeh, Fuzzy (1965)
[74] See Chap. 5
[75] Of course, the other proposed fuzzy operators can also be used; in those cases, correspondingly different fuzzy relations are obtained

6.5 Fuzzifications of Elements of Mathematical Theory

"fuzzifications", the logical OR conjunction is replaced by the maximum operator that is applied to the corresponding membership functions.[76]

The fuzzy relation $T = Q * R$ is therefore defined via Zadeh's "combination rule of max-min combination"[77] for the membership functions:

$$\mu_T(x, y) = \max_{y \in Y} \min\{\mu_Q(x, y); \mu_r(y, z)\}, \qquad y \in Y.$$

Similarity Relations

Zadeh defined the concept of "similarity" as a generalization of the concept of the equivalence relation, since the similarity relation he defined is reflective, symmetrical and transitive, i. e. for $x, y \in X$ the membership function of S has the following properties:

(a) Reflexivity: $\mu_S(x, x) = 1$
(b) Symmetry: $\mu_S(x, y) = \mu_S(y, x)$
(c) Transitivity: $\mu_s(x, y) \geq \max_{y \in Y} \min\{\mu_S(x, y); \mu_S(y, z)\}$.

Fuzzy Ordering Relations

Fuzzy orderings are transitive fuzzy relations. The fuzzy relation P in X is a "fuzzy partial ordering" if it is reflexive, symmetrical and antisymmetrical. Antisymmetry means:

$$(\mu_P(x, y) > 0 \wedge x \neq y) \Rightarrow \mu_P(x, y) = 0.$$

Quantitative Fuzzy Semantics

Likewise in 1971, Zadeh was occupied in his paper *Quantitative Fuzzy Semantics*[78] with the definition of a language as a fuzzy relation between a set of terms $T = \{x\}$ and the universe of discourse $U = \{y\}$.
If a term x of T is given, then the membership function $\mu_L(x, y)$ defines a set $M(x)$ in U with the following membership function:

$$\mu_{M_x}(y)\mu_L(x, y).$$

Zadeh called the fuzzy set $M(x)$ *the meaning of the term x*; x is thus the name of $M(x)$. Here the following example is provided: K is the set of integers from 0 to 100, which indicate the ages of people in a population. The universe of discourse is $U = K$. The term set considered here is $T = $ *young, old, middle-aged, not old, not young, not middle-aged, young or old, not young and not old*. If we look at the term $x = $ *young*, its meaning is the fuzzy set $M(young)$

---

[76] In addition to max operator, there are also other conjunction operations for the "fuzzy or" which then lead to other fuzzy relations

[77] The max-min composition rule is replaced in infinite sets with the sup-min composition rule. However, it is adequate to assume here that all of the sets are finite

[78] [652] Zadeh, Semantics (1971)

6 Fuzzifications

in U. In light of this, the following membership function of a fuzzy set of $M(young)$ can be established subjectively:

$$\mu(y|young) = \begin{cases} 1 & \text{for } y < 25 \\ \left(1 + \left(\dfrac{y-25}{5}\right)^2\right)^{-1} & \text{for } y \geq 25 \end{cases}$$

Similarly, Zadeh established the membership function for $M(old)$:

$$\mu(y|old) = \begin{cases} 0 & \text{for } y < 50 \\ \left(1 + \left(\dfrac{y-50}{5}\right)^{-2}\right)^{-1} & \text{for } y \geq 50 \end{cases},$$

as well as the membership function for $M(middle\text{-}aged)$:

$$\mu(y|middle\text{-}aged) = \begin{cases} 0 & \text{for } 0 \leq y < 35 \\ \left(1 + \left(\dfrac{y-45}{4}\right)^4\right)^{-1} & \text{for } 35 \leq y < 45 \\ \left(1 + \left(\dfrac{y-45}{5}\right)^2\right)^{-1} & \text{for } y \geq 45 \end{cases}.$$

Without having to go into too many details here, it should be noted that the meanings of more complex terms can also be calculated from their simpler terms by, for example, determining that "*very x*" means the square of x, and therefore "*very very x*" means x^4, and so forth. One illustration of this should suffice:

$x = $ *not very small* is the composite term and *small* is defined as follows (universe of discourse is $U = \{1, 2, 3, 4, 5\}$):

$$\textit{small} \quad = \quad (1, \mu_{small}(1) = 1), (2, \mu_{small}(2) = 0.8),$$

$$(3, \mu_{small}(3) = 0.6), (4, \mu_{small}(4) = 0.4),$$

$$(5, \mu_{small}(5) = 0.2)$$

Therefore:

$$\begin{aligned}\text{very small} \quad = \quad &(1, \mu_{small}(1) = 1), (2, \mu_{small}(2) = 0.64),\\ &(3, \mu_{small}(3) = 0.36), (4, \mu_{small}(4) = 0.16),\\ &(5, \mu_{small}(5) = 0.04).\end{aligned}$$

Using the complement formation defined for fuzzy sets, the following is also true:

$$\begin{aligned}\text{not very small} \quad = \quad &\neg \text{very small}\\ &(1, \mu_{small}(1) = 1), (2, \mu_{small}(2) = 0.64),\\ &(3, \mu_{small}(3) = 0.36), (4, \mu_{small}(4) = 0.16),\\ &(5, \mu_{small}(5) = 0.04).\end{aligned}$$

Zadeh later[79] defined the following "modifiers" for fuzzy sets:

– the concentration **CON** $(A) := A^2$
– the dilatation **DIL** $(A) := A^{1/2}$
– the contrast intensification

$$\mathbf{INT}(A) := \begin{pmatrix} 2A^2 & \text{for} & 0 \leq \mu_A(y) \leq 0.5 \\ \neg\, 2(\neg A)^2 & \text{for} & 0{,}5 \leq \mu_A(y) \leq 1 \end{pmatrix}.$$

The use of these modifiers on a term x occurs, for example, according to the following rules (where $A = (x, \mu_A(x))$) is the fuzzy set representing the term x):

– very x $\longrightarrow$ CON (A),
– very very x $\longrightarrow$ CON [CON (A)],
– somewhat x $\longrightarrow$ DIL (A).

6.6 Other Fuzzifications

As early as 1962, Zadeh was already trying to find "the mathematics of fuzzy or cloudy quantities" in order to be able to describe living systems.[80] At that time, even he still had no idea what this math would look like, yet seven years later he presented the ever-widening Fuzzy Set Theory at the "International Symposium on Biocybernetics of the Central Nervous System" in Boston when he spoke on the theme of *Biological Applications of the Theory of Fuzzy Sets and Systems*[81]:

---

[79] [641] Zadeh, Outline (1973), p. 32
[80] [662] Zadeh, System (1962), p. 857. See also Chap. 1
[81] [594] Zadeh, Biological (1969)

The great complexity of biological systems may well prove to be an insuperable block to the achievement of a significant measure of success in the application of conventional mathematical techniques to the analysis of systems. By "conventional mathematical techniques" in this statement, we mean mathematical approaches for which we expect that precise answers to well-chosen precise questions concerning a biological system should have a high degree of relevance to its observed behavior. Indeed, the complexity of biological systems may force us to alter in radical ways our traditional approaches to the analysis of such systems. Thus, we may have to accept as unavoidable a substantial degree of fuzziness in the description of the behavior of biological systems as well as in their characterization.[82]

The program of applying fuzzy sets in the life sciences was carried out successfully in the following years, as the next chapter will show. Yet in 1969, when Zadeh spoke on this subject, there were also already scientists from these ranks who agreed with him. During the discussion which followed Zadeh's address, for example, William L. Kilmer[83] of the University of Michigan, who was working with artificial neuronal networks, developed scenarios in which Zadeh's fuzzy algorithms could be used in pattern recognition.[84]

Additionally, Zadeh found a comrade-in-arms in the fight for Fuzzy Set Theory at his own university in his colleague Hans-Joachim Bremermann[85] of the department of mathematics, for Bremermann had also recognized the great potential this theory possessed as an application in artificial neuronal networks. He spoke about this at an April 1970 meeting called *Character Recognition by Biological and Technical Systems* in Berlin.[86] *What Mathematics Can and Cannot Do for Pattern Recognition* was his subject and in his presentation he discussed the case of deformed prototypes in pattern recognition. His doctoral student Richard Hodges at Berkeley had implemented a method of measuring the degree of deformation of the prototype in letter recognition: The degree of deformation received the value 0 if a letter was identical to the

---

[82] [594] Zadeh, Biological (1969)
[83] William L. Kilmer earned his doctorate at the University of Michigan in 1958 and is now a professor at the Faculty of Electrical and Computer Engineering of the University of Massachusetts, Amherst
[84] [277] Kilmer, Discussion (1969)
[85] Hans-Joachim Bremermann (1926–1996), mathematician born in Bremen, Germany. Bremermann earned his doctorate in Münster in 1951 and then spent two years in the US at Stanford and Harvard. In 1959, he arrived at the mathematics department of the University of California in Berkeley and conducted research on questions of artificial intelligence. In 1979, he moved to Berkeley's department of medical physics, later the division of biophysics and cell physiology in the department of molecular and cell biology
[86] 4th Congress of the German Society for Cybernetics, held at the Technical University of Berlin on April 6–9, 1970. The proceedings were published in 1971: [207] Grüsser et al., Zeichenerkennung (1970)

prototype, but it received a small non-zero value depending on the admissible deviation from the prototype as long as the deformed prototype still belonged to the same pattern. The degree of deformation was given a large value if the deformation deviated completely from the pattern or even fell into another pattern. In Hodge's project, the deformed prototypes that were generated were modeled as fuzzy sets. The deformation function d assumed the value 0 for non-deformed prototypes, and so $(1 - d)/max\ d$ described a fuzzy set.[87] Bremermann also used this opportunity to address the criticism issued by the stochasticians, who found Fuzzy Set Theory superfluous because they did not recognize in it any innovations over probability theory:

> Fuzzy Sets have occasionally been criticized as unnecessary on the grounds that the characteristic function could or should be interpreted as a probability density. Our method shows that the critiques are wrong. It would be quite reasonable to interpret deformations as probabilities. It might be interesting to develop a *theory of groups acting on fuzzy sets*. Such a theory could accommodate deformations of a prototype as described above.[88]

Bremermann was able to inspire enthusiasm for Zadeh's fuzzy sets in yet more students. He managed to harmonize his interests in mathematics and medicine very well in the field of pattern recognition. He considered the identification of an illness as the task of adapting the system complex the patient is presenting to the known pattern of illnesses. In this way, he tied the problem of medical diagnosis into the problem of pattern recognition, which would then be implemented by the computer for processing. He had student Priscilla Wong compile the "nonfuzzy" mathematical-statistical computer diagnosis techniques that were common at the time in her Master's thesis.[89] At the same time, Merle Anne Albin wrote her dissertation under Bremermann about medical diagnostics performed with the aid of fuzzy set applications.[90]

In 1967, Bremermann had succeeded in selling the mathematician Joseph Goguen[91] on Zadeh's theory, which was still quite new at the time. *Categories of Fuzzy Sets* had been the title of Goguen's Ph.D. thesis, for which Zadeh had served as first reviewer and Bremermann as second.[92] In this work, Goguen generalized the fuzzy sets to so-called "L-sets". An L-set is a function that maps the fuzzy set carrier X into a partially ordered set L:

$$A : X \longrightarrow L.$$

---

[87] Bremermann noted that $max\ d$ must be finite; this was not the case in Hodge's work, however
[88] [100] Bremermann, Cybernetic (1971), p. 43
[89] [572] Wong, Techniques (1975)
[90] [17] Albin, Fuzzy (1975). The subject of medical diagnostics in conjunction with Fuzzy Set Theory is reserved for the final chapter
[91] Joseph Goguen, American mathematician, is now a professor in the department of computer science and engineering at the University of California in San Diego
[92] [198] Goguen, Categories (1968)

The partially ordered set L Goguen called the "truth set" of A. The elements of L can thus be interpreted as "truth values"; in this respect, Goguen then also referred to a *Logic of Inexact Concepts*.[93] Since Zadeh's earlier definition had established this truth set as the unit interval, Fuzzy Set Theory was very soon associated with the two multi-valued logics or with probability logic, as well. Goguen's generalization of the set of values to a set L for which the only condition was to be partially ordered cleared up these misunderstandings. Goguen's work was laid out in terms of logical algebra and category theory, and his proof of a representation theorem for L-sets within category theory justified Fuzzy Set Theory as an expansion of set theory.[94]

Fig. 6.6. J. Goguen

Zadeh's efforts to use his fuzzy sets in linguistics led to an interdisciplinary scientific exchange between him and Goguen on the one hand and between the psychologist Eleanor Rosch[95] and the linguist George Lakoff[96] on the other.[97] Rosch had developed her prototype theory on the basis of empirical studies. This theory assumes that people perceive objects in the real world by comparing them to prototypes and then ordering them accordingly. In this way, according to Rosch, word meanings are formed from prototypical details and scenes and then incorporated into lexical contexts depending on the context or situation. It could therefore be assumed that different societies process perceptions differently depending on how they go about solving problems.[98] When Lakoff heard about Rosch's experiments, he was working at the Center

---

[93] [199] Goguen, Inexact (1969)
[94] [200] Goguen, L-fuzzy (1967). See also [199] Goguen, Inexact (1969)
[95] Eleanor Rosch (born 1938), American psychologist, professor at the University of California, Berkeley
[96] George P. Lakoff (born 1941), American linguist, Professor at the University of California, Berkeley
[97] [706] R. S. interview with L. A. Zadeh (2000), [724] R. S. interview with J. Goguen (2002)
[98] [468] Rosch, Categories (1973)

for Advanced Study in Behavioral Sciences at Stanford. During a discussion about prototype theory, someone there mentioned Zadeh's name and his idea of linking English words to membership functions and establishing fuzzy categories in this way. Lakoff and Zadeh met in 1971/72 at Stanford to discuss this idea, after which Lakoff wrote his paper *Hedges: A Study in Meaning Criteria and the Logic of Fuzzy Concepts*.[99] In this work, "hedges" (meaning barriers) were employed to categorize linguistic expressions. The text contained an error, which Lakoff later corrected and likewise discussed with Zadeh[100]:

> Zadeh was not interested in the points that didn't work. Zadeh was interested in the hedges and he was interested in the idea of fuzzy logic. I invented the term fuzzy logic in that paper. Goguen has used "logic of inexact concepts".[101]

Lakoff had indeed used the term "fuzzy logic" in his article and he therefore deserves the credit for first introducing this expression in the scientific literature. Based on his later research, however, Lakoff came to find that fuzzy logic was not an appropriate logic for linguistics: "It doesn't work for real natural languages, in traditional computer systems it works that way."[102]

"Inspired and influenced by many discussions with Professor G. Lakoff concerning the meaning of hedges and their interpretation in terms of fuzzy sets,"[103] Zadeh had also written an article in 1972 in which he contemplated "linguistic operators", which he called "hedges": *A Fuzzy Set-Theoretic Interpretation of Hedges*.

> A basic idea suggested in this paper in that a linguistic hedge such as *very, more, more or less, much, essentially, slightly* etc. may be viewed as an operator which acts on the fuzzy set representing the meaning of its operand.[104]

After this, neither Lakoff nor Zadeh made any further attempts to apply Fuzzy Set Theory to linguistics. In the mean time, however, a multitude of other developments has arisen. Once Zadeh had fuzzified the basic terms of set theory, mathematical theories were next. The first theory into which Zadeh sought to introduce his fuzzy sets was probability theory, since he had already written *Probability Measures of Fuzzy Events* during his guest residency at MIT in 1968. Any competition between probability theory and Fuzzy Set Theory he ignored completely. Instead, he suggested broadening probability theory by adding the concept of fuzzy sets:

---

[99] [305] Lakoff, Hedges (1973), see also [304] Lakoff, Fuzzy (1972)
[100] [305] Lakoff, Hedges (1973), see also [304] Lakoff, Fuzzy (1972)
[101] [729] R. S. interview with G. Lakoff (2002)
[102] [729] R. S. interview with G. Lakoff (2002)
[103] [617] Zadeh, Hedges (1972), p. 4
[104] [617] Zadeh, Hedges (1972)

By using the concept of a fuzzy set, the notions of an event and its probability can be extended in a natural fashion to fuzzy events ... It is possible that such an extension may eventually significantly enlarge the domain of applicability of probability theory, especially in those fields in which fuzziness is a pervasive phenomenon.[105]

Zadeh studied stochastic systems in a fuzzy environment in 1970 in the article *Decision-Making in a Fuzzy Environment*[106], which appeared in the journal *Management Science*. The subject here was decision-making processes in which the conditions and goals were defined as fuzzy sets. The goal function was represented as a fuzzy set of alternatives.

In the early 1970s, Richard Hamming[107] had received a letter from Japan in which the sender posed topologically motivated questions with regard to Zadeh's fuzzy sets. Hamming forwarded a copy of his reply to Zadeh:

And then I showed this to C.-L. Chang. He was a student at that time. And he had good mathematical background and I told him, "Why don't you look into that?" And so then he looked into that and then he wrote a short paper on fuzzy topological spaces, on fuzzy topology, and that paper started this whole field with fuzzy topology.[108]

Edward T. Lee and Samuel C. Lee, two other students, co-authored the first text about *Fuzzy Neurons and Automata* in 1970.[109] This text was also published five years later in the journal *Mathematical Biosciences* under the title *Fuzzy Neural Networks*.[110] Lastly, Enrique H. Ruspini[111] should be mentioned. As a research assistant at the Brain Research Institute at UCLA in 1969, he wrote the article *A New Approach to Clustering*, in which he introduced into

---

[105] [643] Zadeh, Probability (1968), p. 421
[106] [47] Bellman, Decision (1970)
[107] Richard Wesley Hamming (1915–1998), American mathematician, joined the Manhattan Project in 1945 and following the war went to Bell Telephone Laboratories, where he worked with Claude E. Shannon. In 1976, he accepted a professorship for computer science at the Naval Postgraduate School in Monterey, California
[108] [706] R. S. interview with L. A. Zadeh (2000). The cited article is [122] Chang, Spaces (1968). The author in question is Chin-Liang Chang, who had also written his dissertation on pattern recognition and fuzzy sets under Zadeh in 1967; see Sect. 1 in this chapter
[109] [318] Lee et al., Neurons (1970)
[110] [317] Lee et al., Neural (1975)
[111] Enrique Hector Ruspini (born 1942), Argentinean-American mathematician, Licenciado en Ciencias Matemáticas at the university in Buenos Aires, Argentina, Ph.D. in system science at the University of California at Los Angeles. Ruspini works at the Stanford Research Institute (SRI). He earned his doctorate in 1977 with the dissertation *A Theory of Mathematical Classification* at UCLA; third reviewer of this dissertation was Zadeh. [475] Ruspini, Theory (1977), [739] R. S. interview with E. H. Ruspini (2002). See Fig. 6.8

6.6 Other Fuzzifications 265

Fig. 6.7. E. Ruspini

the field of pattern recognition the concept of "fuzzy partition" to represent clusters in data sets:

> The object of cluster analysis is to classify experimental data in a certain number of sets where the elements of each set should be as similar as possible and dissimilar from those of the other sets. This implies the existence of a measure of distance or similarity between

Fig. 6.8. Attendees at the 1st NAFIPS Meeting in Logan, Utah, 1982. (left to right) First row: P. P. Wang, E. Mamdani, K. F. Fu, T. P. Yao, L. Saitta. Second row: M. Roubens, Ph. Smets, J. Efstathiou, R. Tong, R. R. Yager. Top row: P. Bonissone, J. Bezdek, E. H. Ruspini, E. Sanchez

the elements to be classified. The number of such classes may be fixed beforehand or may be a consequence of some constraints imposed on them.[112]

All of these fuzzifications led to further developments, which today represent vast areas of research: fuzzy probability theory, fuzzy statistics, fuzzy algebra, fuzzy topology, fuzzy logic and many more. In 1972, Zadeh could already list a very impressive inventory.

> Since its introduction, the concept of fuzziness has been extended to algorithms, learning theory, automata, formal languages, pattern classification, probability and the decision making process.[113]

Nevertheless, it was not any of these theoretical fuzzifications that have made Zadeh's fuzzy sets popular. It was the principle of fuzzy control which proved to be very successful in the practical implementation of concepts of Fuzzy Set Theory in a small laboratory system and became the model for many other fuzzy control systems.

6.7 Fuzzy Control

In the article cited last, Zadeh and his co-author Sheldon S. L. Chang wanted to highlight an approach to fuzzifying control theory.

> There is another Chang, S. S. L. Chang, who co-authored a paper with me on fuzzy control 1972 and this is the one paper that was written mostly by S. S. L. Chang. I suggested fuzzy control to him and so he wrote something.[114]

Sheldon S. L. Chang[115] had already pursued the fuzzification of Bellman's dynamic programming in 1969.[116] Now he and Zadeh argued that control engineers tended to treat their mathematical models of physical systems as exact and precise though they knew that the models were neither.

> They obtain an optimum solution for the nominal model with possibly an added criterion of minimum sensitivity. However, no one can be sure how the system performs if it deviates from the nominal model

---

[112] [473] Ruspini, Clustering (1969)
[113] [123] Chang et al., Mapping (1972), p. 30
[114] [706] R. S. interview with L. A. Zadeh (2000)
[115] Sheldon S. J. Chang, Chinese-American electrical engineer. Chang studied at Tsinghua University in China and earned a doctorate at Purdue University in Lafayette, Indiana. After working several years in the industry, he went to New York University in 1952. In 1963, he became a professor at New York State University. He wrote the book *Synthesis of Optimum Control Systems*
[116] [124] Chang, Programming (1969)

6.7 Fuzzy Control

in some finite way. It is desirable ... to have fuzzy mathematics which represents exactly the inexact state of knowledge.[117]

To this, Chang and Zadeh added the following concepts of *fineness* and *observation*:

- The *fineness* of a fuzzy set or a fuzzy mapping represents the degree of exactitude in our knowledge about a system: The higher our standard of knowledge becomes, the greater the fineness. If we have exact knowledge about a system, the fuzzy set is a point. (The membership function has the value 1 at this point and the value 0 everywhere else.) In the case of a fuzzy mapping that represents the system, this becomes the ordinary function for exact knowledge about a system.[118]
- An *observation* is represented by an observation operator. Since our knowledge about the state increases by observation, the effect of an observation operator on a fuzzy set is to make our knowledge finer.[119]

For a state space $X = \mathbf{E}^n$ and in it the set $U \subset \mathbf{E}^n$ of all allowed controls, Chang and Zadeh defined the state of a system as the fuzzy set $p(t)$ on X. The dynamic system is then represented by a fuzzy mapping $f : X \times U \longrightarrow X$, which has the following membership function:

$$\mu_{p(t+1)}(x(t+1)) = \mu_f(x(t), u; x(t+1)).$$

They defined the observation of a fuzzy subset q of the fuzzy set p representing the state as a renormalization $\bar{q}$ of q:

$$\mu_q(x) < \mu_p(x) \qquad \mu_{\bar{q}}(x) = \frac{\mu_q(x)}{\sup\limits_{x \in X} \mu_q(x)}.$$

For the observation, Chang and Zadeh assumed an "observation operator" O. The observation operator O and the state p may be given, but the fuzzy set that represents the observed state is not unequivocal. The set of all possible $\bar{q}$ in this case was denoted by the combination $O \circ p$ and therefore $q \in O \circ p$[120]:
Finally, Chang and Zadeh defined "fuzzy *feedback control systems*":
A Fuzzy Feedback Control System consists of the following:

1) a fuzzy mapping f such that $f : X \times U \longrightarrow X$;
2) an observation operator O;
3) a goal set G on X;

---

[117] [123] Chang, Mapping (1972), p. 30
[118] The authors defined fuzzy mappings in this article: A fuzzy mapping f from X to Y is a fuzzy set on $X \times Y$ with membership function $\mu_f(x, y)$. A fuzzy function $f(x)$ is a fuzzy set on Y with membership function $\mu_{f(x)}(y) = \mu_f(x, y)$. ([123] Chang, Mapping (1972), p. 30)
[119] See [123] Chang, Mapping (1972), p. 30
[120] When any confusion could be ruled out, the dash over the $\bar{q}$, which denoted the observed state, was omitted

4) a control policy η which maps the observed state to a control u such that $\eta : Q \to U$, where Q is the set of observed fuzzy sets: $Q = \{q|q \in O \circ p, p = \text{fuzzy set on } X\}$.[121]

The fuzzy feedback control system works like this:

1) The initial state is a fuzzy set $p(0)$ on X.
2) An observation is made and the state of the system becomes $q(0) \in O$ such that $q(0) \in O \circ p(0)$.
3) $U(0) = \eta \circ q(0)$.
4) Since the fuzzy mapping, $\mu_f(x(0), u(0); x(1))$, which represents the dynamic system is given, $\mu_f(q(0), u(0); x(1))$ is obtained from the definition of a fuzzy mapping of a set.[122]
5) $p(1)$ becomes the new initial state and steps 2), 3), 4) are repeated with the increased time variable.

With that, Chang and Zadeh had represented a control problem with the triplet (f, O, G) and a feedback control problem with a triplet (f, O, η). The goal G is considered achievable if an η exists such that for a t which can potentially be determined: $q(t) \subseteq G(t)$. In this paper, Chang and Zadeh proved the following central

Theorem 3. *Let $P_1(f_1, O_1, \eta)$ and $P_2(f_2, O_2, \eta)$ be two control problems such that f_2 is finer than f_1 and O_2 permits better observation than O_1, then G is attainable in P_2 if it is attainable in P_1. For feedback control problems, this means that a precise goal can be attained with rather sloppy control and observation. However, as the goal is approached the observation must be precise, since otherwise one cannot tell whether the goal is attained or not.*

This result yielded the theoretical proof that complex systems can be described and successfully controlled with the help of fuzzy sets. The mathematics of fuzzy sets also resulted in success in cases where precise mathematics failed because the complexity of real systems could not be overcome. In the area of control theory and its technical applications, this discrepancy between theory and practice was keenly felt when mathematical assumptions did not accord with actual conditions or when it was not expedient to produce mathematically exact calculations because it would take too long to obtain the results.

With his article *A Rationale for Fuzzy Control* in the *Journal of Dynamic Systems, Measurement and Control* the same year, Zadeh campaigned among control theoreticians and engineers to employ the theory of fuzzy sets and fuzzy algorithms.[123] The advent of the age of space travel in the year

---

[121] [123] Chang, Mapping (1972), p. 32
[122] If A is a fuzzy set on X, then the fuzzy set $f(A)$ is defined by the following membership function: $\mu_{f(A)}(y) \equiv \sup_{x \in X}(\mu_A(x) \wedge \mu_f(x, y))$
[123] [647] Zadeh, Rationale (1972)

Fig. 6.9. S. L. Chang

1957 had had a dramatic effect on control theory, he wrote. Heretofore the level of mathematization had been relatively low and publications on control theory had only very seldom included theorems and proofs. Since 1960, however, the "level of mathematical sophistication" had grown rapidly, "swinging the pendulum all the way from the low-brow imprecision of the forties to the high-brow mathematical formalism of the seventies".[124] Nowadays, in the early 1970s, it was a "must" for any article in a high-class control engineering periodical to include at least a few theorems and proofs. This was laudable, he said, as it forced the author to make precise statements about his assumptions and conclusions, but this trend also had its disadvantages:

> On the other hand, the question for precision frequently tends to overshadow other, perhaps more important goals, such as the invention of new types of control systems or the discovery of results which, though not of mathematical interests, are of high relevance to real world problems.[125]

A visible consequence of this "excessive concern with precision" was that large portions of the literature about control theory were moving further and further away from reality and contributing less and less to solving problems involving complex and large-scale man-machine systems, which was at the root of the many crises confronting modern society.

> In short, I believe that excessive concern with precision has become a stultifying influence in control and system theory, largely because it tends to focus the research in these fields on those, and only those, problems which are susceptible of exact solutions.[126]

The host of important problems with data, objects or conditions that were too complex or too imprecisely defined to be susceptible to a mathematically

---

[124] [647] Zadeh, Rationale (1972), p. 3
[125] [647] Zadeh, Rationale (1972), p. 3f
[126] [647] Zadeh, Rationale (1972), p. 4

exact analysis were and are ignored due to "mathematical intractability". For problems of this type, this stubborn insistence upon precision had to be abandoned and answers permitted which were *fuzzy* or *uncertain*. Zadeh expressed the hope that his call would be heeded in the future:

> In conclusion, I believe that in the years ahead fuzzy algorithms and control policies will gain increasing though perhaps grudging acceptance. They will have to be accepted and accorded some measure of respectability because the conventional nonfuzzy algorithms cannot, in general, cope with the complexity and ill-definedness of large-scale systems. I also believe that, in order to provide a hospitable environment for the development of fuzzy algorithms, control theory must become less preoccupied with mathematical rigor and precision, and more concerned with the development of qualitative or approximate solutions to pressing real world problems. Such a theory may well turn out to be far richer and far more exciting than control theory today.[127]

It would take about another year before a development began which would prove Zadeh right once again..

6.8 The First "Fuzzy Logic Controller" for Controlling a Steam Engine

The standard methods of designing automatic control systems that were taught until the early 1970s were based on mathematical models that had been established in advance, according to which all numerical calculations were then carried out. A number of newer studies were concerned with self-organizing or adaptive control systems[128] in which the control strategy was not determined in advance but could be adapted to an optimization algorithm. In this way, the experiences of control experts could be taken into account during the optimization process. These control systems displayed several properties that bore a resemblance to human learning; they were therefore also presented as examples of "artificial intelligence" or as "learning machines".
Ebrahim Mamdani[129] at Queen Mary College in London was also interested in the so-called "learning machines". Earlier he had worked in the field of pattern

---

[127] [647] Zadeh, Rationale (1972), p. 4
[128] [573] Yovits et al., Systems (1962), [538] Truxal, Control (1963), [373] Mendel et al., Techniques (1970).
[129] Ebrahim H. Mamdani, Indian-British electrical engineer, studied at the College of Engineering in Poona, India and then went to England, where be became professor of electrical engineering at London's Queen Mary College and at Westfield College of London University (1984–1995). Since 1995, Mamdani has been a professor of electrical engineering at the Imperial College of Science, Technology and Medicine at the University of London. [731] R. S. interview with E. Mamdani (1998)

6.8 The First "Fuzzy Logic Controller" for Controlling a Steam Engine

Fig. 6.10. E. Mamdani

recognition with artificial neuronal networks, in particular the recognition of handwriting and spoken language. Leaning and self-organization were two of the many attributes ascribed to human intelligence. Mamdani now expressed an interest in a third facet of human intelligence: the ability to comprehend commands and develop strategies on the basis of verbal communication rather than experience. Here he fell back on Alexander R. Luria's[130] graphic illustrations, which showed how the development of perceptual-motor skills in small children had a verbal basis.[131] He likewise invoked the cybernetic theory of learning devised by Andrew Gordon Speedie Pask[132], which stressed the linguistic nature of many aspects of intelligence behavior. Lastly, the artificial intelligence researcher Terry Winograd[133] had demonstrated shortly before that a robot arm could be controlled linguistically in order to manipulate various toy blocks.[134] Mamdani now wished to design a control system that could "learn" on the basis of linguistic rules. Naturally, only a fraction of a linguistic structure, such as the one Terry Winograd had shown, was to be considered: "... we were primarily concerned with the translation of semantic expressions into control laws, and not with the recognition of the expressions themselves or their manipulation. To the control engineer quantitative languages supporting arithmetic are the natural ones."[135]

Commands provided and understood by human control experts were to be learned. Normally, an expert observed the sequence of processes and knew based on experience how he should intervene if necessary. If any rules governed how he should proceed, they would include linguistically vague expressions, since he would use worlds like "much", "little", "some", "very" and so forth.

---

[130] Alexander Romanovich Luria (1902–1977), Russian neuropsychologist
[131] [332] Luria, Speech (1960)
[132] [437] Pask, Learning (1963), [438] Pask, Method (1971). Andrew Gordon Speedie-Pask (1928–1996), English cyberneticist. Pask was a professor in the department of cybernetics at Brunel University in England
[133] Terry Winograd (born 1946), American computer scientist. In the late 1960s, he constructed the AI program SHRDLU. Winograd is a professor of computer science at Stanford University
[134] [569] Winograd, Language (1972)
[135] [348] Mamdani, Fuzzy (1975), p. 2

In his January 1973 article *Outline of a New Approach to the Analysis of Complex Systems and Decision Processes,* Zadeh had identified words such as this as linguistic terms or modifiers of linguistic variables.

> The true antecedent of the work described here is an outstanding paper by Zadeh (1973) which lays the foundations of what we have termed linguistic synthesis ... and which had also been described by Zadeh as Approximate Reasoning (AR). In the 1973 paper Zadeh shows how vague logical statements can be used to derive inferences (also vague) from vague data. The paper suggests that this method is useful in the treatment of complex humanistic systems. However, it was realized that this method could equally be applied to "hard" systems such as industrial plant controllers.[136]

Mamdani had read this article and subsequently suggested to his doctoral student Sedrak Assilian[137] that he devise a fuzzy algorithm to control a small model steam engine (Fig. 6.11).[138]

The entire system consisted of the combination of a steam engine and a boiler (see Fig. 6.12). The steam was supposed to reach a certain predetermined pressure within the boiler; this was achieved by regulating the temperature. The engine was to run as consistently as possible at a particular piston speed, for which purpose a throttle was installed. This was therefore a system with

Fig. 6.11. The fuzzy-controlled steam engine

---

[136] [350] Mamdani, Advances (1976), p. 325

[137] No other facts about Sedrak Assilian are available; he also does not appear in later literature about Fuzzy Set Theory and its applications

[138] [731] R. S. interview with E. Mamdani (1998)

6.8 The First "Fuzzy Logic Controller" for Controlling a Steam Engine 273

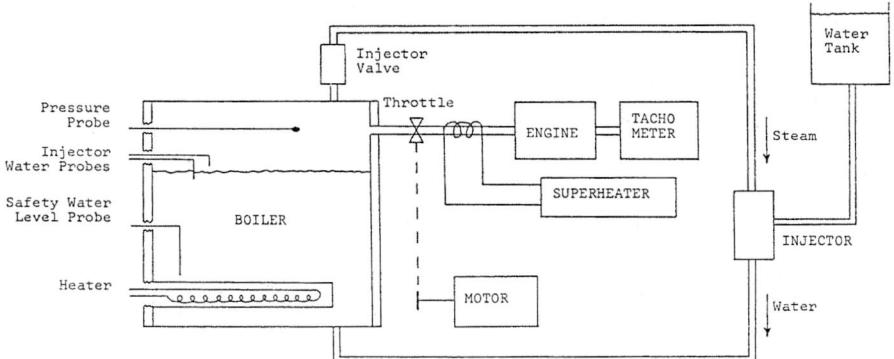

Fig. 6.12. Diagram of the system consisting of a steam engine and a boiler

two inputs (heat supplied to the boiler, engine throttle) and two outputs (pressure in the boiler, engine speed) (see Fig. 6.14).
Sensors constantly monitored the boiler and indicated the current pressure. If the prevailing pressure corresponded to the set point value, then nothing needed be done. If it deviated from the set point, then some action had to be taken, and this task was to be assumed by an automatic fuzzy controller.

Simple identification tests on the plant proved that it is highly nonlinear with both magnitude and polarity of the input variables. Therefore the plant possesses different characteristics at different operating points, so that the direct digital controller implemented for comparison purposes had to be returned (by trial and error) to give the best performance each time the operating point was altered.[139]

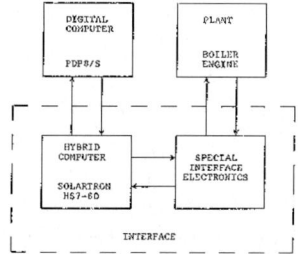

Fig. 6.13. The system of the fuzzy steam engine

---

[139] [348] Mamdani, Fuzzy (1975), p. 2

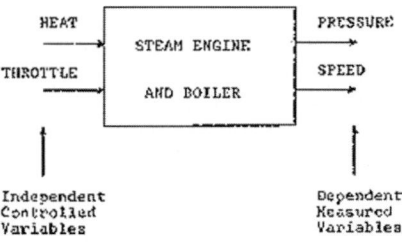

Fig. 6.14. The process variables of the fuzzy steam engine

Assilian and Mamdani defined six linguistic variables (four input and two output variables):

(1) *PE*   *Pressure Error*, defined as the difference between the actual value and the set point of the pressure in the boiler.
(2) *SE*   *Speed Error*, defined as the difference between the actual value and the set point of the of the piston speed.
(3) *CPE* *Change in pressure error*, defined as the difference between the actual value of *PE* and its most recent value.
(4) *CSE* *Change in speed error*, defined as the difference between the actual value of *SE* and its most recent value.
(5) *HC*   *Heat Change* (action variable, as the result of which a command occurs).
(6) *TC*   *Throttle Change* (action variable, as the result of which a command occurs).

Mamdani and Assilian introduced linguistic terms for these variables:

PB -	Positive, Big
PM -	Positive, Mittel
PS -	Positive, Small
P0 -	Positive, Zero
N0 -	Negative, Zero
NS -	Negative, Small
NM -	Negative, Medium
NB -	Negative, Big.

The variables were distributed over a number of points in accordance with the universe of discourse. For the variables *PE* and *SE* there were 13 points, which ranged from the maximum negative error through zero to the maximum positive error, with the zero being divided into a "negative zero error" *NO* and a "positive zero error" *PO* ("*NO* - just below the set point ... *PO* - just above the set point"[140]). Assilian defined fuzzy sets subjectively with the following values (see Fig. 6.15):

---

[140] [348] Mamdani, Fuzzy (1975), p. 7f

6.8 The First "Fuzzy Logic Controller" for Controlling a Steam Engine

	−6	−5	−4	−3	−2	−1	−0	+0	+1	+2	+3	+4	+5	+6
PB	0	0	0	0	0	0	0	0	0	0	0·1	0·4	0·8	1·0
PM	0	0	0	0	0	0	0	0	0	0·2	0·7	1·0	0·7	0·2
PS	0	0	0	0	0	0	0	0·3	0·8	1·0	0·5	0·1	0	0
PO	0	0	0	0	0	0	0	1·0	0·6	0·1	0	0	0	0
NO	0	0	0	0	0·1	0·6	1·0	0	0	0	0	0	0	0
NS	0	0	0·1	0·5	1·0	0·8	0·3	0	0	0	0	0	0	0
NM	0·2	0·7	1·0	0·7	0·2	0	0	0	0	0	0	0	0	0
NB	1·0	0·8	0·4	0·1	0	0	0	0	0	0	0	0	0	0

Fig. 6.15. Fuzzy sets for the variables PE (Pressure Error) and SE (Speed Error)

The variables CPE and CSE have been similarly quantized; the division of zero was of course omitted here. The subjective fuzzy sets were formed as follows (see Fig. 6.16):

	−6	−5	−4	−3	−2	−1	0	+1	+2	+3	+4	+5	+6
PB	0	0	0	0	0	0	0	0	0	0·1	0·4	0·8	1·0
PM	0	0	0	0	0	0	0	0	0·2	0·7	1·0	0·7	0·2
PS	0	0	0	0	0	0	0	0·9	1·0	0·7	0·2	0	0
NO	0	0	0	0	0	0·5	1·0	0·5	0	0	0	0	0
NS	0	0	0·2	0·7	1·0	0·9	0	0	0	0	0	0	0
NM	0·2	0·7	1·0	0·7	0·2	0	0	0	0	0	0	0	0
NB	1·0	0·8	0·4	0·1	0	0	0	0	0	0	0	0	0

Fig. 6.16. Fuzzy sets for the variables CPE (Change in Pressure Error) and CSE (Change in Speed Error)

The variable HC was ultimately quantized over 15 points. The subjective fuzzy sets are the follows (see Fig. 6.17):

	−7	−6	−5	−4	−3	−2	−1	0	+1	+2	+3	+4	+5	+6	+7
PB	0	0	0	0	0	0	0	0	0	0	0	0·1	0·4	0·8	1·0
PM	0	0	0	0	0	0	0	0	0	0·2	0·7	1·0	0·7	0·2	0
PS	0	0	0	0	0	0	0	0·4	1·0	0·8	0·4	0·1	0	0	0
NO	0	0	0	0	0	0	0·2	1·0	0·2	0	0	0	0	0	0
NS	0	0	0	0·1	0·4	0·8	1·0	0·4	0	0	0	0	0	0	0
NM	0	0·2	0·7	1·0	0·7	0·2	0	0	0	0	0	0	0	0	0
NB	1·0	0·8	0·4	0·1	0	0	0	0	0	0	0	0	0	0	0

Fig. 6.17. Fuzzy sets for the variable HC (Heat Change)

Similarly, the variable TC was distributed over five points (see Fig. 6.18): Mamdani and Assilian now defined 24 rules as IF-THEN rules, such as:

IF PE is NB THEN HC is PB,

	−2	−1	0	+1	+2
PB	0	0	0	0·5	1·0
PS	0	0	0·5	1·0	0·5
NO	0	0·5	1·0	0·5	0
NS	0·5	1·0	0·5	0	0
NB	1·0	0·5	0	0	0

Fig. 6.18. Fuzzy sets for the variable TC (Throttle Change)

or to issue three rules according to which the rest of the process should occur[141]:

Rule 1: IF PK and N, THEN NK.

(If the deviation in pressure is small and positive and the deviation in pressure does not change much, then reduce the supply of heat a little.)

Rule 2: IF PO and N, THEN N.

(If the deviation in pressure is approximately zero and the deviation in pressure does not change much, then do not change the supply of heat.)

Rule 3: IF PK and PK, THEN NK.

(If the deviation in pressure is small and positive and the deviation in pressure is slowly increasing, then reduce the supply of heat a little.)

These rule relationships were implemented as fuzzy relations for which Zadeh had already indicated the max-min rule as a composition rule in his first publication on Fuzzy Sets.[142] Additionally, a PDP 8/S digital computer[143] calculated a corresponding fuzzy set as a value for the output variable (Fig. 6.13). This method can be represented graphically in the following way (see Fig. 6.19): The sensors indicate sharp values for the input variables *pressure deviation* and its *change*, whose membership values with respect to the corresponding fuzzy sets can be read on the triangular membership functions. In the illustrated example for rule 1, the membership value with respect to the fuzzy set *pressure deviation* PK is 0.2 and it is 0.4 with respect to the fuzzy set change in *pressure deviation* N. Today this part of the fuzzy control process is known as "fuzzification".

The max-min rule prescribes that the maximum of these two values is computed first. (In the example for rule 1 illustrated above, this value is 0.2).

---

[141] The representation of these control examples stems from [370] McNeill et al., Fuzzy (1993). For the sake of simplicity, the authors of that work did not differentiate between "positive zero" and "negative zero"

[142] See Chap. 5

[143] [26] Assilian, Control (1974), p. 17

6.8 The First "Fuzzy Logic Controller" for Controlling a Steam Engine

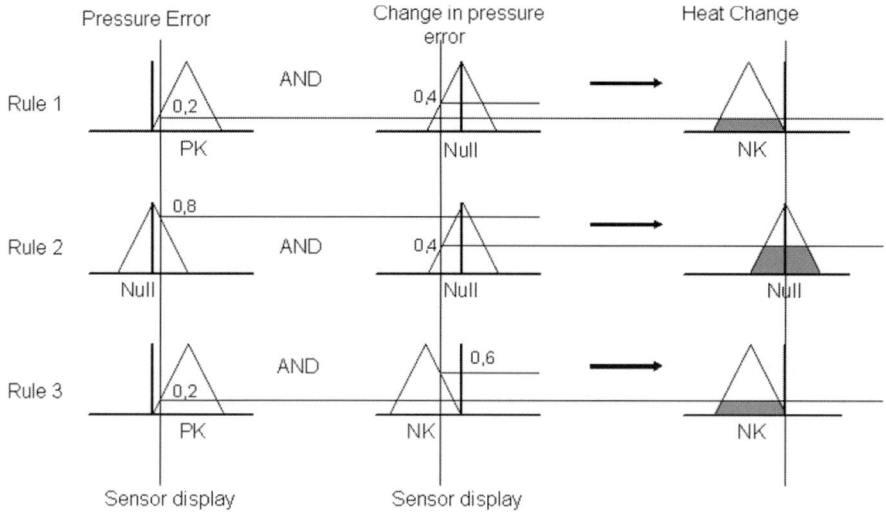

Fig. 6.19. Illustration of the application of the min-max rule

Accordingly, after executing this rule model alone, the output command was *change heat supply NK* and it had a membership value of 0.2! The result of rule 1 thus results in a triangular function that is truncated at the value 0.2 – a trapezoidal membership function.

However, rule 2 and rule 3 have also fired and so they must be evaluated analogously and parallel to rule 1. The final membership function for the fuzzy set as a value of the output variable *change in pressure deviation* is ultimately composed of the trapezoidal membership functions of the individual rule results. This composition occurs according to the max-min rule by forming the maximum of the membership functions of all three output fuzzy sets.

> To recapitulate, two algorithms were implemented in this application: one to compute the "heat change" (HC) control action and the other to compute the "throttle change" (TC) control action. Every rule in these algorithms is a relationship between the input variables PE, CPE, SE, CSE (in that order) and either HC or TC. The control actions are computed by presenting values for the input variables to the two algorithms. The input vectors are of course obtained by sampling the states of the steam engine at the sampling instants. The output of either algorithm is obviously a fuzzy set which assigns grades (of membership) to the possible values of the control fuzzy variable. In order to take a deterministic action one of those values must be chosen, the choice procedure depending on the grades of membership.[144]

---

[144] [348] Mamdani, Fuzzy (1975), p. 5

```
IF PE = (NB OR NM) THEN IF CPE = NS THEN HC = PM
OR
    IF PE = NS THEN IF CPE = PS THEN HC = PM
OR
    IF PE = NO THEN IF CPE = (PB OR PM) THEN HC = PM
OR
    IF PE = NO THEN IF CPE = (NB OR NM) THEN HC = NM
OR
    IF PE = PO OR NO THEN IF CPE = NO THEN HC = NO
OR
    IF PE = PO THEN IF CPE = (NB OR NM) THEN HC = PM
OR
    IF PE = PO THEN IF CPE = (PB OR PM) THEN HC = NM
OR
    IF PE = PS THEN IF CPE = (PS OR NO) THEN HC = NM
OR
    IF PE = (PB OR PM) THEN IF CPE = NS THEN HC = NM
```

Fig. 6.20. Fuzzy control commands for the steam engine designed by Assilian and Mamdani. Abbreviations used for the linguistic values: ZE, zero; PZ, positive zero; PS, positive small; PM, positive medium; PB, positive big; similar for the negative values NZ, NS, NM and NB. Negative deviations signify a movement toward the set point, positive deviations signify a movement away from the set point. Other abbreviations: PE, pressure error; CPE, change in pressure error; HC, heat input change

Just how was the output variable change in pressure deviation supposed to be adjusted, though? For this a sharp value is required. Mamdani and Assilian discussed this question only briefly in their article. For this aspect of the fuzzy control process, which is called "defuzzification", a number of different "defuzzification methods" have been proposed in the intervening years. As this development was beginning, Mamdani and Assilian decided on a simple procedure:

> Various considerations may influence the choice procedure depending on the particular application and in our case effectively that action is taken which has the largest membership grade. It is possible of course that more than one peak of a flat is obtained as illustrated below [see Fig. 6.21]. The particular procedure in our case takes the action indicated by the arrow, which is midway between the two peaks or at the centre of the plateau.[145]

In the dissertation entitled *Artificial Intelligence in the Control of Real Dynamic Systems* that Assilian produced in response to this fuzzy control problem, he wrote that the control strategy they had realized was one that a human operator – in this case, Assilian himself – could use to control a steam engine.

> These control policies were established first by imagining the entire state space ($PE \times CPE \times SE \times CSE$) to be divided into a number of areas, and second, writing down a control policy for each of these

---

[145] [348] Mamdani, Fuzzy (1975), p. 5; Fig. 1, ibid

6.8 The First "Fuzzy Logic Controller" for Controlling a Steam Engine

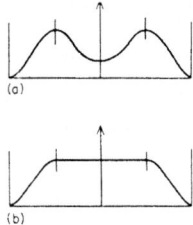

Fig. 6.21. Illustration of the selection of the centroid as a defuzzification method by devised Assilian and Mamdani

areas. Obviously, the first set of rules obtained in this manner does not necessarily produce the best quality of control possible ...[146]

This control algorithm was thus profoundly subjective. Not only the algorithm but also the membership function had been designed subjectively. Yet as Assilian and Mamdani managed to demonstrate, this fuzzy control system exceeded the performance of conventional control systems in several ways (see Fig. 6.22).

- Much less information is required for fuzzy control than for conventional control.
- The verbal knowledge of human experts did not have to be mathematically exact in order to be reprocessed by the automatic control.

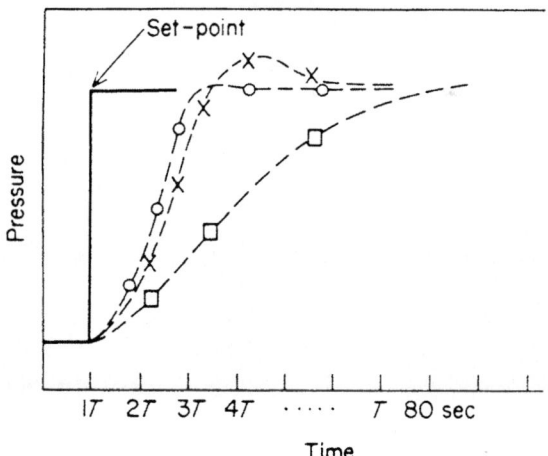

Fig. 6.22. The result of the Assilian-Mamdani Fuzzy Controller (○) compared to a conventional controller. (Dynamic Divergence Caching (DDC) algorithm damped (□) and undamped (x).)

---

[146] [26] Assilian, Control (1974), p. 135

280   6 Fuzzifications

- Errors were reduced little by little until the set point could be reached; digital controllers "overshot" this target instead.
- The fuzzy control system worked faster than a conventional control system; the possibility of processing the parallel firing of several rules at the same time shortened the required control time.

Therefore, the first technically realized fuzzy controller behaved exactly how Chang and Zadeh had predicted a year earlier.

Fig. 6.23. Ebrahim Mamdani (second from left), Lotfi Zadeh (fourth from left) and several of Mamdani's students in London in the late 1970s

With this fuzzy control of a steam engine – or more precisely a combination of a boiler and a steam engine – the essential principles for the construction of

6.8 The First "Fuzzy Logic Controller" for Controlling a Steam Engine

an entire class of fuzzy control systems were established. Today these systems are known as "Mamdani controllers".[147]

Several years later, a first *commercial* fuzzy control system was built according to this model. Lauritz Peter Holmblad, a Danish engineer who joined F. J. Smidth & Co. upon graduation from the Technical University of Copenhagen in 1973, and Jens-Jorgen Østergaard developed a fuzzy control system for the automatic control of a cement kiln.[148]

Attempts to automate cement production had always failed in the past because the process of cement burning is highly complex, ovens do not behave linearly and only a few measurements can be taken during the process. The fuzzy cement kiln Holmblad and Østergaard developed functioned very successfully and reliably, however. It was the first large-scale commercial fuzzy system. Many other products would follow it in the 1980s. Since then, fuzzy systems have been designed, produced, marketed and sold in almost all areas of technology – first in Japan, later in the US and Europe. To describe these developments of fuzzy technology here would go beyond the bounds of this work. This is why the period covered in this book ends before the so-called "fuzzy boom" began in earnest. The fuzzy systems in the field of medicine, which this work will next address, developed in parallel with the fuzzy control systems. The final chapter is dedicated to their story.

---

[147] In light of the fact that Sedrak Assilian had designed this fuzzy controller and that it had also been part of his dissertation, this designation is unjust. Lotfi Zadeh put it this way: "Assilian never got any credit. ... It is a little bit unfortunate because, after all, that was his Ph.D. thesis. So this particular point requires some historical correction." [709] R. S. interview with L. A. Zadeh (2001)

[148] See: [245] Holmblad et al., Fuzzy (1982)

7

The Fuzzification of Medical Diagnostics

Robert S. Ledley[1] and Lee B. Lusted[2], two American scientists who played a definitive role in the development of medical informatics, are the protagonists of the story told in the second part of this chapter – the story of computer-assisted medical diagnostics. It was their wish to use computers to support doctors in the task of drawing conclusions about patients' illnesses based on symptoms, signs and the results of their examinations. This help seemed to be urgently needed, as is shown by this comment from the foreword of a textbook[3]:

> The belief has been expressed that errors in diagnosis are more often errors of omission than of commission.[4]

References to "errors of omission" in medical diagnostics were commonplace in the forewords of the technical literature. In another textbook, the physician Logan Clendening[5] expressed himself on this matter:

---

[1] Robert Steven Ledley (born 1926) received his DDS in dentistry from New York University in 1948 and an MA in mathematical physics from Columbia University in 1949. In 1950, he was commissioned a first lieutenant in the U.S. Army Dental Corps. He was employed as a physicist and mathematician at the National Bureau of Standards and as an instructor and research analyst at Johns Hopkins University. Ledley had worked in a number of research posts in the academic field and the government before accepting a professorship in radiology at the Georgetown University Medical Center in 1970. In addition to his pioneering work in the use of computers in medicine, he also invented the CT diagnostic X-ray scanner, a device for producing three-dimensional images and reconstructions, which has found success in radiation therapy for cancer patients

[2] Lee Browning Lusted (1922–1994), American mathematician and physician, professor of biochemical engineering at the University of Rochester. In 1984, he became a Fellow of the American College of Medical Informatics. Lusted was the first editor of the journal *Medical Decision Making*

[3] [453] Pullen, Diagnosis (1944), p. vii

[4] [333] Lusted et al., Models (1960), p. 214

[5] Logan Clendening (1884–1945), American physician and medical historian. In 1942, Clendening became vice-president of the American Association of the

284   7 The Fuzzification of Medical Diagnostics

> How to guard against incompleteness I do not know. But I do know that, in my judgement, the most brilliant diagnosticians of my acquaintance are the ones who do remember and consider the most possibilities. Even remote ones should be brought up even though they may be immediately rejected.[6]

Several years later, a suggestion was made as to how the source of these errors of "omission" could at least be reined in, if not eliminated entirely:

> What is needed is a device which will answer the question "What are the possible causes of the group of symptoms and signs I have elicited from my patient?" Theoretically a giant table containing hundreds of vertical columns of symptoms and signs and hundreds of horizontal rows of diseases would suffice, but such a table would be as big as the wall of a room and far too unwieldy for practical use.
> The reason why no such device has been found is that the only commonly used fact holder has been the book or card, both of which hold the facts as symbols on a page. The continuous use of the page for thousands of years makes it unnecessary to speak of its advantages for recording. Yet we may conceive that the book, and particularly its unit the page or card, may not be the ultimate form of record. Though some of the defects of the bound book are obviated by a loose-leaf ledger or card system, even cards have the limitations inherent in their structure. In narrative records, side-by-side comparisons of groups of data is impossible, and the table on a single page or folding sheet is a sign that a vague want is felt for devices to display any required relationships. Unfortunately if one makes a table big enough to include as much data as would occupy a whole book of narrative type, the thing becomes utterly unmanageable. For any particular problem we have in mind, the table is cluttered with irrelevant data the obscure the data and their relationships that we are trying to trace.[7]

The author of this 1954 appraisal, the London doctor F. A. Nash, proposed a solution to the problem involving a mechanical apparatus he had designed. It consisted of a frame populated by long rods, over which a kind of "slide" could glide, as in a slide rule, which could assist the doctor in arranging various combinations of 82 signs and symptoms in such a way that the most probable diagnoses were selected from among 337 illnesses (Fig. 7.1)[8]:

> The apparatus allows the free manipulation at will of prefabricated groups of data for the solution of classificatory problems in medicine

---

History of Medicine. When the president of the association, Dr. Jaber Elliott, died, Clendening assumed his office

[6] [130] Clendening et al., Methods (1947), p. 59f., quoted in: [333] Lusted et al., Models (1960), p. 214
[7] [412] Nash, Diagnosis (1954), p. 874
[8] [412] Nash, Diagnosis (1954), p. 874

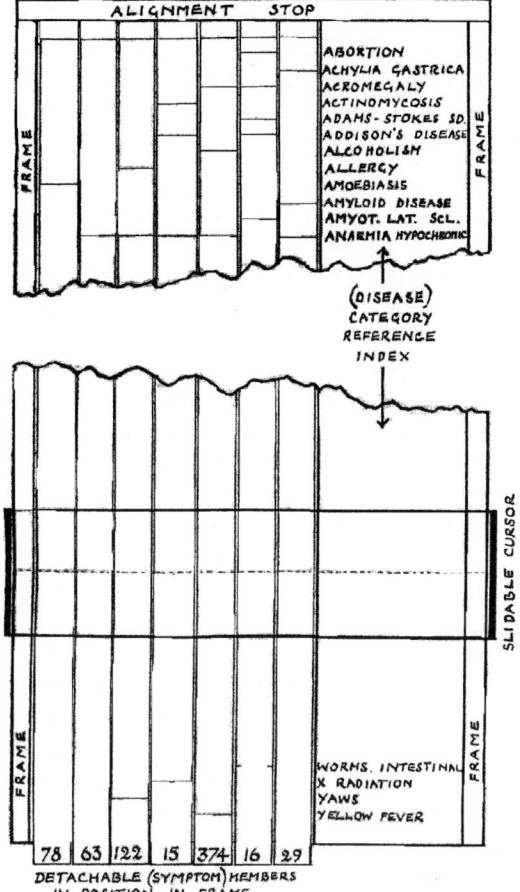

Fig. 7.1. Mechanical table to assist in diagnosis, developed by F. A. Nash, 1954

and other subjects. If the symptoms of a particular patient are put into the device, it will indicate at a glance the range of differential diagnostic possibilities. It is a supplementary aid and in no way a substitute for books, knowledge, experience, or clinical judgement. Like any tool it can be misused; but time may give it a useful, though limited, place in teaching and practice.[9]

The treating physician could indeed use this instrument, but it did not solve the principal problem of diagnostics and differential diagnostics that were always becoming more differentiated. A consulting doctor would strain the limits of his memory here – even if his brain were still in good shape – for as the French ophthalmologist François Paycha had demonstrated in 1955,

---

[9] [412] Nash, Diagnosis (1954), p. 875

doctors in the field of ophthalmology, for example, needed to consider about 1,000 illnesses and some 2,000 signs and symptoms relating to the cornea alone, and developments were continuing apace. Paycha proposed a system for the differential diagnosis of eye diseases that registered all of the symptoms on punch cards.[10] Other medical scientists also made similar suggestions at around the same time.[11]

7.1 Mechanization in Medical Diagnostics

Toward the end of the 1950s, Martin Lipkin and James D. Hardy were inspired by New York Hospital and the achievements of the first mainframe computers and they began to wonder how this new technology could be used in medical research and within the scope of a doctor's activity. In the department of medicine of New York Hospital-Cornell Medical Center, Lipkin and Hardy sought ways to master the flood of information that was constantly growing there, as well. They were well aware of the developments in computer technology, thanks to the writings of Vannevar Bush but also from other publications reaching back to the 1940s and even the 1930s and touching upon "mechanical" computing and sorting machines that used cards and needles or punch cards.[12] Lipkin and Hardy questioned the efficiency of such machines in searching for correlations, for example, between symptoms and laboratory findings, and they referred to corroborating experiments and findings by other researchers.[13]

Saving information on punch cards and sorting it on the basis of particular codes were skills scientists had possessed since the legendary United States census of 1890, which marked the beginning of the success enjoyed by Hollerith and the IBM Corporation. The idea arose of using machines of this type to build collections of data sets that were being accumulated during medical research, to carry out classifications and to develop interconnections among them. It was also thought that it might be possible to use this method to mechanically store data from patients' medical histories and to study whether this technology might be helpful in medical diagnostics. In 1958, Lipkin and Hardy reported on their project in the *Journal of the American Medical Association*, in which they sought to classify all of the diagnosis data from hematological cases by means of a "mechanical apparatus" and to identify relationships between them. How efficient could this ?[14]

---

[10] [440] Paycha, Mémoire (1955), [439] Paycha, Diagnosis (1955)
[11] [38] Baylund et al., Cards (1954), [253] Immich, Ereignisstatistik (1952)
[12] [195] Gibbs et al., Analysis (1947), [496] Schreiber et al., Code (1938), [84] Black-Schaffer et al., Studies (1943), [18] Allen, Card (1943)
[13] [106] Brodman et al., Medical (1951)
[14] [323] Lipkin et al., Corellation (1958)

Having consulted a number of standard textbooks on the subject of hematology, they selected for their study 26 blood disorders and listed all of the characteristic symptoms for each. The saving and sorting of these data was accomplished using a card and needle system that they had developed in 1955. Each card was 8×10.5 inches in size and divided by a numbered scale at the edge into a total of 138 small surface patches. A small hole was punched in each of these surface patches. For each disease, the data given for the characterizing symptoms were used to create a "master code" by transferring said data to the edge of a card. As long as all of the cards being used were identical, any given surface patch on the edge of each card represented the same information. The data from the corresponding case history were entered on one side of the card edge, data regarding the physical examination of the patient appeared on second side. The third side was reserved for data about the peripheral blood examination and the fourth side was for the findings of the bone marrow exam and other laboratory tests.

On each card, a triangular notch was punched in the location provided for the symptoms corresponding to one of the 26 diseases; since a notch of this type would cover the original hole, that hole was then obliterated. On the other hand, for symptoms that were absent, no notch was applied and the original hole remained. Each surface patch was thus a secondary information carrier: either notched or not. So every card bore a conglomerate of binary information units profiling the characteristic symptoms of illnesses.

A set of these cards could then very easily be sorted by a single data item. For example, if the doctor wanted to find the illnesses that are characterized by a single item identified during the physical exam, such as "enlarged spleen", he placed the cards in a stack and passed a metal or plastic wire through the hole indicating this item. When the wire was raised, all of the cards representing a disease characterized by an enlarged spleen would have to drop out of the stack of cards, since the triangular notch had removed the cardstock that would otherwise have held the wire in the card. Cards without this notch were lifted up by the wire.

The cards could also be used to separate out diseases on the basis of multiple entries. For this purpose, hospital records of 80 hematological cases were entered into prepared cards. These cases represented examples of 23 of the 26 disorders mentioned above. All positive results of any individual hospital case were listed and each result was then entered as a code number in the appropriate space on the card.

The 26 cards, each of which contained the data for one disease, were then arranged one after another such that the side containing the historical data faced upward. For each data item obtained from the medical history, a single wire was passed through the holes corresponding to the appropriate spaces on the 26 cards. Many wires were inserted in this way. The wires were then lifted. Cards having a notch for all of the indicated data, meaning that the wire could not hold them, fell out of the row of cards while those with one or more negative results were lifted up. The latter were removed and the remaining cards could then be examined in order to note which disorders included all of the characteristics that were being evaluated.

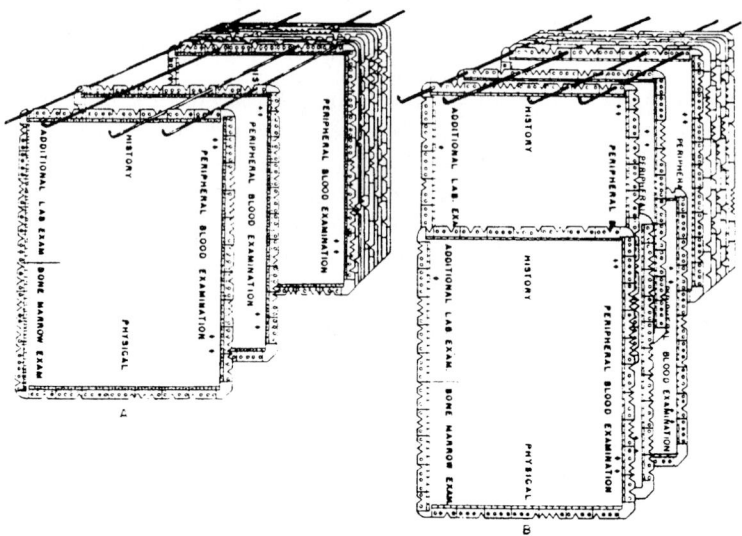

Fig. 7.2. Notch card system developed by Lipkin and Hardy, 1958

The same procedure was followed with the remaining cards using the side bearing the code for the physical examination; those cards lifted out of the groups were once again removed. The remaining cards were then sorted once more in the same way with respect to the peripheral blood examination and finally for the findings of the bone marrow exam and other laboratory tests. Those cards remaining after the fourth and final elimination represented the illnesses for which all of the utilizable findings were positive. The examiner now compared the code numbers of the diagnostic criteria on the cards with those in the hospital data, determined when positive items were positive in the hospital case and noted the illnesses whose criteria were identical to those in the hospital case. Each one of the 80 hematological cases was examined in the manner described, and the diagnoses obtained in this way were compared with the ones on the hospital lists.

7.2 Biomedicine and Digital Computers

Physicians certainly did not rank among the most euphoric users of electronic digital computers in the early 1950s. These machines, functioning as they did according to the principles of mathematical logic, gave rise to powerful misgivings among doctors, who feared that the diagnosis – until now the most exclusive and essential activity a doctor could perform – would eventually be completely usurped by computers.[15] Nevertheless, there were some interested researchers in medical science who had grand visions of the promise this new technology held for the life sciences.

Fig. 7.3. R. Ledley

These pioneers included Robert Steven Ledley and Lee Browning Lusted, two scientists who had been educated both in medicine and in mathematics and technology. Before the war, Lusted had been a graduate student in mathematics and physics at Harvard before then studying medicine at the Harvard Medical School, where he first heard of the new interdisciplinary field of "biophysics".[16] Ledley had been a dentist, served out his military duty in this capacity and subsequently studied mathematics and physics. While the powerful digital computer was blossoming, both researchers demonstrated great interest in this technology though they were being confronted with the collectively perceived "knowledge explosion" at the same time, as Lusted recalled:

> There was concern expressed in medical journals that a knowledge explosion was enveloping the field of medicine and there was speculation that computers could be used to help solve some of the problems. I felt that medical data could be processed by computer and that medical information could be made more useful to physicians by replacing it in a more usable form. I wasn't sure how this could be done but the idea of making information useful by making it more usable stuck with me.[17]

Lusted had worked in radiology in the 1950s. He had entered into this medical specialization in 1951 and by January of 1956 he had been hired for a two-year stint as a radiologist at the Clinical Center of the National Institutes of Health (NIH). Just then, at the start of 1956, he received an invitation from the Airborne Instruments Laboratory (AIL) in Minelao, Long Island, New York, to serve as an advisor for the general application of electronic methods in medicine. Here he occasionally talked with William J. Horvath of the Medical and Biological Physics Department at AIL about the possibilities of performing diagnostics with the aid of a computer. Horvath established

---

[15] See [335] Lusted, Decision (1979), p. 4
[16] See [338] Lusted, Haze (1991), p. 76
[17] [343] Lusted, Roots (1987), p. 173

Lusted's first contact with Ledley, whose article on logic and diagnostics he had read a year earlier.[18] When Lusted returned to the radiology department at NIH following his tenure at AIL, he soon met with Ledley:

> We found that we had been thinking about similar problems and possible solutions. The problems were caused by the large and increasing volume of medical literature. The possible solutions involved mathematical and computational aids for the physician.[19]

Ledley's memories on the subject go further beyond this meeting; he linked the story of his work with Lusted to that of the punch card technique discussed in the previous section:

> The idea of using computers to assist in medicine had, of course, been discussed from time to time by many people, but a specific idea of how to go about it had not yet been published. My attempts in the area were actually first made a few years earlier while I was at the Bureau of Standards where I put together a deck of the McBee key sort cards (the cards with the holes around the margins) for diseases of the tongue. Each card was a disease, and the symptoms related to the disease of the card were punched out to the card margin. Then, if needles were pushed through the edge hole of the deck of cards corresponding to a selection of symptoms, the cards that dropped would be only those corresponding to diseases having these symptoms. I even made a little device for facilitating the shaking and dropping of the cards, and as I carried the deck and my device around the halls of NBS, it didn't take the physicists more than a fraction of a second to say to me, "Oh, you're going to automate medical diagnosis, huh?" Of course, the cards did not truly carry out the logic that was required, and of course no probabilities are involved. But this work led Dr. Lusted and I to our first research accomplishment in medical informatics in 1959.[20]

Interest in recent developments in the quantitative methods, techniques and instruments of electrical engineering had already been growing in the American biomedical scientific community in the 1950s, even though the number of biophysicists who were working with this aspect was quite small; in 1966, Lusted estimated this number at less than one percent of all biophysicists in the '50s. Electrical devices such as amplifiers, Geiger-Müller tubes, photocells and the instruments developed from television technology were utilized in medical laboratories. Blood, nerve and tissue cells could be counted using TV technology, cancer cells could be targeted and irradiated. Yet electronic instrumentation was proliferating in medical diagnostics and therapy, as well: thermal elements to measure body temperature, devices to measure blood

---

[18] [313] Ledley, Logical (1955)
[19] [343] Lusted, Roots (1987), p. 175
[20] [315] Ledley, Personal (1987), p. 34

pressure and flow, electrocardiographs and electroencephalographs to record heart and brain activity, ultrasound and X-ray technology and many other applications of electronics were developed.[21]

When Lusted was still a radiology assistant at the University of California Hospital in San Francisco and an instructor of radiology at the University of California School of Medicine, he wrote the article *Medical Electronics* under the category "Medical Progress" in *The New England Journal of Medicine*.[22] The second sentence stated "Electronic instruments are important medical tools". He then listed those electronic devices which had recently been employed in medical laboratories, for the purpose of diagnosis or therapy, and he anticipated many of the applications that would be developed later on:

Fig. 7.4. L. Lusted

In addition to the devices presented, a host of "ancillary" electronic devices are used in medicine. A few examples illustrate the wide range of uses – radios for hospital patients, digital computers for large-scale data problems, image amplifiers for x-ray fluoroscopy, closed-circuit television for teaching, mass spectrometers, pocket radio receivers for personal paging of doctors in a hospital, tape recorders for teaching purposes and special count-rate meters for work on isotopes.[23]

Finally, Lusted underscored the importance of cooperation between doctors and engineers:

---

[21] For instance, the experiment computer ARC (Average Response Computer) and the TX-O Computer were used, for example, by the neurophysiologists at Massachusetts General Hospital and at the Massachusetts Institute of Technology (MIT) to study brain waves. See: [95] Braizer et al., Applications (1956), p. 325, [34] Barlow, Method (1957), p. 340, [129] Clark, Digital (1961), quoted in: [340] Lusted, Medicine (1966). A thorough description of many electronic devices in the field of "medical electronics" can be found in [337] Lusted, Electronics (1955).

[22] [337] Lusted, Electronics (1955)

[23] [337] Lusted, Electronics (1955), p. 584

> Members of the medical profession and electronic engineers have shown increasing interest in the field of medical electronics, with the result that electronic instrumentation has contributed to recent advances in some fields of medicine. Greater application of electronic instrumentation to medical problems will result from teamwork between the engineer and physician.[24]

The data accumulated in biomedicine had developed rapidly, had become exceedingly elaborate and was growing ever more complex. Retaining and managing all of it had become problematic and the usual methods no longer made it possible to consider and represent all of the interrelations that existed within this mass of information. Many biophysicians realized that they would have to rely upon automated methods of calculating and processing if they were to continue conducting and improving their research. Sooner or later, they were gong to have to turn to the new information and communication technologies and could expect significant improvements as a result of new electronic instrumentation. With that, a shift was heralded in biomedicine from the qualitative and descriptive type of science to one more quantitative, a shift that occurred very quickly; its computerization became a symbol of this development:

> Biology and medicine are becoming more like the other quantitative or exact sciences in that the role of calculating is emerging as an important investigative or research procedure. Because of the organic and integrated aspects of biological and medical sciences it is to be expected that the nature of calculation which may become an effective tool will be somewhat different from that in physical or engineering science and, in fact, will eventually involve the computation of behavior of very much more complicated calculation procedures than are now prevalent in the physical sciences. The computer has provided an essential means of exploring and investigating these calculations.[25]

Lusted did not sense in this development any great pressure from outside of the biomedical community. In 1962, an interest in the scientific history of his own profession had led him to study Thomas Kuhn's[26] book *The Structure of Scientific Revolutions*, published that same year, and when he publicized his personal view of the computer in medicine in 1966, he presupposed a

---
[24] [337] Lusted, Electronics (1955), p. 584
[25] [340] Lusted, Medicine (1966), p. 366
[26] Thomas Samuel Kuhn (1922–1996), earned a doctorate at Harvard University in 1949. He remained there as an assistant professor of general education and history of science until becoming a professor of history of science at the University of California, Berkeley in 1961. In 1964, he was given the M. Taylor Pyne Professorship of Philosophy and History of Science at Princeton University, and in 1979 he became a professor of philosophy and history of science at MIT. *The Structure of Scientific Revolutions* is his most well-known book

NIH ADVISORY COMMITTEE ON COMPUTERS IN RESEARCH – MAY 1962

Seated Left to Right: Mary A. B. Brazier, W. Ross Adey, Otto Schmitt, Ralph W. Stacy, Joseph E. Schenthal, William N. Papian James M. Sakoda and Homer R. Warner.

Standing Left to Right: Fay M. Hemphill, George S. Malindzak, Jr., Rose S. Doying, George A. Sacher, Jr., Max A. Woodbury, Lee B. Lusted, Alston S. Householder, Bruce D. Waxman, Scott Adams and David Garfinkel.

Fig. 7.5. Members of the NIH Advisory Committee on Computers in Research, May 1962

knowledge of this work: "The development of any field of science must be considered in terms of pressures which exist both within and without the scientific community..."[27] Nevertheless, he could not recognize any noteworthy external influences in this regard, such as the nation making some kind of insistent demand for "biomedical computing" in the health care system. He regarded similar pronouncements, such as the call for an "action program for strengthening medical information and communication" by Senator Hubert H. Humphrey's[28] Subcommittee on Reorganization and International Organization[29], as more of an exception.

Lusted considered the internal pressure within the biomedical field at this time to be all the more effective, and he himself was one of the most fervent proponents of computerization there. He had penned an editorial in the *IRE Transactions on Bio-Medical Electronics* in 1962 under the title *Quantification in the Life Sciences*. In it, he explained that he had come into the

---

[27] [342] Lusted, Quantification (1962), p. 1
[28] Hubert Horatio Humphrey (1911–1978), American political scientist and pharmacist. Humphrey was elected US Senator from Minnesota in 1948 and was re-elected in 1954 and 1960. He was the Vice-President of the United States of America under President Lyndon B. Johnson from 1965 to 1969
[29] [250] Humphrey, Memorandum (1962), quoted in [340] Lusted, Medicine (1966), p. 336

possession of a copy of the science history journal in which the proceedings of a conference on *Quantification in the Sciences* had been published: "... I happened to pick up the June, 1961, copy of ISIS, the journal of the History of Science Society."[30] After briefly outlining the content of some of the articles, he argued vehemently in favor of the necessity of "quantification in the life sciences" using predictions about future population growth: In 1962, when his text was published, a doctor came into contact with 750 people, but a ratio of 1,125 patients per doctor was to be expected for the year 1970. If doctors and nurses were still going to have time to see all of these patients, new equipment and procedures were necessary.

> To help the physician use his time more efficiently, we need to learn how to record a patient's history, physical examination and laboratory data in a form which can be processed by electronic computers. All physicians will then be able to have help with difficult diagnostic problems.[31]

Electronic Techniques in Medical Practice was the title of an unpublished memoir by Vladimir K. Zworykin[32], who was director of the Medical Electronic Research Group at the Rockefeller Institute in the 1950s and who was already famous at that time as a pioneer in television technology. It was an early

Fig. 7.6. V. Zworykin

---

[30] [342] Lusted, Quantification (1962), p. 1
[31] [342] Lusted, Quantification (1962), p. 3
[32] Vladimir Kosma Zworykin (1889–1982) was born in Murom, 200 miles east of Moscow. He studied electrical engineering in Russia and together with Boris Rosing conducted experiments using early cathode ray tubes that had been developed by Carl Ferdinand Braun in Germany. In 1910, the two scientists created a TV system with a mechanical scanner and Braun's tubes as receivers. Rosing disappeared during the revolution of 1917; Zworykin fled. He briefly studied X-rays with the French physicist Paul Langevin (1872–1946) in Paris and traveled to the US in 1919, where he was employed by Westinghouse Laboratories in Pittsburgh. At a congress of radio engineers on November 18, 1929, he demonstrated a television receiver in the form of his "cinescope"

concept paper that addressed the issue of how the computer could be helpful in medical practice in the future and in which "computer-assisted diagnosis" was considered. "I don't believe the paper was published but it was circulated privately by Dr. Zworykin",[33] Lusted later wrote. He also recalled that a number of doctors were willing to assign certain tasks in the health care system to computers at that time, but the vast majority did not take this idea seriously and most doctors were critical of it.[34] To illustrate this sentiment, Lusted cited a column entitled *Triple Bromides* by Gerry Feigan, which had appeared in the *San Francisco Medical Society Bulletin* in June of 1965:

> The waiting room of the future will serve eight doctors; it will be completely sound proof. The furniture will consist of contour chairs with built-in gentle massage. Color television will delight the eye, and soft music will allay anxiety. Silent air conditioning will waft delicately scented odors and soft drink dispensers, operating at the touch of a button, will deliver cool libation with or without a Miltown tablet. For those who prefer literature, magazine articles will be softly read by wire. History-taking will be painless. A group of preferential questions will be asked by tape and the answer punched on a card. Nothing will be left to chance, and by cybernetics, the card will be quickly deposited in a slot which will provide the three most probable historic diagnoses.[35]

Medical electronics at first represented a quiet corner of electrical engineering but began to grow continuously and quickly once the vacuum triode developed by Lee de Forest[36] in 1907 made it possible to amplify the very weak electric signals produced by the heart or the central nervous system. The advances recorded here soon made it clear that electronics should be applied in the life sciences, and electrical engineers likewise recognized great potential in successfully using their instruments here. In 1932, the American Institute of Electrical Engineers (AIEE) first sponsored a round of discussions on "electromedical problems", and over the next seven years the AIEE, with the Institute of Radio Engineers (IRE) and the Instrument Society of America, co-sponsored national conferences on electronic technology in medicine and biology. In April 1952, the Professional Group on Medical Electronics (PGME) was founded within the IRE, and three years later it counted 1,000 members, including doctors, biologists, physiologists, physicists and biophysicists.[37]

Toward the end of 1956, the Air Research and Development Command of the U.S. Air Force consulted the Division of Medical Sciences of the National

---

[33] [340] Lusted, Medicine (1966), p. 367
[34] [336] Lusted, Design (1983), p. xi
[35] [171] Feigan, Bromides (1956)
[36] Lee de Forest (1873–1961), American physicist, initially worked with the vacuum diode that had been discovered and invented by the British engineer and natural scientist Sir Ambrose Fleming (1849–1945) in 1904. He then developed it into the triode in 1907
[37] See [337] Lusted, Electronics (1955), p. 581

Research Council with inquiries about all manner of applications of the computer in biology and medicine. A gathering organized at the Harvard Computation Laboratory on October 5, 1956 was the beginning of a series of conferences on this subject. A conference on "Electronic Techniques for Mathematical Operations in Biology and Medicine" was sponsored at the National Institutes of Health (NIH) by NIH and the IRE Professional Group on Medical Electronics on November 29 of the same year. Bolstered by their tremendous success, the Professional Group on Medical Electronics next prepared a Symposium on the Applications of Computers in Biology and Medicine for the annual IRE Convention in March of 1957.

Robert S. Ledley, working at that time as an expert on data processing systems at the National Bureau of Standards, was contracted by the air force and NIH in September 1957 to compile an overview of the computers being used in biological and medical research. This survey, which was published in 1960, cited a greater number of computers being employed in biomedical research projects than the members of one of the advisory committees which assisted in this study had suspected.

> Could an electronic computer with its extensive memory capacity and data-processing facility help the physician avoid some of the diagnostic errors of omission? ... Could the computer be programmed to analyze the patient's signs and symptoms and to give a differential diagnosis? If the computer could be programmed to analyze the patient's data, should logical analysis processes be followed similar to those used by the physician? Diagnosis is, of course, a mental process, but what kind of mental process is it?[38]

Lusted and Ledley had taken that last question and the assertion preceding it from the introductory statements by Logan Clendening and Edward Hageman Hashinger in *Methods of Diagnosis*.[39] It had been these questions that had led them to study those logical processes which were occurring during medical diagnosis and which were given far too little attention in medical textbooks. In another publication from that year, Lusted argued a similar point:

> Dr. Ledley and I started about a year and one half ago to ask ourselves the general question of how electronic computers could aid in medical diagnosis and the first part of this program was to consider the logical analysis of medical diagnosis. What we did was to look at the Clinical Pathological Conference (CPC) which provides data in concise form, although not in a form analogous to the information the physician has on the ward.[40]

Several events in 1959 were decisive in ensuring that biomedical research in the United States would increasingly rely on the abilities of computers: On

---

[38] [333] Lusted et al., Models (1960), p. 214
[39] [130] Clendening et al., Methods (1947), p. 1
[40] [333] Lusted et al., Models (1960), p. 255f

January 14, Zworykin organized the Conference on Diagnostic Data Processing at the Rockefeller Institute.[41] On the 9th and 16th of July, two hearings on the use of automatic data processing in medicine were held before the Subcommittee on Reorganization and International Organization, chaired by Senator Hubert H. Humphrey. They came to the conclusion that corresponding developments ought to be organized and fostered by the government.

Also in 1959, Ledley and Lusted published their article *Reasoning Foundations of Medical Diagnoses*[42], in which they opened the field of application for medical diagnostics not simply to mathematical methods in general, but also to computerization. Ledley, who had meanwhile become an associate professor of electrical engineering at George Washington University in Washington, D.C. and was the mathematician responsible for data processing systems at the National Bureau of Standards, and Lusted, now an associate professor at the University of Rochester School of Medicine in Rochester, New York, authored this article together, and with it and numerous texts like it, they permanently steered biomedical research in a new direction.[43]

Medical diagnoses, they argued, were based on logical conclusions, and these could be inferred from information for which they identified two sources:

1) *Medical knowledge*, which is information about relationships that exist among symptoms and illnesses, and
2) *Symptoms* a patient exhibits, from which other pertinent information can be inferred for this patient.

They sketched out a doctor's answer to the question "How do you make a medical diagnosis?" as follows:

> *First*, I obtain the case facts from the patient's history, physical examination, and laboratory tests.
> *Second*, I evaluate the relative importance of the different signs and symptoms. Some of the data may be of first-order importance and other data of less importance.
> *Third*, to make a differential diagnosis I list all the diseases which the specific case can reasonably resemble.

---

[41] [152] Eden, Proceedings (1960), p. 232

[42] [308] Ledley et al., Foundations (1959). Today this article is considered to mark the beginning of "medical informatics"

[43] At the Third Annual Symposium on Computer Application in Medical Care in approximately 1979, when Lusted looked back on a terrific success story in a text entitled *Twenty Years of Medical Decision Making Studies*, he was able to note that in the period from 1959 to 1968 he and Ledley, working solo or as co-authors, had published some 45 articles in 23 American and nine overseas journals as well as seven proceedings of international conferences, all of them dealing with the subject of computer-assisted medical diagnostics or decision making. [335] Lusted, Decision (1979), p. 4

Then I exclude one disease after another from the list until it becomes apparent that the case can be fitted into a definite disease category, or that it may be one of several possible diseases, or else that its exact nature cannot be determined.[44]

The publication of this article initiated the rise of a new scientific discipline, that of "medical informatics", for the basis had been established here for the use of computers and the computer sciences in medical-scientific procedures, i. e. in research and teaching as well as in diagnosis and therapy. Ledley and Lusted also analyzed symbolic logic, probability theory and game theory, which they explained for their readers who had been trained as doctors and other natural scientists.

Symbolic Logic
The authors highlighted three ingredients of logical concepts inherent in medical diagnosis: (i) medical knowledge, (ii) the signs and symptoms presented by the patient and (iii) the final medical diagnosis itself. Medical knowledge presents certain information about relationships that exist between symptoms and the diseases. The patient's symptoms (i. e. symptoms, signs and laboratory tests) contain further information associated with this patient. Based on these two sources of information and the ability to draw logical conclusions, the doctor makes his diagnosis.

Probability Theory
Since chances, probabilities and statistics are components of medical knowledge, they are also evident in medical diagnosis. Although, as Ledley and Lusted noted here, probabilities can on seldom be stated exactly – "Medical diagnostic textbooks rarely give numerical values, although they may use words such as "frequently", "very often", and "almost always"." – they argue here academically and assume that probabilities could be known or obtained. This is why their text discusses the application of stochastic methods in medical diagnostics.

Values
When making decisions about how to treat the patient, doctors must often contend with complicated conflict situations which not only depend on the established diagnosis but also on therapeutic, moral, ethical, social and economic considerations concerning the patient, his family and the society in which he lives. Similar complicated decision problems frequently arise in military, economic and political situations, and mathematicians had developed them analytically and quantitatively within the framework of a value theory. In arguing for the application of these methods for medical decision making, as well, Ledley and Lusted suggested following John von Neumann's game theory approach.

---

[44] [308] Ledley et al., Foundations (1959)

7.2 Biomedicine and Digital Computers

The last section of the article, under the titles "Learning Device" and "Learning Machine", discusses what at that time were experimental tools for the implementation of the preceding logical and probabilistic analyses. In a realistic application, the two authors surmise, one would expect to deal with circa 300 possible illnesses and some 400 symptoms. The logical basis of such a set of symptoms and diseases would require 2^{700} columns, which is more than 10^{200} columns! A reduced basis should be established, but this is not practicable. The columns that can be eliminated correspond to the disease-symptom relationships that will never occur; the reduced basis corresponds to the complexes that will occur. Here the authors take up the idea of using punch cards; they describe a system of cards and needles similar to that of Lipkin and Hardy. Alongside the hardware, however, it was vital that software also be developed so that the computer could fulfill its true application potential.

The two authors had not been able to interest a medical journal in their manuscript, but luckily the journal *Science* accepted the text and printed it as the lead article in the edition of July 3, 1959. Thus Ledley and Lusted's ideas were guaranteed wide distribution.[45] The article did indeed become very well known; it was translated into Russian in 1961 and published as a monograph in Moscow.[46]

The logical analysis to which Ledley and Lusted had subjected the medical diagnosis process is sketched out here (cf. Fig. 7.7):

- A patient's "attributes", such as the sign "fever" or the disease "pneumonia", they represented with the lower-case letters $x, y, ...$, while statements about these attributes were denoted with the corresponding upper-case letters $X, Y, ...$
- If Y represents the sentence "The patient has the attribute y", then its negation $\neg Y$ is: "The patient does not have the attribute y".
- $X \cdot Y$ represents the combined statement "The patient has the attributes x and y",
- while $X + Y$ represents the following combination: "The patient has attribute x or attribute y or both."
- "If the patient has attribute x, then he has attribute y" is symbolized with $X \Rightarrow Y$.

The authors illustrate these connections by means of Fig. 7.7. The shaded patients in Fig. 7.7 (a) have attribute y, so statement Y is true for them. The patients shaded in the other direction should now have a second attribute x. Figure 7.7 (b) shows all of the patients for whom statement $X \cdot Y$ holds true. Similarly, Fig. 7.7 (c) shows the patients for whom $X + Y$ is

---

[45] See also [338] Lusted, Haze (1991), p. 77
[46] According to Lusted, requests for reprints of this article ran into the thousands. At various times Ledley and Lusted distributed between 5,000 and 8,000 such offprints; the exact number is unknown. See also [338] Lusted, Haze (1991), p. 78 and [335] Lusted, Decision (1979), p. 4

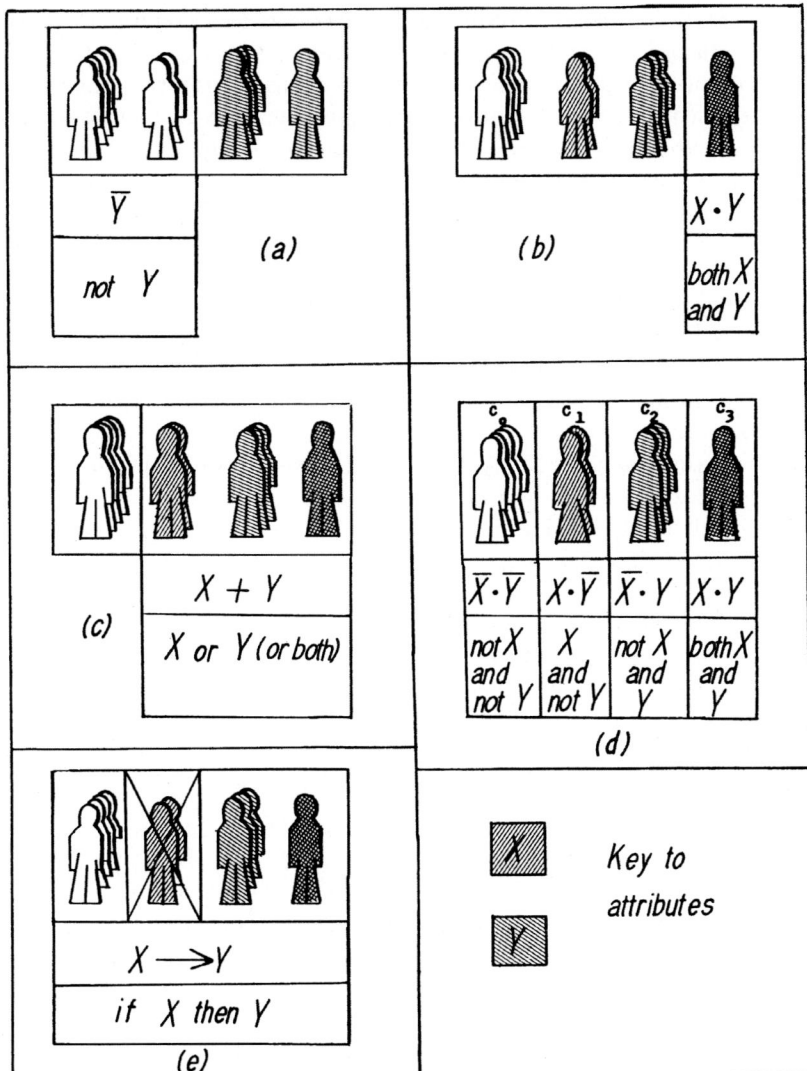

Fig. 7.7. Logical combinations of attributes in a patient population as devised by Ledley and Lusted, 1959

true. Finally, Fig. 7.7 (d) includes the four patient classes that are created by these logical operations, while Fig. 7.7 (e) illustrates a patient population for whom $X \Rightarrow Y$. Since more than two attributes as well as more complicated expressions are generally observed, Ledley and Lusted carried out "Boolean functions" $f(X, Y, ...)$ in order to be able to formulate the appropriate combinations.

For more than two attributes, they classified the patients in a corresponding number of sets C_i: For m attributes, there are 2^m possibilities of being attributed to a patient or not, and there are therefore 2^m sets $C_i : C_0, C_1, ..., C_{2^m-1}$. Let us limit ourselves here to just two attributes, namely symptoms (S) and diseases (D), in which case:

- $S(i)$, i.e.: "The patient has symptom i." $\quad i = 1, ..., n$.
- $D(j)$, i.e.: "The patient has disease j." $\quad j = 1, ..., m$.

Corresponding examples from a diagnosis textbook look like this:
- If a patient has disease 2,
  then he must have symptom 1. $\qquad D(2) \Rightarrow S(1)$
- If a patient has disease 1
  and not disease 2,
  then he must have symptom 2. $\qquad D(1) \cdot \neg D(2) \Rightarrow S(2)$
- If a patient has disease 1
  and not disease 2,
  then he cannot have symptom 2. $\qquad \neg D(1) \cdot D(2) \Rightarrow \neg S(2)$
- If a patient has either or both
  of the symptoms,
  then he must have one or
  both of the diseases. $\qquad S(1) + S(2) \Rightarrow D(1) + D(2)$

From these possibilities, Ledley and Lusted combined a Boolean function E of all of the symptoms and diseases under consideration:

$$E = [D(2) \Rightarrow S(1)] \cdot [D(1) \cdot \neg D(2) \Rightarrow S(2)] \cdot$$
$$\cdot [\neg D(1) \cdot D(2) \Rightarrow \neg S(2)] \cdot [S(1) + S(2) \Rightarrow D(1) + D(2)].$$

They similarly expressed all of the symptoms present in a patient using a Boolean function G, which was much simpler, as the following example demonstrates: A patient might have symptom 2 and not symptom 1, in which case the G-function is:

$$G = \neg S(1) \cdot S(2).$$

The G-function includes symptoms the patient does not have as well as those he does have. If it is unknown whether or not a patient has a symptom, said symptom does not appear in G. The function f is computed from known E and G into $f = D(1) \cdot \neg D(2)$. The logical aspect of a medical diagnosis problem can be formulated as follows: The diseases f can be determined such that:

> If medical knowledge E is known,
> then :
>> If the patient presents symptoms G,
>> then he has diseases f.

The authors then presented the "fundamental formula of medical diagnosis":

$$E \Rightarrow (G \Rightarrow f).$$

In the section entitled "Probabilistic Concepts", the authors address the problem that although statements like "If a patient has disease 2, he must have symptom 2" are components of "medical knowledge", strictly speaking such textbook rules can only be applied if their assumptions are fulfilled with certainty. A doctor must also know with certainty whether a symptom or a disease is present in a patient. In many cases, however, this certainty cannot be established after anamnesis, medical examination and appropriate laboratory tests, and so textbook rules like the above have to be modified for practical implementation. For example: "If a patient has disease 2, then there is only a certain chance that he will have symptom 2 – that is, say, approximately 75 out of 100 patients will have symptom 2".[47] This, at least, was the argument put forth by Ledley and Lusted in the section following the logical analysis sketched out above; they also drew their conclusions right away: "Since "chance" or "probabilities" enter into "medical knowledge", then chance, or probabilities, enter into the diagnosis itself".[48] The fact that in 3/4 of cases, a particular symptom proves a particular disease does not allow the diagnosis to seem very precise, and for a patient whose health is in question, it is not very helpful. This may be why six years later Lusted emphasized the "inadequacy of conventional mathematical methods for dealing with biological problems".[49]

In my opinion, however, there is no indication that Lusted himself ever considered giving up the "conventional mathematical methods" in favor of "new", "unconventional methods". In his 1962 editorial *Quantification in the Life Sciences* for the *IRE Transactions on Bio-Medical Electronics*, he argued vehemently for the mathematization of the life sciences, following the pattern of the tremendously successful mathematization of physics from a century earlier:

> The importance of "mathematization of the biological sciences" can not be underestimated if we are to clarify and formulate precisely the important problems in these fields.[50]

He even closed his text by quoting Lord Kelvin:

> We recognize Lord Kelvin's famous dictum: "If you cannot measure, your knowledge is meager and unsatisfactory", may be as true for the life sciences as for the physical sciences.[51]

---

[47] [308] Ledley et al., Foundations (1959), p. 13
[48] [308] Ledley et al., Foundations (1959), p. 13
[49] [334] Lusted, Computer (1965), p. 321
[50] [342] Lusted, Qualifications (1962), p. 2
[51] [342] Lusted, Qualifications (1962), p. 2

Three years later, Lusted was expressing much different opinions about "meager" knowledge; in his contribution to a multi-volume work about computers in biomedical research, he wrote that new mathematical methods were needed in biomedicine in order to collect and code all of the information being acquired, and here he closes not with the words of Kelvin, but rather those of Lotfi Zadeh, which he had written three years earlier:

> In a recent statement, Professor L. A. Zadeh (1962) summed up the situation as follows: "In fact, there is a fairly wide gap between what might be regarded as "animate" system theorists and "inanimate" system theorists at the present time, and it is not at all certain that this gap will be narrowed, much less closed, in the near future. There are some who feel that this gap reflects the fundamental inadequacy of the conventional mathematics – the mathematics of precisely-defined points, functions, sets, probability measures, etc. – for coping with the analysis of biological systems, and that to deal effectively with such systems, which are generally orders of magnitude more complex than man-made systems, we need a radically different kind of mathematics, the mathematics of fuzzy or cloudy quantities which are not describable in terms of probability distributions. Indeed, the need for such mathematics is becoming increasingly apparent even in the realm of inanimate systems, for in most practical cases the *a priori* data as well as the criteria by which the performance of a man-made system is judged are far from being precisely specified or having accurately-known probability distributions."[52]

Lusted had expressed his opinion of the inadequacy of "conventional" mathematics in medical applications in the very same year in which Zadeh inaugurated his new "unconventional" fuzzy mathematics.

7.3 Computer Systems in Medicine

In periods of great euphoria following initial success with "electron brains", the scientific seriousness behind the visions of computerization was often left behind, and not just in popular scientific literature! So predictions about the future role of computers in medicine also sometimes crossed the line into science fiction: "Some articles did give the reader the impression that computers would be able "to do medical diagnosis" in the future", Lusted said, through unfortunately without indicating the source of his quote[53], in an attempt to put to rest right away any idea that the writers could have possibly gleaned any such ideas from the text he and Ledley had published in 1959. Lusted repeats from the text he had co-authored with Ledley twenty years earlier:

---

[52] [334] Lusted, Computer (1965), p. 321
[53] But see the column by G. Feigan which cited [340] Lusted, Medicine (1966), p. 368

The Reasoning Foundations article talked about "the mathematical techniques that we have discussed and the associated use of computers are intended to be an aid to the physician. This method in no way implies that a computer can take over the physician's duties. Quite the reverse; it implies that the physician's task may become more complicated. The physician may have to learn more; in addition to the knowledge he presently needs, he may also have to know the methods and techniques under consideration in this paper."[54]

Ledley became the founding president of the National Biomedical Research Foundation (NBRF) in 1960 and that same year he published his book *Digital Computer and Control Engineering*, in the introduction of which he drafted scenarios for the application potential of digital computers. There he anticipated developments that would not become reality until many years later, and in the section entitled *Aids to Medical Diagnosis*, he linked the beginnings of two research directions that were at best on the far horizon at the time but that have become reality today and are in fact growing together in the way Ledley had envisioned: "knowledge-based systems" and "telemedicine".

Consider a modern high-speed electronic digital computer, and suppose that it has been programmed to aid medical diagnosis. Let us assume further that the physician can directly communicate with the computer by telephone, teletype, radio, etc. The value of such a computer interrogation arises from three factors: (1) the ability of the computer to formulate a treatment plan that will maximize the chance of curing the patient; (2) the ability to determine the minimum number of necessary medical laboratory tests or other diagnostic procedures for the particular patient; and (3) the ability to evaluate more accurately diagnostic-test results for a particular patient based upon his previously recorded health records. A network of such computers could form a hypothetical health-computing system (see Fig. 7.8). Here each computer can communicate with individual physicians and hospitals within its area, receiving, transmitting, and computing medical information as required. However, the area computers must be capable of communicating with each other as well, since approximately 20 per cent of Americans change addresses each year, and probably most of us go on at least one trip a year. Also all the area computers could communicate with a special research computer that could sample data as required for various research and public health-control investigations. The great significance and importance of such a health-computer network cannot be overestimated as an aid to increasing individual good health and longevity and as a vast new source of medical information concerning mankind.[55]

---

[54] [335] Lusted, Decision (1979), p. 4
[55] [312] Ledley, Digital (1960), p. 21–23. The numbering of the figures was adapted to the order of the figures in this book

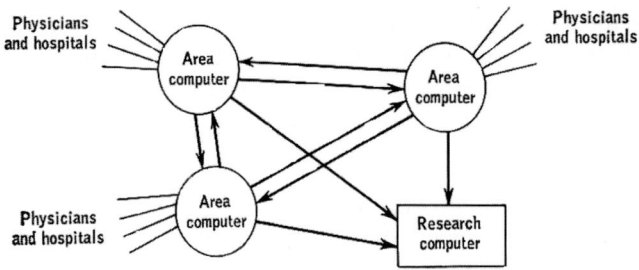

Fig. 7.8. Ledley's vision of a computer network in the health care system of the 1960s

After the grandiose expectations of computers in general as the "Great Problem Solvers", "artificial intelligence" and "electronic doctors" had been dampened at least by the time Minsky and Papert's work had been published, the euphoria gave way to resignation. The research field of artificial intelligence did not gain momentum again until the 1980s.[56]

The use of computer assistance in medicine developed in another way. At first, scientists tried to determine what assistance they could obtain from the exact sciences and technology without necessarily having to rely on computers. Logic and mathematics were standing by with decision trees and matrices, stochastics and operations research offered the theorem by Bayes and the cost-benefit relationships for applications in medical decision making problems and, based on the stochastic decision theory, the so-called ROC (Receiver Operating Characteristic) analysis had emerged in the 1960s and become indispensable to various laboratory techniques.

ROC analysis was developed from the theory of signal detection in electrical engineering, such as during the analysis of radar signals. However, this process was very soon also being used for the scientific examination of human sensory perception, in psychophysics[57] and eventually, in the 1970s, for medical diagnostic systems, as well.

Lusted's thoughts led him to study decision making problems in medical technology. In 1962, he considered the question of how the doctor could exploit medical data to its fullest to aid in his decision making and how he could overcome observer variation, such as that noted in the interpretation of X-rays. To this end, he also studied literature that had only a peripheral connection to medicine, as he recalled in 1979:

> During 1962–1964 from conversations with Ward Edwards, who was then at the University of Michigan, and from various articles I learned about the subjects of personal probability, Bayesian statistical interference for psychological research and signal detection theory applied to human observers. Ideas from these areas influenced my thinking

---

[56] See Chap. 3
[57] [205] Green et al., Theory (1966)

about medical decision making studies and culminated in my demonstration that signal detection theory and operating characteristic analysis helps to explain observer variation in radiology and offers a strong vantage point for consideration of decision outcome tradeoffs in all areas of medicine.[58]

In his 1968 book *Introduction to Medical Decision Making*[59], Lusted had recommended ROC analysis in addition to the aforementioned methods, characterizing it then, in 1979, as a method fitting between the strict statistical decision making matrix and the dynamic cost-benefit relationship. Unhappy that most doctors did not understand this analytical method and that it had yet to find a place in medical decision analysis, Lusted was nevertheless vaguely confident in 1979 that this situation could change, for he recognized a number of indicators that it was going to flourish in medical decision theory:

– Good timing: The costs of the health system were of great interest both nationally and internationally.
– Good technical literature: Journal articles and textbooks that were well-suited to introduce the topic were now available.[60]
– Great interest: The number of young doctors and medical students who were interested in studies on medical decision theory was growing and similar, very popular programs of study had been launched.
– Scientific community and professional journal: The Society for Medical Decision Making met for the first time from the 11th to the 13th of September 1979 in Cincinnati, Ohio; their journal *Medical Decision Making* appeared soon thereafter.

However, in illustrating the success story medical decision making methods had enjoyed, Lusted did not mention any single such computer system in his contribution for the proceedings of the Third Annual Symposium on Computer Application in Medical Care – not even the Vienna system of computer-assisted diagnosis, which had existed for ten years and had expressly been modeled in the tradition of Lusted and Ledley's article.[61]

The oldest vision of this Vienna system, the CADIAG (Computer-Assisted Diagnosis), resulted in 1968 from the cooperation by Walter Spindelberger[62] and Georg Grabner[63] at the Institute for Medical Computer Sciences at the

---

[58] [335] Lusted, Decision (1979), p. 6
[59] [339] Lusted, Introduction (1968)
[60] Lusted cited specifically [248] Houle, Bayesians (1979) and [132] Cutler, Problem (1979)
[61] [516] Spindelberger et al., Computerverfahren (1968)
[62] Walter Spindelberger, Austrian mathematician at IBM Austria
[63] Georg Grabner (1922–2006), Austrian doctor. At the time, Grabner was the chief physician at the Second University Clinic for Gastroenterology and Hepatology at Vienna General Hospital and a director of the Institute for Medical Computer Sciences at the medical school of the University of Vienna

University of Vienna's medical school.[64] The differential diagnosis of liver diseases presented itself as a first field of application.[65] The myriad possible relationships between medical entities (symptoms, diagnoses, diseases, patients) were represented by formulae from the calculus of first-order predicate logic.[66] Inspired by the theoretical model by Ledley and Lusted, which had of course been preceded by Lipkin and Hardy's suggestion to employ "mechanical apparatus" to aid in the differential diagnosis of blood disorders[67], a strictly logical computer program was created to support the physician in making differential diagnoses within a narrowly limited medical area predefined by the doctor.

7.4 Computer-Assisted Diagnosis at Vienna General Hospital

The Medical Computing Center was founded at the University of Vienna in 1967. A first computer system had been financed by the "Austrian People's Radio Donation – Fight Against Cancer – 1965" campaign as well as by the federal government[68]: An IBM 360/30 with a 16 kilobyte core memory was acquired and personnel positions created (Fig. 7.11).

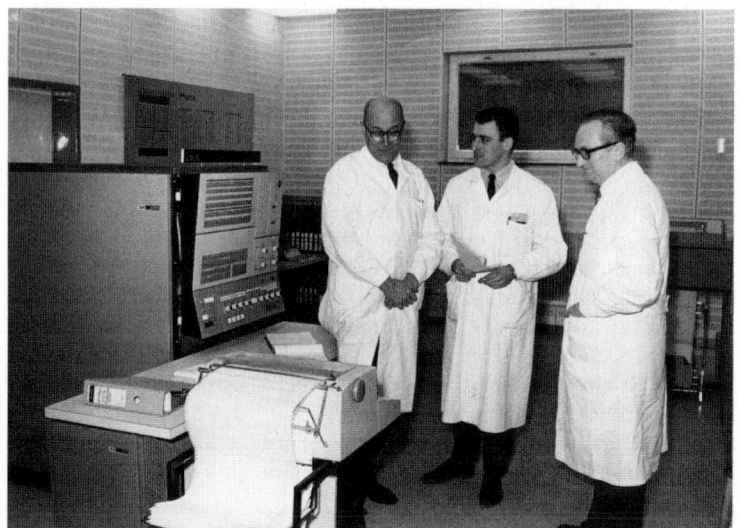

Fig. 7.9. Karl Fellinger, Peter Bauer, Georg Grabner (left to right)

---

[64] [516] Spindelberger et al., Computerverfahren (1968)
[65] [192] Gangl et al., Erfahrungen (1969)
[66] [337] Lusted, Electronics (1955), p. 580 (a review with 53 references), [314] Ledley, Methods (1954), p. 3, [311] Ledley, Digital (1955), p. 7
[67] [323] Lipkin et al., Correlation (1958)
[68] Karl Fellinger (1904–2000), Austrian physician, was a university professor in Vienna from 1945 and chancellor of Vienna University in 1964–65. From 1946 to

From the very start, the scientific work carried out at the Medical Computing Center was characterized by two distinctive features: On the one hand, highly sensitive patient-related information was being processed, and on the other hand, there was no model for this kind of procedure itself. For this reason, research and development projects were just as necessary as comprehensive organizational measures.[69]

By 1970, the system's core memory had been increased to 32 kilobytes, then to 64 kilobytes; in 1973, the computer system was replaced by a larger one. The increasing interest in medical informatics doctors and students were demonstrating and the need to carry out their own research in the Medical Computing Center resulted the same year in the founding of the Institute for Medical Computer Sciences, which received its own assignments in research and teaching from the medical school. It was led in dual role with the Second University Clinic for Gastroenterology and Hepatology. This clinic began the test phase of the research program WAMIS (Wiener Allgemeines Medizinisches Informationssystem), a medical database and information system.

Fig. 7.10. Karl Fellinger, Peter Bauer, Georg Grabner (left to right)

After a two-year trial run, WAMIS was made available in 1975 to other clinics via terminals.[70] WAMIS was designed modularly; it included *Operational components*:

1975, he was director of the Second Medical University Clinic and Georg Grabner's predecessor in this position

[69] [107] Bund, Rechenzentrum (1982), p. 9
[70] [107] Bund, Rechenzentrum (1982), see also [204] Grabner, WAMIS (1985)

7.4 Computer-Assisted Diagnosis at Vienna General Hospital

Fig. 7.11. IBM/2260 computer screen unit and IBM/1053 printer for the WAMIS system

- A medical documentation system recorded the entered medical data on the screen and the WIELAB laboratory system recorded laboratory data completely automatically.

Informational components:

- An *information* function could provide saved patient-related data, and a statistical evaluation system (WAMAS) was used for the scientific evaluation of the database.

Components from research projects:

- An instructing and teaching system was introduced for student training.
- Possibilities for improvement with the aid of computers were studied in the area of biosignal processing, especially in ultrasonography.
- A computer-assisted diagnosis system (CADIAG, Computer Assisted Diagnosis) was intended to suggest diagnoses based on the symptoms saved in the database.

The CADIAG module that was integrated into WAMIS had the following properties at the time:

- On the basis of medical knowledge, the system suggested all of the possible diagnoses that fit with given patterns of symptoms.
- All diagnoses were treated equally. The doctor could analyze the results and influence the computer's actions.

– Suggestions were made for further diagnostic steps, ordered by the efficiency with which they would confirm the diagnosis.
– The reasons for all diagnostic decisions were shown upon request.

The next sections will trace the development of this CADIAG system, above all its fuzzification.

7.4.1 CADIAG

In 1968, Georg Grabner and Walter Spindelberger of IBM Austria had first publicized the computer technique for diagnostic support which they had developed and which had been motivated by the work of Ledley and Lusted.[71] This technique was soon being applied in the differential diagnosis of liver diseases.[72] In this diagnosis system, a number – albeit a small number – of sensible relations between symptoms and illnesses were interpreted as operators in Boolean logic and were automatically processed using the PL/1 programming language. The Viennese scientists were tracking American efforts to "make diagnoses with the help of an electronic data processing system" as well as subsequent developments of a "series of computer methods" they had been lacking, since they were not compatible, "it is not possible to combine them and their application must in practice be followed by a differential diagnosis that outlines the specialist area".[73] They considered a second disadvantage of these methods to be the fact that, along with the individual diseases, considerations during diagnosis also had to include the probabilities of occurrence or even subjective weightings that had been formed on the basis of personal experience but that had nothing to do with the definition of the ailment, and methods such this therefore did not fulfill the demand for universality.

The new consultation system in Vienna was supposed to be an active partner to the doctor in the diagnostic process, to provide expert medical knowledge of internal medicine and to deduce and justify medical diagnoses logically based on patient symptom patterns it was given. In this way, the clinicians hoped to increase the quality of their diagnoses, provide quicker diagnoses and prevent false diagnoses.

In this system, many different possible relationships could be represented among symptoms, diagnoses, diseases and patients using formulae from the calculus of first-order predicate logic. Probability theory and statistics were intentionally and pointedly not used because:

> The aim of the program is not to establish a single probable diagnosis, but instead all of those diseases should be considered that are at all possible in conjunction with a particular constellation of symptoms.[74]

---

[71] [516] Spindelberger et al., Computerverfahren (1968).
[72] [192] Gangl et al., Erfahrungen (1969).
[73] [516] Spindelberger et al., Computerverfahren (1968), p. 189
[74] [192] Gangl et al., Erfahrungen (1969), p. 585

7.4 Computer-Assisted Diagnosis at Vienna General Hospital

Foregoing statistical methods and thus limiting themselves to logical methods points to a strategy that was maintained throughout the entire subsequent CADIAG line. Before this aspect is discussed, however, we should take a quick look at the technical equipment.

Hardware

The program of the first Vienna system ran on an IBM system/360 model 30 with a minimum configuration of a 32 kilobyte core memory and a harddisk unit. Peripheral devices included a card puncher and reader, a writing console, a high-speed printer and four magnetic tape units (Fig. 7.12). The present and absent symptoms and diagnoses were entered via punch cards. Additionally, the number obtained from the index of all symptoms and diagnoses was punched into the card, supplemented by a numeral 1 or 0, and entered into the computer by means of the punch card reader. The process then followed the program sections described below. Once the process was finished, the diagnoses that had been identified or eliminated could be printed out.

In order for all of the relationships representing the medical knowledge to be documented in a way the computer could process, a system for submitting the punch cards was designed in which the diagnoses and symptoms were entered (Fig. 7.13).

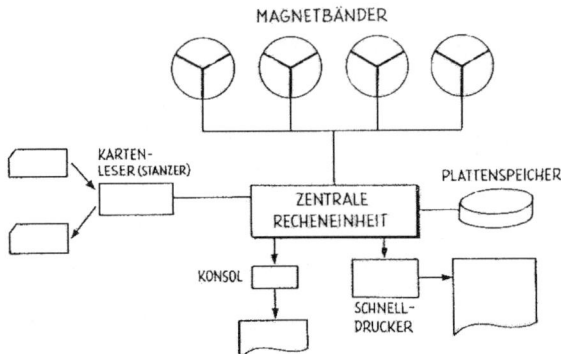

Fig. 7.12. Diagram of the computer for the WAMIS system. Kartenleser (Stanzer): card reader (puncher); Magnetbänder: magnetic tape; Zentrale Recheneinheit: mainframe computer; Konsol: console; Schnelldrucker: chain printer; Plattenspeicher: disk memory

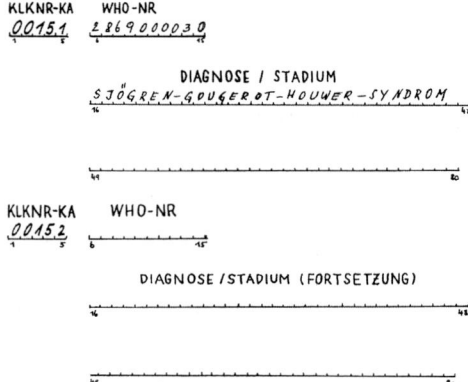

Fig. 7.13. Punch cards No. 1 and No. 2 for the Vienna diagnosis system

- Columns 1 to 5 on the card were reserved for information about the clinic and clinic type.
- Columns 6 to 15 contained the WHO's[75] diagnosis key[76] and additional supplemental numbers from the clinic.
- Columns 16 to 80 contained the diagnosis names and the stage, if more than one stage existed.

A second card was (Fig. 7.13) provided in case the entries on card 1 needed to be continued and space had run out. A third card (Fig. 7.14) and any possible cards after that contained as thorough information as possible on all of the symptoms associated with the clinical picture.

- Columns 1 to 15 were copies of those on the first card.
- Column 16 contained the page number of the form.
- Columns 17 to 18 contained the line number.
- Column 19 contained the code W, A, Z or was left empty.
- Columns 20 to 74 contained the text of the symptoms.
- Column 75 contained the codes 1, 2, 5, 6, 7 and 8.
- Columns 76 to 80 contained the categories in which the symptom was entered.

A single symptom was written in each line; in addition to the name of the symptom, the terms were entered in categories that had been marked at the edge (columns 76–80) of the form. These included:

---

[75] World Health Organisation
[76] The WHO key (WHO Key, 8th edition) was expanded to 10 digits for the internal use of the clinic. The 9th digit, for example, served to identify the degree of severity of the illnesses (from eliminated to mortal danger) and the 10th digit was reserved for terms such as "condition after ..." and "suicide by ..."

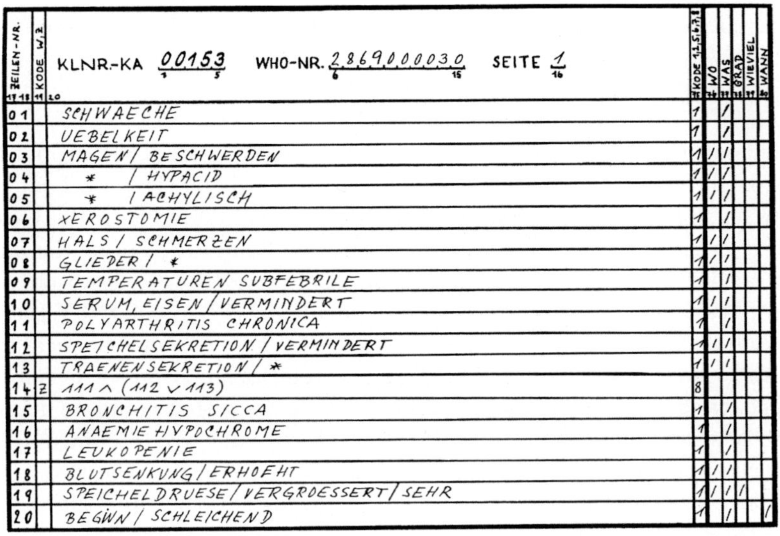

7.4 Computer-Assisted Diagnosis at Vienna General Hospital

ZEILEN-NR.	KODE W/2	KLNR-KA 0.0.1.5.3 WHO-NR. 2.8.6.9.0.0.0.0.3.0 SEITE 1	KODE 1,5,7,1	WO	WAS	GRAD	WIEVIEL	WANN
01		SCHWAECHE	1	/				
02		UEBELKEIT	1	/				
03		MAGEN / BESCHWERDEN	1	/	/			
04		* / HYPACID	1	/	/			
05		* / ACHYLISCH	1	/	/			
06		XEROSTOMIE	1	/				
07		HALS / SCHMERZEN	1	/	/			
08		GLIEDER / *	1	/	/			
09		TEMPERATUREN SUBFEBRILE	1	/				
10		SERUM, EISEN / VERMINDERT	1	/	/			
11		POLYARTHRITIS CHRONICA	1	/				
12		SPEICHELSEKRETION / VERMINDERT	1	/	/			
13		TRAENENSEKRETION / *	1	/	/			
14	2	111 ∧ (112 ∨ 113)	8					
15		BRONCHITIS SICCA	1	/				
16		ANAEMIE HYPOCHROME	1	/				
17		LEUKOPENIE	1	/				
18		BLUTSENKUNG / ERHOEHT	1	/	/			
19		SPEICHELDRUESE / VERGROESSERT / SEHR	1	/	/	/		
20		BEGWN / SCHLEICHEND	1	/				/

Fig. 7.14. Punch card No. 3 for the Vienna diagnosis system. Columns 76–80: Wo: Where; Was: What; Grad: Degree; Wieviel: How Much; Wann: When

WHERE: pathological changes;
WHAT: pathological changes (and diagnoses) that occur independently or in conjunction with category WHERE;
DEGREE: extent of the pathological change; degree of severity of illness; terms like "very", "moderate", "excessive" should be used here, not numbers!
HOW MUCH: numbers indicating sizes and frequencies;
WHEN: times of occurrence and durations of symptoms.

Column 19 is especially important because codes were employed here which indicated how each respective symptom was to be evaluated in light of its relationship to the diseases being considered. The following codes were selected arbitrarily and had become conventional:

Symptom is facultative and not proof of illness	Code 1
Symptom is facultative and proof of illness	Code 2
Symptom occurs obligatorily with this illness but does not prove it with certainty	Code 7
Symptom is obligatory for this illness and also proves it	Code 8

After these symptom data had been punched and scanned into the system, they were transferred to a magnetic tape and could then be resorted into any desired individual categories; the lists created in this way could pass through various correction procedures, and so a relatively error-free diagnosis symptom tape was then available.

Software

Spindelberger and Grabner first used their system to realize the propositional logic-based model of computer-assisted medical diagnosis that Ledley and Lusted had proposed ten years earlier. "It was made famous in the German-speaking countries by the Vienna internist Josef Schmid."[77] In it, the familiar operators for conjunction (AND, $\wedge$), disjunction (OR, $\vee$), negation (NON, $\neg$), implication (IF ... THEN, $\Rightarrow$) and equivalence (IF AND ONLY IF ..., $\Leftrightarrow$) represented the appropriate combinations of statements about medical knowledge and about information regarding the relationships between symptoms that occurred in patients, and corresponding diseases that could be diagnosed on the basis of these symptoms. Symptoms and diseases assumed the role of logical variables whose measurement is given by one of the two admissible truth values ("true" or "false", 0 or 1). Alongside the propositional logic operators of conjunction and disjunction, Spindelberger also provided their system with application-motivated combinations, which have the equivalents from propositional logic given alongside them:

Symptom S obligatory and proof of disease D $S \Leftrightarrow D$.
Symptom S facultative and proof of disease D $S \Rightarrow D$.
Symptom S obligatory and not proof of disease D $S \Leftarrow D$.
Symptom S rules out disease D $S \Rightarrow \neg D$.

The system with the predicate logic-based method described above soon found success as an application the fields of internal medicine, hepatology[78] and rheumatology.[79] The hepatological knowledge base consisted of 82 liver diseases, 323 symptoms, signs, laboratory tests and other information such as patient history, the patient's physical condition, results of biopsies, histologies, X-ray and other possibly specialized examinations. In 20 test cases, the computer system offered the clinically confirmed diagnosis at least as a diagnostic hypothesis.

7.4.2 CADIAG-I

The Computer Assisted DIAGnosis system in Vienna was changed dramatically in just its second year of existence, and the computer system was also given a new design and the name CADIAG-I. The limits of two-valued logic

---

[77] [490] Schneider, Grundlagen (1970), p. 161. In 1967, Dr. Josef Schmid was the first private doctor to put an electronic data processing system into operation, an IBM 1130
[78] [192] Gangl et al., Erfahrungen (1969)
[79] [36] Bauer et al., Computerdiagnostik (1968), [247] Horak et al., Computerdiagnostik (1968)

had also presaged the limits of representing "medical knowledge" for processing in a computer. Statements from classical propositional logic can only be either "true" or "false", and accordingly the first version of the Vienna computer-assisted diagnosis systems could characterize symptoms only as either "present" or "not present" and diagnoses only as either "applicable" or "not applicable". However, doctors and clinics were (and are) certainly not always able to collect all of the possible data or to examine the patient for all possible symptoms, and so in many cases the term "not examined" was noted in the evaluation of a symptom. Additionally, it was (and is) often appropriate during the diagnosis process not to immediately rule out a diagnosis or consider it absolutely certain but instead to say merely that it is "possible" because, in this way, it is possible to leave open whether the disease in question is actually present or not. It was thus plausible particularly in the medical-diagnostic context to contemplate using a three-valued logic in computer-assisted medical diagnosis, and in just the second year of the Vienna system, the underlying two-valued logical calculus was replaced by the three-valued logical calculus[80] Steven Kleene had introduced: In addition to the two values 1 and 0 for symptoms and diagnoses, meaning "present" and "not present", there was now also a third value 1/2 for "possible" or "not examined". A small table demonstrates this:

Logical Value	Symptom S	Diagnosis D
0	Not present	Not present (eliminated)
1	Present	Present (proven)
1/2	Not examined	Possible (hypothesis)

Of course, the CADIAG-I relations can also be represented by first-order predicate logic or by IF-THEN rules. Relations can exist between individual symptoms; here the symptom-disease relations OB, FB, ON and A are defined. Similarly, relations OB, FB, ON and A can also exist between the diseases.

OB   **obligatory and proof**
If the relation OB exists between a symptom S and a disease D, then the symptom S must occur with this disease and is proof of this disease D. Conversely, if a symptom S is not present, disease D can be eliminated.

Propositional logic operator:
Equivalence $S \Leftrightarrow D$.

D \ S	0	1	1/2
0	1	0	1/2
1	0	1	1/2
1/2	1/2	1/2	1/2

---
[80] For more on Steven Kleene's three-valued logic, see Chaps. 4 and 5

FB facultative and proof

If the relation FB exists between a symptom S and a disease D, then the symptom S does not necessarily occur with this disease, but if it does, then it is proof of this disease D.

Propositional logic operator:
Implication $S \Rightarrow D$.

S \ D	0	1	1/2
0	1	1	1
1	0	1	1/2
1/2	1/2	1	1/2

ON obligatory and not proof

Symptom S must occur with disease D, but it does not prove it. Conversely, D is eliminated when symptom S is absent.

Propositional logic interpretation:
$S \Leftarrow D$ or $\neg S \Rightarrow \neg D$.

S \ D	0	1	1/2
0	1	0	1/2
1	1	1	1
1/2	1	1/2	1/2

FN facultative and not proof

Symptom S does not have to occur with disease D; when it does occur, it does not prove disease D. Therefore, the strength of this relationship cannot be defined with certainty; this relation is thus selected relatively often in cases of uncertainty.

Propositional logic interpretation:
$(S \Rightarrow D)$ or $(S \Rightarrow \neg D)$.

S \ D	0	1	1/2
0	1	1	1
1	1	1	1
2	1	1	1

A eliminated

If symptom S is present, then disease D is eliminated.

Propositional logic interpretation:
$S \Rightarrow \neg D$ or $\neg(S \wedge \neg D)$.

S \ D	0	1	1/2
0	1	1	1
1	1	0	1/2
1/2	1	1/2	1/2

H   **indication**
Symptom S indicates a secondary diagnosis.

Propositional logic interpretation:
$S \wedge D_1 \Rightarrow D_1 \vee D_2 \vee D_3 \vee \ldots \vee D_N$.

D \ S	0	1	1/2
0	1	0	1/2
1	0	1	1/2
1/2	1/2	1/2	1/2

'_'   **unknown or unspecific**
There is either no known relationship or only an unspecific relationship between symptom S and disease D.

7.4.3 From CADIAG-I to CADIAG-II

About a decade after the advent of CADIAG-I, the system once again underwent significant changes, which also justified a new version number.[81] The new CADIAG-II was characterized by expansions undertaken by Klaus-Peter Adlassnig[82], who had gone to the Institute for Medical Computer Sciences in Vienna in September 1976 and was very soon engaged with the work done on computer-assisted diagnosis by Peter Bauer[83], Walter Spindelberger and Georg Grabner. He wanted to write a dissertation on consistency checks in three-valued logic.[84]

In conjunction with the WAMIS, expectations of scientific and practical success with this system were very high and within the circle of doctors at Vienna General Hospital they were eager to discover whether computer assistance would actually be able to improve conditions such that

---

[81] The first CADIAG system originally had no version number at all, though

[82] Klaus-Peter Adlassnig (born 1950), Austrian computer scientist. Adlassnig studied computer sciences from 1970 to 1974 at the Technical University in Dresden, where he graduated in 1974. From 1977 to 1983, he studied computer sciences at the Technical University in Vienna, where he received a doctorate in 1983. Today he is a professor at the Medical University of Vienna in the Core Unit for Medical Statistics and Informatics, previously known as the Institute for Medical Computer Sciences at the medical school of the University of Vienna

[83] Peter Bauer (born 1942), Austrian mathematician and statistician, was an instructor at the Institute for Medical Informatics at the University of Vienna from 1967 to 1970, and from 1970 to 1985 an instructor at the Institute for Medical Statistics and Documentation there. From 1985 to 1993, he was head of the Institute for Medical Documentation and Statistics at the University of Cologne and since 1994 has been the director of the Institute for Medical Statistics at the medical school of the University of Vienna, which has been part of the Medical University of Vienna since 2002

[84] [714] R. S. interview with K.-P. Adlassnig (2000)

- the quality of diagnostics was increased,
- false diagnoses were avoided,
- diagnoses were made more quickly.

In October of 1976, Grabner suggested that Adlassnig offer a presentation on Vienna's CADIAG diagnosis system for the Second World Conference on Medical Informatics (MEDINFO '77), which took place from August 8 to 12, 1977 in Toronto, Canada.[85] The English translation of Adlassnig's text was accepted for this congress[86], and so in August he delivered a lecture at the conference in Toronto on the subject of this computer-assisted Vienna diagnosis system. In another session of the conference, two papers were presented that also addressed fuzzy methods in medical diagnostics: In *Bayesian Probability of Fuzzy Diagnosis*, Philippe Smets[87], H. Vainel, R. Berbard and F. Kornreich emphasized the imprecision of diagnostic terms:

> As an example consider a group of patients suffering from arteriosclerosis: can we really decide for every patient whether or not he belongs to that set, despite knowing everything about him? One approach consists in deciding some supposedly well-defined borders between the initially poorly defined set and its complement. But this is often a procrustean approach dictated more by mathematical conveniences and open to strong criticism as these borders are often arbitrary, oversimplified and far from generally accepted.[88]

So just as the sets of the *large* numbers or the *tall* men – examples Lotfi Zadeh had used in his text – could sensibly be interpreted as fuzzy sets, the same could also be said for "the set of patients suffering from arteriosclerosis". To cite another example: In the statement "angina pectoris is usually related to arteriosclerosis and observed among old and obese patients", four fuzzy sets are mentioned – *angina pectoris*, *arteriosclerosis*, *old people* and *obese people* – which a practitioner could handle or treat very well.

In their paper *Medical Diagnostic System with Human-Like Reasoning Capability*, Richard Moon, S. Jordanov, Alonso Perez-Ojeda and I. Burhan Türksen[89] also referred to the fact that a great potential for application existed for Fuzzy Set Theory in medical diagnostics.[90] Adlassnig missed these two presentations in Toronto but he soon read both papers in the proceedings. These were likely the first texts about Fuzzy Set Theory Adlassnig had

---

[85] [714] R. S. interview with K.-P. Adlassnig (2000)
[86] [4] Adlassnig et al., WAMIS (1977)
[87] Philippe Smets was a professor at the Université Libre de Bruxelles (USB) and from 1985 to 1999 the director there at the Institut de Recherches Interdisciplinaires et de Développements en Intelligence Artificielle (IRIDIA). See Fig. 6.8
[88] [515] Smets et al., Diagnosis (1977)
[89] I. Burhan Türksen (born 1937) has been director of the Knowledge/Intelligence Systems Laboratory in the department of mechanical and industrial engineers at the University of Toronto since 1987. See Fig. 7.19
[90] [406] Moon et al., Diagnostic (1977)

encountered, since even when he was a student in East Germany he had not acquired any knowledge of the "theory of unsharp amounts" in lectures or other settings.[91] Almost 25 years later, Adlassnig recalled at least having read the text by Smets relatively soon.[92] Adlassnig's intensive study of Fuzzy Set Theory began with the publication of the review article *The Process of Medical Diagnosis: Routes of Mathematical Investigations* by Petre Tautu and Gustav Wagner in *Methods of Information in Medicine* the following January (Wagner being the editor-in-chief of this journal) as well as the bibliography *Problems of Medical Diagnosis* by Gustav Wagner, Petre Tautu and U. Wolber in the same issue, and this was followed soon thereafter by his first research and development projects on applying Zadeh's theory in medical informatics. He recognized in Fuzzy Set Theory "a concept that takes into account the great complexity and imprecision of definitions in these areas".[93] He assessed the situation in his new domain as follows:

> In medical diagnostics, "there are mostly no crisp boundaries between the individual diseases, the occurrence of several diseases in a patient blurs the symptom profile and complicates diagnostic and therapeutic decision making, classifying the acquired diagnostic findings as 'normal' or 'pathological' is often arbitrary in borderline cases, describing the intensity of pain can only be done verbally and thus depends on the subjective evaluation of the patient and descriptions of illnesses can only very seldom include exact correlations between symptoms and diseases."[94]

Fig. 7.15. K.-P. Adlassnig

Adlassnig saw a fundamental difference between the high degree of precision in the exact natural and engineering sciences on the one hand and

---

[91] [714] R. S. interview with K.-P. Adlassnig (2000). In the GDR, fuzzy sets were referred to as "unsharp amounts"
[92] [714] R. S. interview with K.-P. Adlassnig (2000)
[93] [10] Adlassnig, Modell (1982), p. 12
[94] [10] Adlassnig, Modell (1982), p. 12

the"inexactness on the other hand which is found in descriptions from sociology, psychology, medicine, linguistics, literature, art, philosophy and so on".[95] Motivated by the comprehensive literature research on fuzzy systems that can be used in the medical field, Adlassnig came to the decision to fuzzify the Vienna CADIAG-System system:

> Fuzzy Set Theory with its capability of defining inexact medical entities as fuzzy sets, with its linguistic approach providing an excellent approximation to medical texts as well as its power of approximate reasoning, seems to be perfectly appropriate for designing and developing computer assisted diagnostic, prognostic and treatment recommendation systems.[96]

For the new version of the system, he pursued the following goals:

- Storing medical knowledge in the form of logical relationships between symptoms and diseases, symptoms among one another and diseases among one another.
- The logical relationships do not need to satisfy the two- or three-valued logical calculus; they "can be "fuzzy" ".[97]
- Several diagnoses, including rare diagnoses, are offered for each symptom pattern presented.
- The diagnosis process occurs iteratively.
- Reasons are given for diagnostic decisions.

Before an exact description of this system is provided, however, we should take a look at the beginnings of the medical applications of Fuzzy Set Theory.

7.5 Fuzziness in Medicine

Lotfi Zadeh had had absolutely no idea that there would end up being so many applications of fuzzy sets, fuzzy logic and fuzzy algorithms – in short, of fuzzy systems in the technical industry. The conclusion of Chap. 5 notes that he had suspected that his new theory could lead to new developments in the humanities, such as in linguistics, philosophy or economics. As has already been shown, he had argued at first that fuzzy systems were predestined to represent living systems in the life sciences, although in his first journal publication on *Fuzzy Sets* in 1965 he mentioned only "pattern recognition, communication of information, and abstraction" as possible fields of application.[98] Four years later, he explicitly suggested using his new math in biology because classical mathematics might be inadequate: "... the possibility that

---

[95] [10] Adlassnig, Modell (1982), p. 12
[96] [11] Adlassnig, Survey (1982), p. 205
[97] [10] Adlassnig, Modell (1982), p. 13
[98] [612] Zadeh, Fuzzy (1965)

classical mathematics – with its insistence on rigor and precision – may never be able to provide totally satisfying answers to the basic questions relating to the behavior of animate systems".[99] Living organisms were many times more complex than inanimate systems constructed or observed by humans, he argued here, and in doing so linked seamlessly to sentences he had written before he had even conceived of his Fuzzy Set Theory. That he believed a "new mathematics" necessary in order to describe living systems and to examine them scientifically is evident – as has already been stated[100] – in *From Circuit Theory to System Theory* from 1962:

> ... we need a radically different kind of mathematics, the mathematics of fuzzy or cloudy quantities which are not describable in terms of probability distributions.

This "new mathematics" was intended to differ appropriately from its "conventional" sister discipline, "the mathematics of precisely-defined points, functions, sets, probability measures, etc." Just what this would look like, though, Zadeh had not yet figured out. He did, however, describe all the more clearly the settings in which the fuzzy sets could be used: "... for coping with the analysis of biological systems, and that to deal effectively with such systems, which are generally orders of magnitude more complex than man-made systems".[101]

When Zadeh presented on the subject of *A New View of System Theory* in Brooklyn in April 1965, his article about fuzzy sets was in print. Zadeh's "new view of system theory" consisted in the generalization of systems to "fuzzy systems": "S is a fuzzy system, if $u(t)$ or $y(t)$ or $x(t)$ or any combination of them ranges over fuzzy sets".[102]

Fuzziness was the price to pay for the impracticality of exact mathematics when systems are to be analyzed that consist of a large number of interactive elements or in which a large number of variables must be taken into account. His theory of the "so-called fuzzy sets and systems", however, was well-suited to deal with such quantitative or quasi-quantitative fuzziness.[103]

Yet by 1969, Zadeh found that describing biological systems was not the most promising possible field of application for his new mathematical theory. Instead, he was clearly thinking about the field of medical diagnosis.[104] It was here that possibly the most natural applications of his fuzzy concepts lay, and his statements from that time point in the direction of later developments:

– Specifically, from the point of view of fuzzy set theory, a human disease, e. g., diabetes, may be regarded as a fuzzy set in the following sense. Let $X = x$ denote the collection of human beings. Then diabetes is a fuzzy

---

[99] [594] Zadeh, Biological (1969), p. 199
[100] See also Chap. 1 and Chap. 4
[101] [662] Zadeh, System (1962)
[102] [664] Zadeh, Systems (1965), p. 33. See also Chap. 5
[103] [594] Zadeh, Biological (1969), p. 200
[104] [594] Zadeh, Biological (1969), p. 200

set, say D, in X, characterized by a membership function $\mu_D(x)$ which associates with each human being x his grade of membership in the fuzzy set of diabetes.[105]

– In some cases, it may be more convenient to characterize a fuzzy set representing a disease not by its membership function but by its relation to various symptoms which in themselves are fuzzy in nature. For example, in the case of diabetes a fuzzy symptom may be, say, a hardening of the arteries. If this fuzzy set in X is denoted by A, then we can speak of the fuzzy inclusion relation between D and A and assign a number in the interval $[0,1]$ to represent the "degree of containment" of A in D. In this way, we can provide a partial characterization of D by specifying the degrees of containment of various fuzzy symptoms $A_1, ..., A_k$ in D. When arranged in a tabular form, the degrees of containment constitute what might be called a *containment table*.[106]

Apparently unaware of Zadeh's suggestions of these applications of Fuzzy Set Theory in medical diagnosis, Merle Anne Albin wrote a dissertation on this subject in 1975 under Berkeley mathematics professor Hans-J. Bremermann, a friend of Zadeh's. The doctoral committee, of which Zadeh was also a member, accepted Albin's work entitled *Fuzzy Sets and Their Applications to Medical Diagnosis and Pattern Recognition* on June 14, 1975; its bibliography does not include any of the texts by Zadeh cited above and it begins with the statement that both the mathematical theory of medical diagnosis and the theory of fuzzy sets were very new areas, which had gained most of their impetus in the preceding decade. Albin wrote: "It is not too surprising, then, that the application of fuzzy sets to medical diagnosis has never before been seriously attempted."[107]

Zadeh's aforementioned works with comments about possible applications of fuzzy sets in medicine likewise went unnoticed both by Harry Wechsler in his 1976 article *Applications of Fuzzy Logic to Medical Diagnosis*[108] and by Alonso Perez-Ojeda in his M.S. thesis *Medical Knowledge Network. A Database for Computer Aided Diagnosis*[109], which was submitted in Toronto in 1976 and which led to the MEDINFO '77 paper that he had presented with Moon, Jordanov and Türksen. Petre Tautu and G. Wagner, with their aforementioned review article[110], and Richard C. Elder and Augustine O. Esogbue also failed to mention them in the two parts of their paper[111] from 1979 and 1980 about a fuzzy model for medical decision making processes, which traced back to Elder's M.S. thesis *Fuzzy Systems Theory and Medical Decision Making*.[112]

---

[105] [594] Zadeh, Biological (1969), p. 203
[106] [594] Zadeh, Biological (1969), p. 205
[107] [17] Albin, Fuzzy (1975)
[108] [549] Wechsler, Fuzzy (1975)
[109] [442] Perez-Ojeda, Knowledge (1976)
[110] [532] Tautu et al., Diagnosis (1978)
[111] [165] Esogbue et al., Fuzzy I (1979/1980), [166] Esogbue et al., Fuzzy II (1980)
[112] [161] Elder, Fuzzy (1976)

All of these works dealt with the application potential of Fuzzy Set Theory in the medical field without making any mention of Zadeh's explicit comments on this subject; instead, they reinvented the wheel independently of one another:

– Medical laboratory findings that were classified either as *normal* or as *pathological* could be fuzzified (Fig. 7.16, 7.17); a continuous transition, for instance, from *shortened* to *normal* to *prolonged* bleeding times or from *slightly elevated*
to *severely elevated* cholesterol level corresponded more to a medical mindset. Using membership functions, values were established to state the degree to which the test results belonged to the respective fuzzy sets.[113] Thus Moon, Jordanov, Perez-Ojeda and Türksen[114] had suggested, as had Esogbue and Elder[115] two years later, that the painfulness or severity of symptoms such as headaches or cyanosis could be represented by membership functions or by the degree of abnormality of a clinical or diagnostic test result

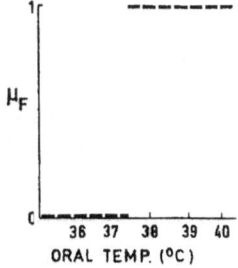

Fig. 7.16. Classical indicator function for "fever"

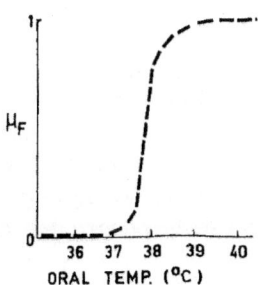

Fig. 7.17. Membership function for "fever"

---

[113] [17] Albin, Fuzzy (1975), [532] Tautu et al., Diagnosis (1978), [165] Esogbue et al., Fuzzy I (1979/1980), [166] Esogbue et al., Fuzzy II (1980), [406] Moon et al., Diagnostic (1977)
[114] [406] Moon et al., Diagnostic (1977).
[115] [165] Esogbue et al., Fuzzy I (1979/1980)

- Esogbue and Elder had proposed a nearly linear membership function for a fuzzy set *abnormal cholesterol C*, expressed in mg/100 ml of serum:

$$\mu_c(x) = \begin{cases} 0, & \text{for} \quad x < 260 \\ \frac{x}{340} - \frac{26}{34}, & \text{for} \quad 260 \leq x \leq 600 \\ 1, & \text{for} \quad x > 600 \end{cases}$$

- Moon, Jordanov, Perez-Ojeda and Türksen[116] used the modifiers[117] Zadeh had introduced to calculate the degree of membership of a test result in a fuzzy set S_2 of the degree of membership of the test result in fuzzy set S_1. In this way, for example, the membership value of the result x from a urine sodium concentration test to the fuzzy set (*very high urine sodium concentration*, $\mu_{S_2}(x)$) can be obtained from the membership value to the fuzzy set high urine sodium concentration $\mu_{S_1}(x)$ by the following modification:

$$\mu_{S_2}(x) = \mu_{S_1}(x)^2.$$

Another example was the modification of the fuzzy set "low blood pressure", as illustrated in Fig. 7.18.
- Esogbue and Elder also used fuzzy sets to enable them to insert missing or misdiagnosed diseases into a patient's medical history.
- Smets and his co-authors[118] also emphasized the fuzziness of diagnostic terms such as *arteriosclerosis* or *angina pectoris*. Such diseases are not clearly or sharply defined, which is why it was often not possible to determine precisely the symptoms that clearly stand for a disease. However,

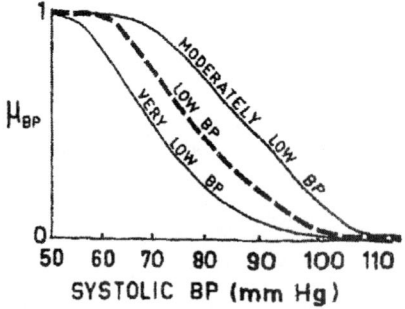

Fig. 7.18. Modifiers for "low blood pressure",(BP: blood pressure)

---

[116] [406] Moon et al., Diagnostic (1977)
[117] See also Chap. 6
[118] [515] Smets et al., Diagnosis (1977)

diagnoses could also be defined as fuzzy sets whose elements are symptoms. These are assigned a membership value that indicates the intensity with which the symptom belongs to the fuzzy set representing the disease in question.
- Moon, Jordanov, Perez-Ojeda and Türksen[119] had also attempted – though with little success – to represent symptom combinations by means of AND and OR conjunctions of fuzzy sets. A short time later Elie Sanchez, working in France, chose for this purpose the concept of the fuzzy relation $R \subset S \times D$ between the symptom set S and the diagnosis set D. In doing so, he assumed that a doctor translates his knowledge and his experience into *degrees of association* between symptoms and diagnoses.

This suggestion by Elie Sanchez resulted in the successful application of Fuzzy Set Theory in the field of medical diagnosis. It should therefore be examined more closely.

Fig. 7.19. Attendees of the IIZUKA-88 International Video-Session, KIT-NASA at the Kyushu Institute of Technology in Iizuka, Japan, 1988. Left to right: First row: I. B. Türksen, T. Teramo, L. Kocy, unknown, B. Kosko, unknown, T. Yamakawa, unknown, M. M. Gupta, unknown, H. J. Zimmermann; upper row: unknown, unknown, unknown, E. Sanchez, R. R. Yager, K. Azai, M. Sugeno

7.5.1 "Medical Knowledge"

Perez-Ojeda had already suggested in his Master's thesis at the University of Toronto in 1976 that medical knowledge could be represented as a network in which symptoms and diseases were linked to one another by relations.[120] He

---
[119] [406] Moon et al., Diagnostic (1977)
[120] [442] Perez-Ojeda, Knowledge (1976)

thus tied into the works by AI researchers like Marvin Minsky[121] and M. Ross Quillian[122], who for several years had been favoring "semantic networks" for graphic representations of the structure and storage of human knowledge.

The semantic network of medical knowledge Perez-Ojeda constructed was to be formed from various types of nodes:

- Disease complex nodes
- Disease nodes
- Symptom complex nodes
- Symptom nodes
- Statement nodes (relations)
- Laboratory test nodes.

The "network of medical knowledge" (Fig. 7.21) he had in mind could then be constructed graphically by elementary nodes and edges: Symptoms and diseases were symbolized by nodes, while the edges lying between them as the fibers of the network are created by logical relations between the symptoms and diseases; examples of typical elements of "medical knowledge" included (Fig. 7.20)[123]

- "Acute Pyelonephritis *usually* presents bladder irritation and infection."
- "Acute Pyelonephritis presents *occasionally* fever, or chills, and malaise."
- "A runny nose is *almost always* present in a common cold."

The diseases "acute pyelonephritis" and "common cold" he abbreviated with D_1 and D_2 and the symptoms "runny nose", "fever", "bladder irritation" as well as "infection", "chills" and "malaise" with S_1 - S_6.

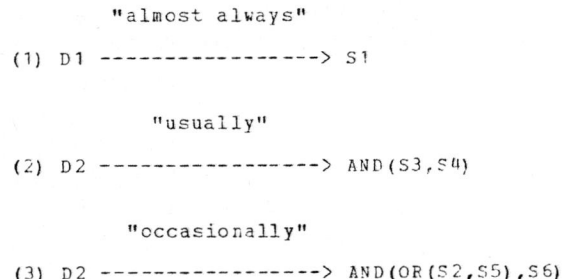

Fig. 7.20. Relationships between symptoms and diagnoses according to Perez-Ojeda

---

[121] See e. g., [405] Minsky, Semantic (1968)
[122] M. Ross Quillian, American cognitive psychologist. Quillian is a professor in the department of political science at the University of California. [455] Quillian, Memory (1968), [454] Quillian, Language (1969)
[123] [442] Perez-Ojeda, Knowledge (1976), 3.2

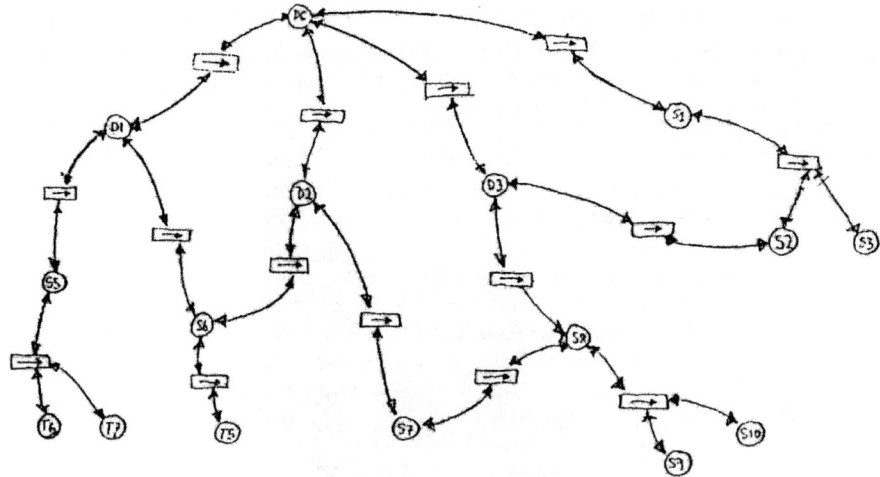

Fig. 7.21. Semantic network of symptoms (S_i) and diseases (D_i), disease complexes (DC) and statements (T_i) according to Perez-Ojeda

In order to model the "relation strength" (in the example "usually", "occasionally" and "almost always") mathematically, he identified them with probability modifiers interpreted using frequency theory.

Recording such relationships between symptoms and diseases mathematically was thus suggested by Elie Sanchez[124], who had studied at the Laboratoire de Biomathématiques, Statistique et Informatique Médicale of the Faculté de Médecine in Marseille in the 1970s, where he submitted a Master's thesis in human biology in 1974 entitled *Equations de Relations Floues*.[125] He later expanded the theory of fuzzy relations in publications including the journal *Information and Control*, where Zadeh was an editor, and he expected interesting results from the application of fuzzy relations "in transportation problems and in belief systems".[126]

He himself pursued a different direction: "We plan to investigate *medical*[127] aspects of fuzzy relations at some future time."[128]

The shape this plan would take can be gleaned from a volume about advances in Fuzzy Set Theory and its applications, which was published in 1979 by

---

[124] Elie Sanchez (born 1944) earned the Ph.D. in mathematics and in human biology at the University of Méditeranée (University of Aix-Marseille II). Today he is a professor at the Institut Méditeranéen de Technologie in Marseille. See Fig. 7.19
[125] [481] Sanchez, Equations (1974)
[126] [483] Sanchez, Resolution (1976), p. 47
[127] Italic emphasis not in original (R. S.)
[128] [483] Sanchez, Resolution (1976), p. 47

Madan M. Gupta, Rommohan K. Ragade and Ronald R. Yager.[129] For this anthology he had written two closely related papers in which he demonstrated how the max-min composition rules Zadeh had introduced could be used as a rule of inference, in particular in medical diagnostics. *Compositions of Fuzzy Relations*[130] is the first text, an abstract mathematical foundation for the second: *Medical Diagnosis and Composite Fuzzy Relations.*[131]

Fig. 7.22. E. Sanchez

In this second article, which was oriented to medical applications, Sanchez referred to the fact that medical diagnoses often had to be made without any precise analysis being possible. One or more illnesses then had to be inferred from a patient's symptoms, which most often be cannot be described in any exact way. In so doing, neither the set of diseases taken into consideration nor the conclusion about the disease(s) drawn from the symptoms can be precise. Likely without any knowledge of Perez-Ojeda's work, Sanchez, who had meanwhile joined the department of biomathematics and informatics of the medical school in Marseille, introduced the relationships between the set of symptoms and the set of diseases as fuzzy relations; these fuzzy relations, he found, represented the corpus of medical knowledge.

> In a given pathology, we denote by S a set of symptoms, D a set of diagnosis and P a set of patients. What we call "medical knowledge" is a fuzzy relation, generally denoted by R, from S to D expressing associations between symptoms, or syndromes, and diagnosis, or groups of diagnosis.[132]

The important new idea in Sanchez's work was his suggestion to use Zadeh's max-min composition rule as an inference rule to develop diagnoses. Given symptom and diagnosis sets S and D and an existing fuzzy relation

---

[129] [218] Gupta et al., Fuzzy (1979).
[130] [479] Sanchez, Compositions (1979).
[131] [480] Sanchez, Diagnosis (1979).
[132] [480] Sanchez, Diagnosis (1979), p. 438

$R \subset S \times D$ between them, the max-min composition can serve as an "inference rule", which makes it possible to deduce imprecise descriptions of a patient's illnesses (fuzzy sets of D) from imprecise symptom descriptions (fuzzy sets of S). With this inference rule, medical diagnoses D_j about a patient's disease can be derived by fuzzy logic from symptoms S_i with the help of the medical knowledge represented by the fuzzy relation R. The membership function is then computed as follows:

$$\mu_{D_i}(d) = \max_{s \in S} \min\{\mu_{S_i}(s); \mu_R(s,d)\}, \qquad \text{where} \quad s \in S, \ d \in D.$$

By taking into account a set P of all patients considered and a fuzzy relation Q between P and the symptom set S, it was now possible with the aid of the max-min composition rule to obtain a fuzzy relation $T = Q * R$ with the membership function $T(p,d)$:

$$\mu_T(p,d) = \max_{s \in S} \min\{\mu_Q(p,s); \mu_R(s,d)\}, \qquad s \in S, \ d \in D, \ p \in P.$$

The membership function of the fuzzy relation R is denoted with $\mu_R(s,d)$. The fuzzy relation R can be expressed as a matrix, the entries of which can be made after interviewing doctors about their diagnostic experiences. This expert medical knowledge must additionally be translated into *degrees of association* between symptoms and diagnoses.

Sanchez interpreted this equation in this way: If the condition of a patient p is described with the help of a fuzzy set A of symptoms from S, then diagnoses from D can be associated with this patient p with the help of a fuzzy set B, specifically by means of the fuzzy relation R between S and D. Given fuzzy subsets A of S and B of D, the max-min composition $B = A \circ R$ describes the condition of the patient with respect to the symptoms he is experiencing and the diseases from which he may be suffering. The membership function below defined the fuzzy subset B in D.

$$\mu_B(d) = \max_{s \in S}(\mu_A(s), \mu_R(s,d)), \qquad d \in D.$$

Simultaneously studying an entire set P of patients p led Sanchez to the definition of the fuzzy relation $Q \subset P \times S$ to characterize the relationship between these patients and their possible symptoms. Finally, the newly composed fuzzy relation T on $P \times D$ can be composed from the fuzzy relations Q and R: $T = Q \circ R$ with the membership function

$$\mu_T(p,d) = \max_{s \in S} \min(\mu_Q(p,s), \mu_R(s,d)), \quad s \in S, \ d \in D, \ p \in P.$$

Zadeh had devised this max-min rule in 1965 as a composition rule for fuzzy relations.[133] Assilian and Mamdani used it in 1972 to calculate inference rule relationships once they had implemented the fuzzy IF-THEN rules for

---

[133] See also Chap. 5

their fuzzy control algorithm as fuzzy relations.[134] Sanchez now interpreted it directly as a "fuzzy inference rule":

$$B = A \circ R : \qquad \text{IF } A \text{ Then } B \text{ by } R.$$

Using this inference rule, it is possible to logically infer medical diagnoses B of a patient's ailment from symptoms A with the aid of the "medical knowledge" represented as fuzzy relation R. The fuzzy relation can be depicted as a matrix; the values entered into this matrix must be provided by medical experts.

In 1980, Joly, Sanchez, Gouvernet and Valty had employed the max-min composition to diagnose diseases of the heart. The relation matrix R for this purpose was established by cardiologists who in this way established associations between five cardiological symptoms and three diagnoses, namely *normal, left ventricular hypertrophy* and *valvular cardiopathy*, by degrees of association. Test runs were carried out on 21 patients.

A year earlier, Sanchez had turned the problem around: Instead of calculating the fuzzy relation $T = Q \circ R$ from the available fuzzy relations Q and R, it was much more interesting and closer to clinical practice, at least in terms of diagnostics, to determine the fuzzy relation R from the known fuzzy relations T and Q. This would mean, of course, acquiring "medical knowledge" on the basis of knowledge about symptoms and diseases (clinical discharge diagnosis) from a patient set, in other words, to discover the relationships between

Fig. 7.23. L. A. Zadeh and E. Sanchez at a dinner at the IEEE Conference on Decision and Control, New Orleans, Louisiana, December 7–9, 1977

---

[134] See also Chap. 6

symptoms and diseases. In his two (interrelated) texts in 1979, Sanchez then highlights how it is possible to determine the largest relation matrix R for which: $T = Q \circ R$.[135] In these publications, Sanchez also examined the remaining possibility, that of calculating the fuzzy relation Q from a given T and R. His question was this: To what range of variation can a patient's symptoms vary without the patient dropping out of the diagnosis cluster for the respective disease? This meant determining the largest relation matrix Q for which $T = Q \circ R$.

7.5.2 CADIAG-II

In the field of medical diagnostics, vaguely formulated statements are much more common than statements that are certainly true or certainly not true. A "normal" finding can scarcely be differentiated from a "pathological" finding with any certainty, and the doctor very often hears the patient say things that are accompanied by words like "kind of", "almost", "very", "less". The doctor himself also uses expressions such as these when he must describe clinical pictures where the results of examinations do not provide conclusive evidence of a disease or any certain exclusion criteria. The three-valued logic used in CADIAG-I still appeared to be too strict, since tendencies can also often be quite incisive such that, in both the area of symptoms and diagnoses, a lot more can be said than simply "possible" or "not examined". Words like "more", "very", "somewhat" are often crucial in determining the doctor's next move.

CADIAG-II was now supposed to assist in proving, excluding, deriving and substantiating diagnoses or offer suggestions for further examinations based on the symptoms present and absent in a patient. Characterizing features of CADIAG-II were above all the fuzzy relation methods applied in order to represent the relationship between the medical entities, since in addition to definitional, statistical and experience-based knowledge, the knowledge base of CADIAG-II also contained medical-diagnostic knowledge that was represented by fuzzy relations, namely:

- between symptoms and diseases (S_i, D_j),
- between symptoms (S_i, S_j),
- between diseases (D_i, D_j) and
- between combinations of symptoms and diseases (SC_i, D_j).

To be able to use these fuzzy relations for diagnoses, Adlassnig was particularly interested in the assessments relating to the *occurrence* of each respective symptom and its *strength of confirmation* for the corresponding diseases. He measured the first aspect by the *frequency* with which the symptom had heretofore occurred in patients who had suffered from the disease. The second aspect characterized the "logical strength" with which it could be concluded that the disease was occurring. Adlassnig defined the variables

---
[135] [479] Sanchez, Compositions (1979), [480] Sanchez, Diagnosis (1979)

- X_1: the $S_i - D_j$ occurrence (the occurrence of symptom S_i with disease D_j) and
- X_2: the $S_i - D_j$ strength of confirmation (the strength of confirmation for the symptom S_i for disease D_j) $(i = 1, ...M, j = 1, ..., N)$.

Naturally, the two variables assume precise values taken from the universes U_1 and U_2. Adlassnig gave the following examples: U_1 and U_2 are each the interval [0, 100]. In the first case, this means the presence of between zero and one hundred times per 100 possibilities and in the case of U_2 a strength of confirmation of between 0 and 100, where 0 indicates the smallest possible and 100 the greatest possible conclusiveness of proof.

Adlassnig saw two principal ways of determining the membership values: Medical databases could be evaluated, for example, by enumerating the occurrences and the strengths of confirmation or by asking the doctors. Adlassnig chose the "linguistic documentation by medical experts" and questioned physicians whose answers he arranged in a schema of previously defined linguistic concepts and then considered as fuzzy subsets. For the following questions:

- *How often does symptom S_i occur with disease D_j?*
  (occurrence A)
  and
- *How certainly does S_i prove disease D_j?*
  (strength of confirmation B)

ten possible answers were provided: *always, almost always, very often, often, unspecific, seldom, very seldom, almost never, never, unknown*.

After this process, the following relationships from CADIAG-I remained:
- occurrence: "always" corresponds to relationship "obligatory"
- strength of confirmation: "always" corresponds to relationship "proof"
- occurrence   "never" corresponds to relationship "eliminated"
- strength of confirmation: "never" corresponds to relationship "eliminated".

Adlassnig selected appropriate overlapping subsets of U_1 and U_2 to "fuzzify". Following Zadeh's suggestion, Adlassnig defined two linguistic variables:

- $T(X_1)$ = *always, almost always, very very often, very often, fairly often, often, more or less often, unknown, more or less seldom, seldom, fairly seldom, very seldom, very very seldom, almost never, never*.
- $T(X_2)$ = *always, almost always, very very often, very often, fairly often, often, more or less often, unknown, more or less seldom, seldom, fairly seldom, very seldom, very very seldom, almost never, never*.

Proceeding from the primary elements (*always, often, unknown, seldom, never*), secondary and tertiary elements were formed by the "linguistic modifiers" according to Zadeh's model. In his article *A Fuzzy Logical Model of Computer-Assisted Medical Diagnosis*[136], which was written later but published first, Adlassnig used a section entitled "Review of Published Methods" to survey all of the essential publications on the application of fuzzy sets in the field of medical diagnosis. Then, following the suggestion made by Sanchez, he applied Zadeh's max-min composition rule: Given $\Sigma = S_1, S_2, ..., S_m$ as the system set and $\Delta = D_1, D_2, ..., D_n$ as the set of diagnoses taken into consideration; m and n represent the cardinality of these sharp sets.

- Every symptom $S_i \in \Sigma, 1 \leq i \leq m$ is a fuzzy set of the reference set $\mathcal{R} = x_1, x_2, ...$, which contains all possible assumable values of S_i. S_i is characterized by a membership function $\mu_{S_i}(x)$ which defines the membership intensity of $x \in \mathcal{R}$ in S_i.
- Every diagnosis $D_j \in \Delta, 1 \leq i \leq n$ is a fuzzy set of the reference set $\mathcal{P} = p_1, p_2, ...$, characterized by a membership function $\mu_{D_j}(p)$ where the set $\mathcal{P}$ contains all of the patients being considered and assigns each patient his membership value in D_j.

Adlassnig now represented the relationships "occurrence" and "strength of confirmation" that were considered between symptoms and diagnoses as fuzzy relations with the aid of his devised mathematical ammunition. In addition, he defined the "occurrence" of a symptom S_i in the case of a diagnosis D_j as the frequency with which it occurs, while he established the "strength of confirmation" of the symptom as the importance it has for the diagnosis. With these two aspects that a doctor takes into account when making a diagnosis, Adlassnig defined the two variables that had been used two years earlier:

- The "occurrence" of a symptom S_i in the case of a diagnosis D_j is a fuzzy set P of the reference set $U_p(x) = 0, 1, 2, ..., 100$, where x means: S_i occurs x times in 100 cases of D_j.
- The "strength of confirmation" of a symptom S_i for a diagnosis D_j is a fuzzy set C of the reference set $U_c(y) = 0, 1, 2, ..., 100$, where y means: S_i has proven D_j in y of 100 cases.

The membership functions of P and C were formed in a very simple way:

$$\mu_p(x) = f_1(x; 2, 5, 98), \qquad x \in U_p$$
$$\mu_c(x) = f_1(x; 2, 5, 98), \qquad x \in U_c$$

---

[136] [9] Adlassnig, Fuzzy (1980)

where $f_1(x)$ denotes the standard function indicated by Zadeh[137]:

$$f_1(x;\alpha,\beta,\gamma) = \begin{cases} 0 & \text{for } x \leq \alpha \\ 2(\dfrac{x-\alpha}{y-\alpha})^2 & \text{for } \alpha < x \leq \beta \\ 1 - 2(\dfrac{x-\alpha}{y-\alpha})^2 & \text{for } \beta < x \leq y \\ 1 & \text{for } x > y \end{cases}$$

Altogether Adlassnig defined the following membership functions for the two aspects of "medical knowledge" by utilizing the standard functions f_1 and f_2 Zadeh had introduced in 1976[138]:

$$\begin{aligned}
A_1 = B_1 &= \mu_{\text{always}}(x) & &= f_1(x; 97, 98, 99) \\
A_2 = B_2 &= \mu_{\text{almost always}}(x) & &= f_1(x; 80, 85, 90) \\
A_3 = B_3 &= \mu_{\text{very often}}(x) & &= \mu_{\text{often}}(x)^4 \\
A_4 = B_4 &= \mu_{\text{often}}(x) & &= f_1(x; 40, 60, 80) \\
A_5 = B_5 &= \mu_{\text{unspecific}}(x) & &= f_2(x; 20, 50) \\
A_6 = B_6 &= \mu_{\text{seldom}}(x) & &= 1 - f_1(x; 20, 40, 60) \\
A_7 = B_7 &= \mu_{\text{very seldom}}(x) & &= \mu_{\text{seldom}}(x)^4 \\
A_8 = B_8 &= \mu_{\text{almost never}}(x) & &= 1 - f_1(x; 10, 15, 20) \\
A_9 = B_9 &= \mu_{\text{never}}(x) & &= 1 - f_1(x; 1, 2, 3)
\end{aligned}$$

Adlassnig next studied the binary fuzzy relation R_p, a fuzzy set of the Cartesian product $\Sigma \times \Delta$, characterized by the two parameter membership functions $\mu_{R_p}(S_i, D_j)$ in the interval $[0,1]$. $\mu_{R_p}(S_i, D_j)$ is identical to $\mu_p(x)$, which was documented for S_i with respect to D_j as $S_i D_j$ occurrence relationships (Fig. 7.24).

The $S_i D_j$ confirmation strength relationships form the elements of the binary fuzzy relation $R_c \subset \Sigma \times \Delta$, defined by $\mu_{R_c}(S_i, D_j)$, which in turn is identical to $\mu_c(x)$ (Fig. 7.24).

Finally, Adlassnig introduced the binary fuzzy relation $R_S \subset \mathcal{P} \times \Sigma$. The sets $\mathcal{P}$ and Σ are given with the symptom patterns of the patients examined. Symptoms that were not studied but were considered in the doctors' documentation could be ignored in the relation matrices R_p and R_c. All of the functions $\mu_{R_S}(p, S_i)$, which characterize the fuzzy relation R_S, are identical to $\mu_{S_i}(x)$ calculated for the patient corresponding to the defined fuzzy set S_i.

As a composition of these fuzzy relations, four different fuzzy indicators and their membership functions can now be calculated:

- ΣD_j occurrence condition $\qquad R_1 = R_S \circ R_p$,
$$\mu_{R_1}(p, D_j) = \max_{S_i} \min\{\mu_{R_S}(p, S_i); \mu_{R_p}(S_i, D_j)\},$$

- ΣD_j confirmation strength condition $\qquad R_2 = R_S \circ R_c$
$$\mu_{R_2}(p, D_j) = \max_{S_i} \min\{\mu_{R_S}(p, S_i); \mu_{R_c}(S_i, D_j)\},$$

---

[137] [598] Zadeh, Complex (1976)
[138] [598] Zadeh, Complex (1976)

7.5 Fuzziness in Medicine

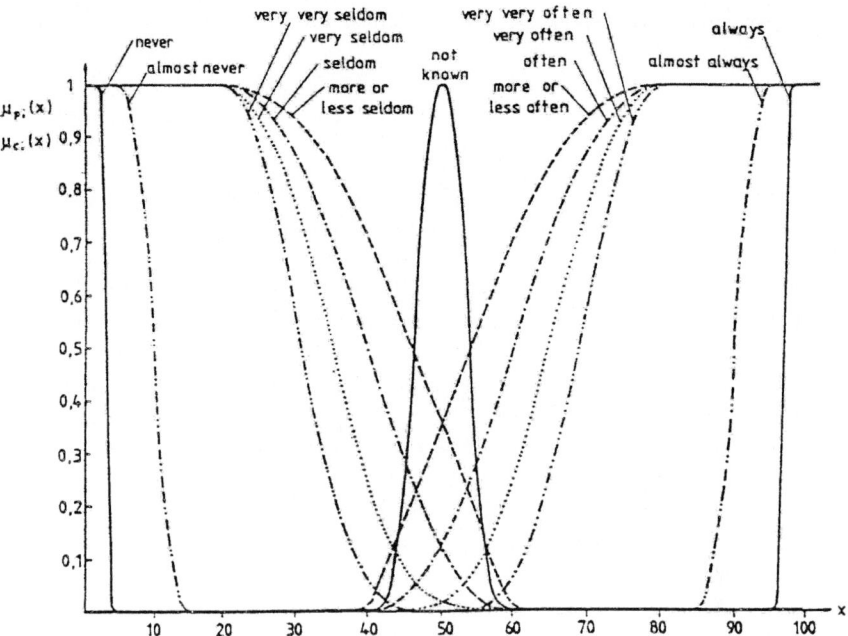

Fig. 7.24. Membership functions for the fuzzy sets "occurrence" and "strength of confirmation"

- ΣD_j non-occurrence condition $\quad R_3 = R_S \circ (1 - R_p)$
  $$\mu_{R_3}(p, D_j) = \max_{S_i} \min\{\mu_{R_S}(p, S_i); 1 - \mu_{R_p}(S_i, D_j)\},$$
- Not ΣD_j confirmation strength condition $\quad R_4 = (1 - R_S) \circ R_p$
  $$\mu_{R_4}(p, D_j) = \max_{S_i} \min\{1 - \mu_{R_S}(p, S_i); \mu_{R_p}(S_i, D_j)\},$$

where: $p \in \mathcal{P}, S_i \in \Sigma, 1 \leq i \leq m$ und $D_j \in \Delta, 1 \leq i \leq n$.

The computer-assisted diagnostics in CADIAG-II functioned by the following steps:
1. Each symptom observed in patient p is evaluated numerically as either 0 or 1. The definition of the symptom fuzzy sets permits the calculation of the degree of membership $\mu_{R_s}(p, S_i)$.
2. The degrees of membership $\mu_{R_1}(p, D_j), \mu_{R_2}(p, D_j), \mu_{R_3}(p, D_j)$ and $\mu_{R_4}(p, D_j)$ for patient p are calculated. The system then displays the following results:

   - All D_j are given as proven diagnoses (DP) if:
     - $0.98 \leq \mu_{R_2}(p, D_j) \leq 1$.
   - All D_j are given as eliminated diagnoses (DE) if:
     - either $0.98 \leq \mu_{R_3}(p, D_j) \leq 1$
     - or $0.98 \leq \mu_{R_4}(p, D_j) \leq 1$.

– All diagnoses D_j for whose degree of membership $\mu_H(p, D_j)$ the following is true:

$$-0.5 < \mu_H(p, D_j) = \min\{\mu_{R_1}(p, D); \mu_{R_1}(p, D_j)\}$$

– are designated reference diagnoses.

CADIAG-II (Fig. 7.25) was installed at Vienna General Hospital as an "electronic reference work for possible diagnoses" and as a "complex on-line consultation system for special cases for the detailed and complete differential diagnostic clarification of the patient's ailment". In 1993, the knowledge base was characterized by "disease profiles and symptom combinations for a total of 262 diseases, including 185 rheumatological ailments (69 arthropathies, 12 diseases of the spinal column, 38 diseases of the connecting and supporting tissue, 45 bone diseases and 21 regional pain syndromes) and 77 diseases from the area of gastroenterology (24 diseases of the gall bladder and biliary tract, 10 pancreatic diseases, 37 diseases of the large intestine) and hepatology (6 hepititides)". In over 600 clinical cases, its suggestions were compared to the clinical diagnoses and its accuracy was rated at between 80 and 95%. The false diagnoses were usually the result of the fact that therapies were already underway to fight symptoms, the disease was still initially or already stabile or the facts of the anamnesis did not seem significant. When the mainframe on which CADIAG-II had been running was shut down in 2004, this also spelled the end for the CADIAG-II system. A resumption of computer-assisted diagnosis within a MEDFRAME medical expert system platform has been in the works for years.

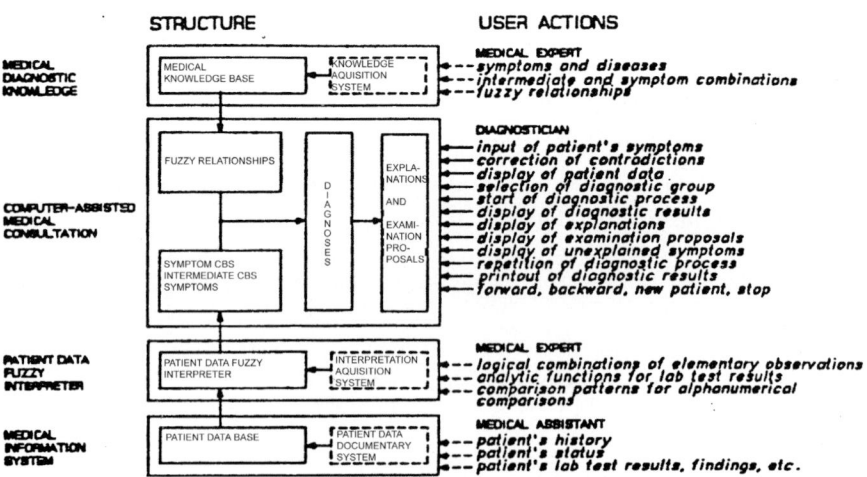

Fig. 7.25. Structure of the computer-assisted diagnosis system (dashed lines: components that take effect before the diagnosis process commences), p. 95, partially relabeled for legibility

8
Conclusion

The 20th century is history. It is still very much with us, though, and our memories of it are still fresh. Our knowledge of this past century is great; the events of the last hundred years changed our environment irreversibly; political and social circumstances have had a lasting effect; literature, music and art all shaped the 20th century; discoveries, inventions and developments have left a lasting impression and – last, but certainly not least – the fruits of scientific and technical research have contributed so much to modern life that it scarcely resembles the world at the turn of the 19th to the 20th century.

Many of the technological innovations we can't imagine doing without exist today thanks to the high degree of mathematization the engineering sciences achieved during the 20th century. Because of this, the middle of the century brought an enormous field of new opportunities for computerization: The solutions to mathematically sophisticated and complex calculations could be formulated as algorithms and programmed in machine languages or assemblers, and they could later be translated into higher languages interpretable by computers or simply written directly in such "programming languages". The computer became a ubiquitous tool for performing mathematical, computable problem solving processes.

Mathematics, on the other hand, also continued to develop dramatically in the 20th century, but without being of any particular use in solving the problems of engineering, since it was drifting ever further from a role as an adequate aid in the scientifically exact description of real systems. Even in the 19th century, the deliberations of mathematicians had led to results that could not be aligned with the world as perceived by people. The non-Euclidean geometries developed by Carl Friedrich Gauß[1] and Bernhard Riemann[2] and the discontinuous and non-differentiable functions first demonstrated by Peter G. L. Dirichlet[3] – "monster functions", as Herbert Mehrtens called them, borrowing a term from Jules Henri Poincaré, who, more polemically, also saw in them a "host of freaks" – were the first signs of a crisis in mathematics.[4]

In addition, Georg Cantor had begun during the second half of the 19th century to examine the concept of the actual infinite and gradually to introduce different concepts of the infinite. In 1874, he published the paper *On a Property of the Embodiment of All Real Algebraic Numbers*[5], in which he postulated "a one-to-one correspondence between the set of natural numbers and the set of positive rational numbers" and proved the non-countability of the set of real numbers. These considerations eventually led him to introduce set theory in his *Contributions to the Transfinite Set Theory*[6] in 1895 and 1897. With the success of these papers, however, antinomies and paradoxes in the logical foundations of mathematics also became apparent.[7] The basis for what was supposed to be the uncontradicted "Queen of the Sciences", which could in principle settle all of the problems in science, was severely

---

[1] Carl Friedrich Gauß (1777–1855), German mathematician, first studied languages and philosophy from 1792 to 1795 at the Collegium Carolinum in Braunschweig, then transferred at the age of eighteen to the University of Göttingen to study mathematics. He concluded his studies in 1799 with a doctoral dissertation at the Academia Julia in Helmstedt. In 1807, Gauß became a professor of mathematics and director of Göttingen's observatory

[2] Bernhard Riemann (1826–1866), German mathematician, began his study of mathematics in 1846 in Berlin and Göttingen under the tutelage of C. F. Gauß and P. G. L. Dirichlet, among others. Riemann earned his doctorate in Göttingen in 1854 and became an extraordinary professor there in 1857. In 1859 he became an ordinary professor and assumed the chair vacated by the deceased Dirichlet, who himself had been the successor to Gauß

[3] Peter Gustav Lejeune Dirichlet (1805–1895), German mathematician, studied mathematics in Paris and became a private lecturer in Breslau (Wroclaw). After next working as a tesher at the General War School in Berlin, he retained that position while also serving as a professor at Berlin University. In 1885, he was appointed a professor in Göttingen

[4] See also [372] Mehrtens, Moderne (1990), p. 84f, [447] Poincaré, Intuition (1902), [449] Poincaré, Methode (1908), p. 111f

[5] [110] Cantor, Zahlen (1874)

[6] [111] Cantor, Mengenlehre (1895/97)

[7] It suffices here to describe the antinomy identified by Bertrand Russell, probably in the spring of 1901: Let M be the "set of all sets that do not contain themselves as members". So what is this set M? Is it a set that does *not* contain itself as a member, or not? If the answer is yes, then M – as a member of the set M – would belong to all of the sets that do not contain themselves as members. Yet it would not be allowed to contain itself! If the answer is no, then the set M would not include itself as a member, but in that case it would have to be a member of M and thus necessarily include itself as a member

and permanently jolted by unresolved contradictions, leaving in its place the "pathology of mathematics"[8], the early high points of which would come in 1931 and 1936 with the publication of Kurt Gödel's paper about formally undecidable propositions[9] and Alan Turing's paper on computable numbers and an application to the *Entscheidungsproblem*.[10]

Investigating the "diseases of mathematics" turned out to be a very constructive reaction. The scientific community of researchers pursuing *pure mathematics* perceived the quaking of these foundations as a serious threat, but also as a challenge to redouble their efforts to understand the fundamentals and the philosophy of mathematics. They established different "schools", which sought a way *out* of the crisis – and which were enormously successful *in* the crisis, so much so that the philosophy of mathematics was able to make great strides over the course of the 20th century.[11]

In the scientific community of *applied mathematicians*, and especially among natural scientists and engineers, the crisis afflicting mathematics went largely unnoticed, although it is possible that the foundations of their sciences, too, were and are resting upon rotten pillars. The conventional calculations that practitioners still considered tried – many thousands of times over – and true were too far removed from the pedantically exact and polished mental acrobatics required by metamathematics for the emergence of paradoxes to have any effect.[12]

What conclusions should electrical engineers of the day have drawn from the academic sophistry of the mathematicians? If mathematics had been shaken to its core, then the validity of all theorems inferred from it were theoretically no longer certain. Should scientists now ignore all of their experiences confirming the theoretical calculations and their practical laboratory work? They were actually better off continuing to assume that the theorems they had been using thus far were still valid. Predictions about the results

---

[8] "Pathology of mathematics" is how Nicolas Bourbaki referred to this phase in [94] Bourbaki, Élements (1960). See also [372] Mehrtens, Moderne (1990), [241] Hobsbawm, Zeitalter (2004), pp. 305–327

[9] [197] Gödel, Sätze (1931)

[10] [539] Turing, Entscheidung (1936)

[11] Of particular note here are *Logicism* and its principle advocates Gottlob Frege, Bertrand Russell and Alfred North Whitehead, the *Formalism* founded by David Hilbert and Luitzen E. Brouwer's *Intuitionism*. See [372] Mehrtens, Moderne (1990), [289] Körner, Philosophie (1960), [534] Thiel, Philosophie (1995). Logicism made it possible to reduce the large system of mathematical concepts to a relatively small number of basic set theory concepts, mathematical theories were consistently axiomatized as a result of the axiomatization program proclaimed by David Hilbert and during the Intuitionism process it was possible to carry out many constructive proofs that did not utilize the principle of contradiction or the law of the excluded middle (*tertium non datur*), but Gödel's law of incompleteness can be considered a success in this crisis

[12] See [534] Thiel, Philosophie (1995), p. 330f, [241] Hobsbawm, Zeitalter (2004), p. 309f

of what were now being called "classical" experiments, predictions that had been successful for decades thanks to mathematical calculations, and the experiments that were being newly conceived spoke in favor of continuing with this method of operation. Within the framework of the conventional margin of error which they had learned to accept, the mathematically founded natural and engineering sciences functioned excellently. In any case, the idea did not occur to them that the "inaccuracies" that were permitted here were not all the result of statistical or systematic errors or mistakes in measurement but rather could also occur as an expression of the inability of the mathematical theory to represent the processes of real systems.

During the process of scientific development, new theoretical elements are sometimes needed in mathematics to provide a way to describe objects and systems and their "behavior" in a more mathematically exact way. This fact was evident in a further "foundational crisis" that arose in the early decades of the 20th century. It pertained to the field of physics, for with the "scientific revolutions" that came about with the popularization and acceptance of the theory of relativity and quantum mechanics, a shift also took place in the relationships between the exact natural science of physics and its mathematical-conceptual foundation. The objects and theoretical quantities established by Albert Einstein and Hermann Minkowski[13] in the case of relativity and by Werner Heisenberg, Niels Bohr and others for quantum mechanics differ fundamentally from those previously known in physics. They have properties which had never before played a role in physics. So the objects in the theory of relativity are not accorded the quantities *shape*, *mass* and *volume* as properties – as is the case in classical physics; in relativity theory, these are relations between the objects and the reference frame.

In quantum mechanics, which was developed as a new mathematical theory to describe real subatomic objects and processes, the theoretical objects cannot be identified either as particles or as waves and they thus differ substantially from the objects of mechanics and electrodynamics that had previously been considered by physicists. For instance, the "state" of an object in the classical theory – a particle – is determined definitively by means of the pair of physical "state variables" *position* and *momentum*, with which its properties are bound, to be at a particular location and to have a particular velocity.[14] In quantum mechanics, however, the "state function" is an abstract mathematical function whose interpretation as a "summarization" of

---

[13] Hermann Minkowski (1864–1909), Russian-born (today Lithuania) German mathematician, studied at universities in Berlin and Königsberg, where he earned the doctorate in 1885. He taught at the universities in Bonn, Königsberg and Zurich, where Albert Einstein attended his lectures. In 1902, Minkowski became a professor in Göttingen

[14] In classical mechanics, the momentum is defined as the product of the two variables velocity and mass of the particle

physical state variables with a real correspondence is exceptionally difficult.[15] Erwin Schrödinger had proposed interpreting the quantum mechanical "state function" as a material or charge density function[16], and Werner Heisenberg attempted to interpret it as matrices.[17] Eventually, what became known as the "Copenhagen interpretation of quantum mechanics" followed Max Born's suggestion of viewing them as "probability amplitudes", that a particle can be measured at a particular location or else that the frequency of a wave can be determined.[18] These two functions, both of which lead to the probability density in the particle or wave image, can be transformed into each other by the Fourier transformation and it becomes apparent here that the Heisenberg uncertainty principle is essential on the abstract mathematical level: A position distribution is created by superimposing frequencies from a band of frequencies and vice versa. It is not possible to use the representation of the quantum object as a wave with a precisely determined frequency in the wave image to determine the exact position to represent the quantum object as a particle in the particle image, and vice versa. In other words: That's as precise as you can get with these tools! Although the mathematical formalisms can be combined with each other for the purposes of mechanics and electrodynamics, the resulting formalism is not well-suited for an optimal representation of real subatomic objects and their interactions. Quantum mechanics can do it better! – It was Johann (John) von Neumann who identified the quantum mechanical state functions in his "axiomatization of quantum mechanics" as vectors in the abstract Hilbert space[19] and thus clarified on the theoretical level: Quanta are not the objects of classical mechanics (particles), nor are they the objects of classical electrodynamics (waves). Quanta are objects of quantum mechanics, and this mathematical theory includes mathematical quantities other than those. This theory makes it possible to achieve a better mathematical-theoretical representation of subatomic reality. Whether it provides an exact representation of what is actually happening, however, is something that cannot be answered – a "real magic formula" is how Albert Einstein described it at the time.

Quantum mechanics lacked the features previously found in physics, such as the vividness and perceptibility as well as the measurability of state variables. A function in the Hilbert space, which is infinitely-dimensional by definition, is not descriptive and interpreting it as "probability amplitude" does not offer any real possibility of perceiving a classical physical property of the

---

[15] Michael Drieschner defined the state of a quantum mechanical object in connection with Carl Friedrich von Weizsäcker's studies in natural philosophy as the "summarization of all contingent properties of this object". [146] Drieschner, Voraussage (1979), p. 95. See also [552] Weizsäcker, Aufbau (1985), p. 517f and [499] Seising, Strukturen (1996), p. 78

[16] [498] Schrödinger, Quantisierung (1926)

[17] [88] Born et al., Quantenmechanik (1926)

[18] [89] Born, Stossvorgänge (1926), [90] Born, Adiabatenprinzip (1926)

[19] [417] Neumann, Quantenmechanik (1932)

real object represented in this way or of measuring the corresponding classical physical quantity. On the other hand, though, tests such as the double-slit experiment, which is important in quantum theory, presuppose that the quanta appear as classical physical objects and have the properties of physical objects and that they are perceived or measured by an observer. This forced creation of a quantum object in an experiment, such as in the representation of the particle image, is also called the "collapse of the wave function". In so doing, the quantum mechanical character of the object description is lost.

Since experiments are only meaningful if they lead to measurement results or perceptions for an observer, the dilemma facing the quantum mechanics revolution lies in the fact that the state function of "quantum mechanics" is an abstract mathematical construct without any concrete physical meaning, without any exact correspondence in reality. For any given state function of an object at the current time, it is possible to make exact predictions about the state function of this object at a future point in time thanks to the Schrödinger equation. Predictions about the implementation of this state function in the particle image cannot be exact, however; according to the Heisenberg uncertainty relation they are uncertain.

Theory and experiments are intertwined areas of modern science. Scientific theories should succeed or fail on the basis of experience. The findings of the experiments carried out to this end must be perceptible. Experiments must therefore be designed and adapted for the perceptual possibilities that humans experience and their results will then fit into the framework of an image of the real object, not necessarily an image of the mathematical nature of the theoretical object. I have so far attempted in this final chapter to demonstrate that the gulf dividing theory from reality in the science of the 20th century has always been present. The foundational crises in mathematics and physics showed unmistakably that mathematics does not rest upon solid ground and that mathematical theories do not portray reality in a certain and exact way. Even the 20th century's new mathematical theory *quantum mechanics*, though it continues to serve very well to this day, displays a fundamental inability to represent real objects and their interactions, since reducing the quantum mechanical state function to a classical representation as a particle or wave eliminates the quantum character the theory provides.[20] These deficiencies or uncertainties remained topics for specialists, fundamental researchers and philosophers while users could largely ignore them.

It was only in the second half of the 20th century that the mathematical theory's inability to describe real systems and their "behavior" was addressed. The mathematically-oriented electrical engineer Lotfi Zadeh made this problem the central focus of his scientific efforts. The story of the genesis of Fuzzy

---

[20] See also [673] Zeilinger, Schleier (2003). The Vienna physicist Anton Zeilinger treats this subject in a way that is intelligible to the layperson and subsequently also explains his experiments on "quantum teleportation", which in deference to the TV series *Star Trek* is also called "beaming". See also fn. 116 in Chap. 3

8 Conclusion

Set Theory that is told in this book concentrates primarily on the work Zadeh carried out in the 1950s and 1960s. His attempts to identify, characterize and classify real electrotechnical systems using mathematically precise descriptions had foundered and pointed to an even greater complexity in real systems. A great distance gaped between the mathematical descriptiveness of real systems on the one hand and of theoretical systems on the other. Real systems could not be described as precisely as theoretical systems, and Zadeh had noted this difference as far back as his discussions with Ernst Adolph Guillemin in the 1940s. In his work on electrical filters, he highlighted the differences between ideal and optimal filters and his efforts prompted him to conclude in 1962 that a different mathematics was needed – a mathematics of fuzzy quantities – in order to describe and account for large complex systems. He eventually turned away from "hard" mathematics and, in 1964, developed the Fuzzy Set Theory, which in the 1990s became the core theory of what would later be known as "soft computing", a collection of theories, methods and techniques for use with highly complex systems. Zadeh looked back on this phase in 1994:

> In traditional – hard – computing, the prime desiderata are precision, certainty and rigor. By contrast, the point of departure in soft computing is the thesis that precision and certainty carry a cost and that computation, reasoning and decision making should exploit – wherever possible – the tolerance for imprecision and uncertainty.[21]

During the Industrial Liaison Program Conference at Berkeley in March of 1991, he founded the *Berkeley Initiative in Soft Computing* (BISC)[22], which since then has represented an Internet-based worldwide community of people and organizations united by their interest in soft computing and applications thereof.

When Zadeh had the opportunity to recall his personal view of the history of Fuzzy Set Theory once again in 1999, he explained this more precisely[23]:

> As systems become more varied and more complex, we find that no single methodology suffices to deal with them. This is particularly true of what may be called information/intelligent systems – systems which form the core of modern technology. To conceive, design, analyze and use such systems we frequently have to employ the totality of tools that are available. Among such tools are the techniques centered on fuzzy logic, neurocomputing, evolutionary computing, probabilistic computing and related methodologies. It is this conclusion that

---
[21] [655] Zadeh, Soft Computing (1994), p. 77
[22] See also the information available at http://www-bisc.cs.berkeley.edu/
[23] [705] Zadeh, Evolution (1999). This text consists of two parts. Zadeh wrote the first part upon receipt of the Honda Prize, which he was awarded by the Honda Foundation in 1989. The second part emerged ten years later when Zadeh was reflecting upon the development of Fuzzy Set Theory over the preceding decade. The section quoted here is taken from the second part

formed the genesis of the concept of soft computing. There are two aspects of SC (soft computing) that stand out in importance. First SC is not a single methodology; rather it is a consortium of methodologies which are aimed at exploiting the tolerance for imprecision, uncertainty and partial truth to achieve tractability, robustness and low solution cost. Second, the constituent methodologies in SC are for the most part complementary and synergistic rather than competitive. What this means is that in many applications it is advantageous to employ the constituent methodologies of SC in combination rather than in a stand-alone mode.[24]

Fig. 8.1. Honda Prize presentation ceremony, awarded to Lotfi Zadeh in 1989. Center: Lotfi Zadeh; to his right, his wife Fay and company founder Soichiro Honda (1906–1991)

As this book has shown, Zadeh saw in his Fuzzy Set Theory a "general system theory" by means of which real systems could be described the way they were perceived by the observer, and the two early application systems discussed in this book, for controlling a steam engine and for the computerized support of medical diagnostics, were really just the beginning of a brilliant success story for fuzzy systems. The composition rule of fuzzy relations proved to be central here. This max-min composition was employed as an inference rule both in the area of fuzzy control and in computer-aided medical diagnostics. In both fields of application, linguistically formulated expert knowledge was represented in the form of linguistic variables with the help of fuzzy sets.

---

[24] [705] Zadeh, Evolution (1999), p. 899f

For the fuzzy control system in Assilian and Mamdani's steam engine, fuzzy IF-THEN rules were formed and then combined as fuzzy relations following the precepts of the max-min rule, thus resulting in an output quantity of the control system. In the CADIAG-II medical diagnostic system, the relationships between patients' symptoms and diseases were represented as fuzzy relations; they can likewise be combined via the max-min rule, whereupon the system makes suggestions about diagnoses. Fuzzy systems are obviously extremely well suited to process expert knowledge when it is presented in the form of language!

Can Expert Systems Be Designed Without Using Fuzzy Logic? was the title of Zadeh's contribution to the *Seventeenth Annual Conference on Information Sciences and Systems* in 1983.[25] He answered this question by referencing the expert systems that had been designed and were functioning successfully at that time without the use of fuzzy logic.[26] He then pointed out, however, that the methods of dealing with uncertainty, imprecision and incomplete data that were used in these expert systems and that were taken or derived from probability theory did not possess a satisfactory methodological structure that allowed them to make inferences from a knowledge base that was built upon practical human experience. If the information in a data base is inexact, incomplete or not entirely certain, then the systematic use of fuzzy logic becomes practically indispensable.

> This does not mean, however, that the employment of fuzzy logic is both necessary and sufficient for the design of a useful expert system. Rather it suggests that fuzzy logic provides a more appropriate conceptual framework for the management of uncertainty than classical probability-based methods.[27]

Other authors also emphasized the advantages of Fuzzy Set Theory over probability theory. For the book *Fuzzy Expert Systems*, published by Abraham Kandel in 1991, Kandel and Lawrence O. Hall wrote in their introductory essay:

> The failings of probability in situations where little or no a priori information is known provide an arena for the use of fuzzy expert systems. The rest of this book will show that their use can provide an enhancement for systems which must do reasoning with much uncertainty and imprecision.

Earlier these authors referred to the "biggest weakness of nonfuzzy methods of dealing with imprecision and uncertainty", namely that of contending with linguistic terms; this by contrast is the greatest strength of fuzzy methods:

---

[25] [677] Zadeh, Expert Systems (1983)

[26] In an extensive article on this subject the following year, Zadeh mentioned by name the medical expert systems MYCIN, PROSPECTOR, CASNET and INTERNIST/CADUCEUS. [685] Zadeh, Making (1984), p. 30

[27] [677] Zadeh, Expert Systems (1983), p. 490

Fuzzy set theory provides a natural method for dealing with linguistic terms by which an expert system will describe a domain. An imprecise numeric term can be effectively described by a fuzzy number. Other terms are simply mapped to and from fuzzy sets. Hence, the use of fuzzy set theory in expert systems has caused an evolution of systems.[28]

On the basis of Fuzzy Set Theory and beginning with his 1973 article *Outline of a New Approach to the Analysis of Complex Systems and Decision Processes*[29], Zadeh has spent the intervening thirty-plus years developing a new conceptual and methodical framework which makes it possible to use linguistic terms and linguistic variables alongside numbers and numerical variables in science.

I suggested that fuzzy logic, in its wide sense, is in essence a body of concepts and techniques which open the door to the use of natural languages in scientific theories.[30]

Since the second half of the 1990s, Zadeh has been talking about this idea as *Computing with Words* and in May of 1996, the *IEEE Transactions on Fuzzy Systems* published his article *Fuzzy Logic = Computing with Words*[31], which begins with the following statements:

Fuzzy logic has come of age. Its foundations have become firmer, its applications have grown in number and variety, and its influence within the basic sciences – especially in mathematical and physical sciences – has become more visible and more substantive. Yet, there are two questions that are still frequently raised: a) what is fuzzy logic and b) what can be done with fuzzy logic that cannot be done equally well with other methodologies, e.g., predicate logic, probability theory, neural network theory, Bayesian networks, and classical control? The title of this note is intended to suggest a succinct answer: the main contribution of fuzzy logic is a methodology for computing with words. No other methodology serves this purpose.[32]

---

[28] [224] Hall et al., Evolution (1991), p. 19
[29] [641] Zadeh, Outline (1973)
[30] [705] Zadeh, Evolution (1999), p. 14
[31] [672] Zadeh, Words (1996)
[32] [672] Zadeh, Words (1996), p. 103. By placing the equal sign in the title, Zadeh was stressing the role of fuzzy logic as a methodology for *Computing with Words* and also that there was no alternative methodology. Nevertheless, the stylistic exaggeration was not intended to lessen the importance of fuzzy logic: "Needless to say, there is more to fuzzy logic than a methodology for CW [Computing with Words, R. S.]. Thus, strictly speaking, the equality in the title of this note should be an inclusion; using the equality serves to accentuate the importance of computing with words as a branch of fuzzy logic"

What is *Computing with Words* (CW) and what is it for? – Zadeh contrasts "computing" as the manipulation of numbers and symbols in a computer, with the use of words in a natural language and he compares perceptions achieved while calculating to conclusions drawn by a human. Human calculations thus result once again in words or mental perceptions, and people also require words in order to pass information along. The meanings of words are always blurred, however. For *Computing with Words* with the computer, the blurred meanings of the words are represented by fuzzy sets. Linguistic description of perceptions can then be formulated in natural (or artificial) language and, for the purpose of relaying information, certain variables are restricted to one value, but one that in *Computing with Words* can be a fuzzy value and not a number.[33]

In 1999, Zadeh built upon *Computing with Words* to sketch out a theory of the arithmetic processing of perceptions, the *Computational Theory of Perceptions*, with which he wants to contribute to the field of "artificial intelligence". The brain can process perceptions – in contrast to a computer. Zadeh thus proposed having the computer process the labels of perceptions – words. *Computing with Words* is therefore the basis for the *computational theory of perceptions*.

> The computational theory of perceptions, or CTP for short, is based on the methodology of CW. In CTP, words play the role of labels of perceptions and, more generally, perceptions are expressed as propositions in natural language.[34]

As he had done in the 1960s and '70s, Zadeh mentioned the fact that although computers are superior to people when it comes to intensive mathematical calculations, the human brain is better than a computer at handling many tasks for which no measurements and numerical computations are necessary.

> The computational theory of perceptions – which is based on CW – is inspired by the remarkable human capability to perform a wide variety of physical and mental tasks without any measurements and any computations. Everyday examples of such tasks are parking a car, playing golf, deciphering sloppy handwriting and summarizing a story. Underlying this capability is the brain's crucial ability to reason with perceptions – perceptions of time, distance, speed, force, direction, shape, intent, likelihood, truth and other attributes of physical and mental objects.[35]

In contrast to before, though, at the end of the 20th century this argument no longer served simply to extol the virtues of Fuzzy Set Theory for application systems, but also as a bridge to the field of artificial intelligence research.

---

[33] As explained above, the necessity for a variable to take on a particular numerical value leads to "state function collapse" in quantum mechanics

[34] [599] Zadeh, Computing (1999), p. 105

[35] [606] Zadeh, Evolution (1999), p. 903

> The principal aim of the computational theory of perceptions is the development of an automated capability to reason with perception-based information. Existing theories do not have this capability and rely instead on conversions of perceptions into measurements – a process which in many cases is infeasible, unrealistic or counterproductive. In this perspective, addition of the machinery of the computational theory of perceptions to existing theories may eventually lead to theories which have a superior capability to deal with real-world problems and make it possible to conceive and design systems with a much higher MIQ (Machine IQ) than those we have today.[36]

Zadeh adopted the concept of machine intelligence quotient (MIQ) in 1994 as a measure for the "intelligence" of man-made systems, which are becoming ever more plentiful:

> In retrospect, the year 1990 may well be viewed as the beginning of a new trend in the design of household appliances, consumer electronics, cameras and other types of widely used consumer products. The trend in question relates to a marked increase in what might be called the Machine Intelligence Quotient (MIQ) of such products compared to what it was before 1990. Today, we have microwave ovens and washing machines that can figure out on their own what settings to use to perform their tasks optimally; cameras that come close to professional photographers in picture-taking ability; and many other products that manifest an impressive capability to reason, make intelligent decisions and learn from experience.[37]

Five years later, Zadeh even contrasted this observation with what was being praised at that time as the "information revolution", which was evident in the Internet, World Wide Web, mobile telephones, fax machines and personal computers, all of which had become permanent fixtures of everyday life. Not so readily apparent but perhaps even more important, in Zadeh's view, was the revolution of "intelligent systems":

> The artifacts of this revolution are man-made systems which exhibit an ability to reason, learn from experience and make rational decisions without human intervention. I coined the term MIQ (Machine Intelligence Quotient) to describe a measure of intelligence of man-made systems. In this perspective, an intelligent system is a system which has a high MIQ.[38]

Of course, the human IQ and the MIQ are not the same thing. Zadeh stressed not only that the MIQ is product specific and has completely different dimensions than the IQ; he also referred to the fact that the MIQ changes over

---

[36] [606] Zadeh, Evolution (1999), p. 904
[37] [655] Zadeh, Soft Computing (1994), p. 77
[38] [705] Zadeh, Evolution (1999), p. 899

time, whereas the IQ for people remains more or less constant. An important dimension for the MIQ was speech recognition, for example, which is an ability that is assumed to be a matter of course in the case of IQ. The MIQ for a camera from the year 1990 can logically only be compared to that of another camera from the same era and these values will surely be lower than the values for cameras from the year 2005.

Fig. 8.2. Lotfi Zadeh and the author of this book in the lecture hall of the Medical University of Vienna, Core Unit for Medical Statistics and Informatics, after a lecture Zadeh delivered there on March 2, 2005

With *Soft Computing, Computing with Words, The Computational Theory of Perceptions* and the Machine Intelligence Quotient, Zadeh revisited subjects a half-century later which represented work from the beginning of his scientific career – thinking machines and complex systems – and which were now growing together. We find ourselves at the very start of the age of (artificial) intelligent systems, he stated at the end of the 20th century, and he asked: Why, despite the research into artificial intelligence, did it take so long? – Because its essential tools come from the fields of symbol manipulation and predicate logic, and we know today that their potential for solving real problems is very limited. These limits ultimately paved the way for fuzzy sets and soft computing – even into the field of artificial intelligence – for in the year 2001 Lotfi Zadeh – at last – could also publish an article in *AI Magazine* bearing the following title: *A New Direction in AI. Toward a Computational Theory of Perceptions.*[39]

---

[39] [605] Zadeh, Direction (2001)

Afterword by Lotfi A. Zadeh

For a historian of science, writing a treatise on the genesis and evolution of fuzzy set theory and fuzzy logic is a challenging task. The challenge relates to the fact that fuzzy logic has long been, and to some degree still is, an object of controversy and debate. On reading Dr. Seising's treatise *The Fuzzification of Systems* a conclusion which emerges is that Dr. Seising has met the challenge with impressive success. His work is accurate, informative, illuminating and insightful. He and the publisher, Springer-Verlag, deserve our thanks and congratulations.

In part, many of the skeptical views of fuzzy set theory and fuzzy logic reflect the fact that "fuzzy" is a word which is usually used in a pejorative sense. When I was writing my first paper on fuzzy sets, a problem which I faced was that of finding an appropriate name for sets with unsharp boundaries. A term which I could have used was "vague", a term drawn from the literature of logic. But, basically, "vague" means insufficiently specific rather than unsharp. To illustrate, "I will be back sometime", is vague, but "I will be back in a few minutes", is fuzzy but not vague. Another example: Consider the shadow of a sphere – a shadow which has unsharp boundaries. Such a shadow would be called "fuzzy" but not "vague". After a great deal of internal debate and consultation with others, I settled on the name "fuzzy", realizing of course that "fuzzy" will be a handicap. More fundamentally, the controversies circling around fuzzy set theory and fuzzy logic are rooted in misconceptions.

To begin with, fuzzy logic is not fuzzy. In large measure, fuzzy logic is precise. Another source of confusion is the duality of meaning of fuzzy logic. In a narrow sense, fuzzy logic is a logical system. But in much broader sense that is in dominant use today, fuzzy logic, or FL for short, is much more than a logical system. More specifically, fuzzy logic has many facets. There are four principal facets: (a) the fuzzy-set-theoretic facet, FLs; (b) the logical facet, FLl; (c) the epistemic facet, FLe; and (d) the relational facet, FLr. In this perspective, fuzzy logic, FL, subsumes fuzzy set theory. The basic concepts of graduation and granulation form the core of fuzzy logic. More specifically, in fuzzy logic everything is or is allowed to be graduated, that is, be a matter of degree or, equivalently, fuzzy. Furthermore, in fuzzy logic everything is or is allowed to be granulated, with a granule being a clump of attribute-values drawn together by indistinguishability, similarity, proximity or functionality. Graduation and granulation are the principal distinguishing features of fuzzy logic.

In his scholarly treatise, Dr. Seising analyzes the genesis of fuzzy set theory and fuzzy logic not in isolation but in the much broader context of set theory, logic, systems analysis and philosophy of languages. The much greater generality of fuzzy logic suggests that many theories can be enriched through employment of concepts and techniques drawn from fuzzy logic. Dr. Seising's discussion of the controversial aspects of fuzzy set theory and fuzzy logic cast much light on the underlying issues. A basic question is: What does fuzzy logic have to offer that is not offered by other theories? What is widely unrecognized at this juncture is that fuzzy logic can add to any theory what may be called NL-capability, that is, the capability to compute and reason with information described in natural languages. The importance of this capability derives from the fact that much of human knowledge is described in natural language. In this connection, what should be noted is that a natural language is basically a system for describing perceptions. Thus, perception-based information may be equated to information described in natural language. What this implies is that NL-capability is concomitant with the capability to operate on perception-based information. This capability plays a particularly important role in probability theory, robotics, decision analysis and economics. In summary, Dr. Seising's treatise is the first work that traces and analyzes the genesis and evolution of fuzzy set theory and fuzzy logic in a broad historical perspective. His work contributes in many important ways to a better understanding of what fuzzy set theory and fuzzy logic are and how they relate to earlier theories which have a position of centrality in human thought and cognition. His scholar treatise and its publisher, Springer-Verlag, deserve our thanks and congratulations.

Lotfi A. Zadeh
Berkeley, CA
January 10, 2007

References

1. Aatre, Vasudev K.: *Network Theory and Filter Design.* (2nd edition), New York: John Wiley & Sons, 1986.
2. Adlassnig, Klaus-Peter, Wolfgang Horak, Gernot Kolarz, Wolfgang Dorda, K. A. Fröschl, Helmut Grabner und Georg Grabner: Expertensysteme in der Medizin. In: *Österreichische Krankenhaus-Zeitung,* 32, pp. 361–384, 1991.
3. Adlassnig, Klaus-Peter, Wolfgang Horak, Gernot Kolarz, Werner Scheithauer, P. Peichl, Ursula Hay, Helmut Grabner, Alois Marksteiner, Peter Sachs, Ilse Gröger, Barbara Schneider, Wolfgang Dorda und Brigitte Haidl: Medizinische Experten- und Konsultationssysteme: Überblick und Anwendungsbeispiel. In: *Acta medica Austriaca,* 14, 5, pp. 136–143, 1987.
4. Adlassnig, Klaus-Peter, Thomas Gergely, Helmut Grabner und Georg Grabner: *A Computer Assisted System for Diagnostic Decision Making – Online Usage of the Database of the Medical Information System WAMIS.* In: [512], pp. 213–218, 1977.
5. Adlassnig, Klaus-Peter: *Ein computergestütztes medizinisches Diagnosesystem unter Verwendung von fuzzy Teilmengen.* Dissertation. Technische Universität Vienna, May 1983.
6. Adlassnig, Klaus-Peter: *Wissensbasierte Entscheidungsunterstützung in der Medizin.* In: [242], 1993.
7. Adlassnig, Klaus-Peter: *Exemplarische Erfahrungen and Projekte. Medizinische Expertensysteme in der klinischen and Laboratoriumsdiagnostik.* In: [249], pp. 37–67, 1993. (Abridged and revised edition in: [2], pp. 361–384, 1991.)
8. Adlassnig, Klaus-Peter: *Medizinische Expertensysteme in der klinischen und Laboratoriumsdiagnostik.* In: [249], pp. 37–67, 1993.
9. Adlassnig, Klaus-Peter: A Fuzzy Logical Model of Computer-Assisted Medical Diagnosis. In: *Methods of Information in Medicine,* Vol. 19, pp. 141–148, 1980.
10. Adlassnig, Klaus-Peter: Ein einfaches Modell zur medizinischen Diagnostik mit fuzzy Teilmengen. In: *EDV in Medizin und Biologie,* 13, 1, pp. 12–16, 1982.
11. Adlassnig, Klaus-Peter: *A Survey on Medical Diagnosis and Fuzzy Subsets.* In: [220], pp. 203–217, 1982.
12. Ahrweiler, Petra: Künstliche Intelligenz-Forschung in Deutschland – Die Etablierung eines Hochtechnologie-Fachs. In: *Internationale Hochschulschriften,* 141. Münster, New York: Waxmann 1995.

13. Aizerman M. A., E. M. Braverman and L. I. Rozonoer: Theoretical Foundations of the Potential Function Method in Pattern Recognition Learning. In: *Automation and Remote Control*, Vol. 25, No. 6, pp. 821–837, 1964. (English translation of *Avtomatika i Telemekhanika*, Vol. 25, No. 6, pp. 917–936, June 1964).
14. Aizerman M. A., E. M. Braverman and L. I. Rozonoer: The Method of Potential Function for the Problem of Restoring the Characteristic of a Function Converter from Randomly Observed Points. In: *Automation and Remote Control*, Vol. 25, No. 12, pp. 1546–1556, December 1964. (English translation of *Avtomatika i Telemekhanika*, Vol. 25, No. 12, pp. 1705–1714, December 1964).
15. Aizerman M. A., E. M. Braverman and L. I. Rozonoer: The Probability Problem of Pattern Recognition Learning and the Method of Potential Functions. In: *Automation and Remote Control*, Vol. 25, No. 9, pp. 1175–1190, September 1964. (English translation of *Avtomatika i Telemekhanika*, Vol. 25, No. 9, pp. 1307–1323, September 1964).
16. Akademie der Wissenschaften der DDR.: *Akademie der Wissenschaften der DDR*, Jahrbuch 1977, Berlin: Akademie-Verlag, 1977.
17. Albin, Merle Anne: *Fuzzy Sets and Their Application to Medical Diagnosis and Pattern Recognition*. Ph.D. Thesis, University of California Berkeley, Berkeley, California 1975.
18. Allen, E. P.: Punch Card for Neoplastic Diseases. In: *New Zealand Journal of Mathematics*, 42, pp. 121–125, June 1943.
19. Altrock, Constantin von: *Fuzzy Logic*. Band 1: Technologie, (2nd edition), Munich, Vienna: Oldenbourg, 1995.
20. Ampère, André-Marie: *Essai sur la philosophie des sciences*. Paris: Bachelier, 1834.
21. Arbib, Michael A.: A Partial Survey of Cybernetics in Eastern Europe and the Soviet Union. In: *Behavioral Science*, 11, pp. 193-216, 1966.
22. Arnold, H. D.: An Introduction to Research in the Communication Field. In: [432], pp. 25–48, 1932.
23. Ashby, William Ross: *An Introduction to Cybernetics*. London: Chapman et Hall, 1956.
24. Ashby, William Ross: *Design for a Brain*. London: Chapman et Hall, 1952. (2nd edition) 1960.
25. Aspray, William: The Scientific Conceptualization of Information: A Survey. In: *Annals of the History of Computing*, 7, No. 2, pp. 117–140, 1985.
26. Assilian, Sedrak.: *Artificial Intelligence in the Control of Real Dynamic Systems*. Ph.D. Thesis, No. DX193553, University London, August 1974.
27. Aubin, Jean-Pierre, Donald Saari and Karl Sigmand (Eds.): Dynamics of Macrosystems. *Proceedings of a Workshop on the Dynamic of Microsystems*. Held at the International Institute for Applied System Analysis (IIASA), Laxenburg, Austria, September 3–7, 1984. (= Lecture Notes in Economics and Mathematical Systems, 257). Berlin, Heidelberg, New York, Tokyo: Springer, 1985.
28. " The Author". In: *Columbia Engineering Quarterly*, 13, January 1950.
29. Badura, Bernhard: *Sprachbarrieren. Zur Soziologie der Kommunikation*. Stuttgart: Frommann-Holzboog, 1971.
30. Balabanian, Norman: *Network Synthesis*. Englewood Cliffs, N.J.: Prentice Hall Inc., 1958.

31. Balakrishnan, A. V. (Ed.): *Advances in Communication Systems. Theory and Applications*, Vol. 1. New York, London: Academic Press, 1965.
32. Ball, G. H. and D. J. Hall: *ISODATA, a Novel Technique for Data Analysis and Pattern Classification*. Stanford Research Institute, Menlo Park, Calif., 1965.
33. Barbaraschi S. and E. Gatti: Modern Methods of Analysis for Active Electrical Networks and Particular Regard to Feedback Systems. Part I. In: *Energia Nucleare*, Vol. 2, pp. 105–119, December 1954.
34. Barlow, J. S.: An Electronic Method for Detecting Evoked Responses of the Brain and for Reproducing their Average Waveforms. In: *Electroencephalography and Clinical Neurophysiology*, 9, pp. 340, 1957.
35. Bauer, Peter, F. Dorau, A. Gangl und Georg Grabner: Entwurf für ein Computer-Lehrprogramm im Bereich medizinischer Diagnostik. In: Sonderdruck aus *Diagramm*, Fachzeitschrift für Datenverarbeitung und Organisation, pp. 1–8, 1972.
36. Bauer, Peter, P. Brunner, G. Michalek, H. Paumgartner, H. Richter, and G. Stöger: *Computerdiagnostik in der Hepatologie*. In: [173], pp. 222–231, 1968.
37. Baylis, Thomas A.: *The Technical Intelligentsia and the East-German Elite*. Berkeley: University of California Press, 1974.
38. Bayland, E. and G. Bayland: Use of Record Cards in Practice, Prescription and Diagnostic Records. In: *Ugeskrift Laeger*, 116, 3, 1954.
39. Beardon, Colin: Computers, Postmodernism and the Culture of the Artificial. In: *AI & Soc*, 8, pp. 1–16, 1994.
40. Beaulieu, Norman C.: Introduction to Certain Topics in Telegraph Transmission Theory. In: *Proceedings of the IEEE*, Vol. 90, No. 2, pp. 276–279, February 2002.
41. Becker, Oskar: *Grundlagen der Mathematik*. Frankfurt am Main: Suhrkamp, 1975.(Freiburg: Alber, 1964).
42. Becker, Barbara: *Künstliche Intelligenz – Konzepte, Systeme, Verheissungen*. Frankfurt am Main, New York: Campus Verlag, 1992.
43. Belevitch, V.: Summary of the History of Circuit Theory. In: *Proceedings of the IEEE*, pp. 848–855, May 1963.
44. Bellman, Richard E., Robert Kalaba and Lotfi A. Zadeh: Abstraction and Pattern Classification. In: *Journal of Mathematical Analysis and Applications*, 13, pp. 1–7, 1966.
45. Bellman, Richard E. and Robert Kalaba: Dynamic Programming and Adaptive Control Processes: Mathematical Foundation. In: *IRE Transactions on Automatic Control*, AC-5, pp. 5–10, January 1960.
46. Bellman, Richard E. and Robert Kalaba: On the Role of Dynamic Programming in Statistical Communication Theory. In: *IRE Transactions on Information Theory*, pp. 197–203, September 1957.
47. Bellman Richard E. and Lotfi A. Zadeh: Decision-Making in a Fuzzy Environment. In: *Management Science*, Vol. 7, No. 4, pp. B-141, B-164, December 1970.
48. Bellman, Richard E.: *Dynamic Programming*. Princeton, NJ: University Press, 1957.
49. Bellman, Richard E. and Stanley, Lee: History and Development of Dynamic Programming, A Memorial to Richard Bellman. In: *IEEE Control Systems Magazine*, 4, pp. 24–28, 1984.
50. Bellman, Richard E.: *Eye of the Hurricane. An Autobiography*. Singapore: World Scientific Publishing, 1984.

51. Bellman, Richard E. and Lotfi A. Zadeh: Local and Fuzzy Logics. In: [148], pp. 105–165, 1977.
52. Bellman, Richard E.: Local Logics. In: *IEEE Logic*, p. 175, 1975.
53. Bellman, Richard E.: *Mathematical Methods in Medicine*. Singapore: World Scientific, 1983.
54. Bellman, Richard E. and Robert Kalaba: A Note on Interrupted Stochastic Control Processes. In: *Information and Control*, 4, pp. 346–349, 1961.
55. Bellman, Richard E. and Robert Kalaba (Eds.): *Selected Papers on Mathematical Trends in Control Theory*. New York: Dover, 1964.
56. Bellman, Richard E. and Charlene Paule Smith: Simulation in Human Systems. In: *Decision Making in Psychotherapy*. New York [et al.]: Wiley & Sons, 1973.
57. Bellman, Richard E.: On the Theory of Dynamic Programming. In: *Proceedings of the National Academy of Sciences of the United States of America*, 38, pp. 716–719, 1983.
58. Bemmel, Jan H. van, Marion J. Ball and Ove Wigertz (Eds.): *MEDINFO 83. Proceedings of the Fourth World Conference on Medical Informatics*. Amsterdam, pp. XI-XV, August 22–27, 1983.
59. Beneking, Heinz: *Praxis des elektronischen Rauschens*. Mannheim, Vienna, Zürich: Bibliographisches Institut, 1971.
60. Bennett, William R.: Applications of the Fourier Transforms in Circuit Theory and Circuit Problems. In: *IRE Transactions on Circuit Theory*, Vol. CT-2, No. 3, pp. 237–243, September 1955.
61. Bennett, Stuart: The Emergence of a Discipline: Automatic Control 1940–1960. In: *Automatica*, Vol. 12, pp. 113–121, 1976.
62. Bennett, William R.: *Data Communications Consultant, Bell Telephone Laboratories, Incorporated: Electrical Noise*. New York, Toronto, London: McGraw Hill Book Comp., 1960.
63. Bentley, Raymond: *Research and Technology in the Former German Democratic Republic*. Boulder [et al.]: Westview Press, 1992.
64. Bergier, J.: Un plan général d'automatisation des industries. In: *Les Lettres françaises*, p. 7, 15, avril 1948.
65. Berkeley, Edmund Callis: *Giant Brains or Machines that Think*. New York, London: John Wiley & Sons, Chapman & Hall, 1949.
66. Berner, Peter, Emil Brix and Wolfgang Mantl (Ed.): *Wien um 1900. Aufbruch in die Moderne*. Munich: Oldenbourg, 1986.
67. Bertalanffy, Ludwig von and Anatol Rapoport (Eds.): *General Systems (= Yearbook of the Society for the Advancement of General Systems Theory*. Vol. I.). Tunbridge Wells: Abacus Press, 1965.
68. Bertalanffy, Ludwig von: *Theoretische Biologie*. Band 1. Berlin: Gebrüder Bornträger, 1928.
69. Bertalanffy, Ludwig von: *... aber vom Menschen wissen wir nichts*. Düsseldorf: Econ, 1970.
70. Bertalanffy, Ludwig von: *Perspectives on General System Theory. A collection of essays gathered together and published two years after his death in 1972*, edited by Edgar Taschdjian, New York: George Braziller, 1975.
71. Bertalanffy, Ludwig von.: *Robots, Men and Minds. Psychology in the Modern World*. New York: George Braziller, 1967.
72. Bertalanffy, Ludwig von: General System Theory. In: [67], pp. 1–10, 1965.
73. Bertalanffy, Ludwig von: *General System Theory. Foundations, Development, Applications*. New York: George Braziller, 1968.

74. Bertalanffy, Ludwig von: Allgemeine Systemtheorie. Wege zu einer neuen Mathesis Universalis. In: *Deutsche Universitätszeitung*, 5/6, 1957.
75. Bertalanffy, Ludwig von: *Kritische Theorie der Formbildung (= Schaxels Abhandlungen,* 27). Berlin: Gebrüder Bornträger, 1928.
76. Bertalanffy, Ludwig von: *Vorläufer und Begründer der Systemtheorie.* In: [303], pp. 17–27, 1972.
77. Bezdek, James C.: *Fuzzy Mathematics in Pattern Classification.* Ph.D. Thesis, Center for Applied Mathematics, Cornell University, Ithaca, New York, 1973.
78. Bezdek, James C.: *Pattern Recognition with Fuzzy Objective Function Algorithms.* New York, London: Plenum Press, 1981.
79. Biorci, Giuseppe (Ed.): *Network and Switching Theory. A NATO Advanced Study Institute.* New York, London: Academic Press, 1968.
80. Birkhoff, Garrett: *Lattice Theory.* (= American Mathematical Society Colloquium Publications, Vol. 25). New York, 1948.
81. Black, Max: Reasoning with Loose Concepts. In: *Dialogue*, 2, pp. 1–12, 1963.
82. Black, Max: Some Aversive Responses to a Would-Be Reinforcer. In: *Wheeler Society*, pp. 125–134, 1973.
83. Black, Max: Vagueness. An Exercise in Logical Analysis. In: *Philosophy of Science*, 4, pp. 427–455, 1937.
84. Black-Schaffer, B. and P. D. Rosahn: Studies in Syphilis: Methods of Analysis of Yale Autopsy Protocols, Including Code for Punch Card Study of Syphilis. In: *The Yale Journal of Biology and Medicine*, 15, pp. 575–586, March 1943.
85. Blaydon, Colin C.: On a Pattern Classification Result of Aizerman, Braverman, and Rozonoer. In: *IEEE Transactions on Information Theory*, Vol. 12, No. 1, pp. 82–83, January 1966.
86. Blumtritt, Oscar: *Nachrichtentechnik. Sender, Empfänger, Übertragung, Vermittlung.* (2nd edition), Munich: Verlag 1997.
87. Bock, Hans Hermann: *Automatische Klassifikation. Theoretische und praktische Methoden zur Gruppierung und Strukturierung von Daten.* (= Studia Mathematica/Mathematische Lehrbücher Band XXIV). Göttingen: Vandenhoek und Ruprecht, 1974.
88. Born, Max, Werner Heisenberg and Pascqal Jordan: Zur Quantenmechanik II. In: *Zeitschrift für Physik*, 35, pp. 557–615, 1926.
89. Born, Max: Zur Quantenmechanik der Stossvorgänge. In: *Zeitschrift für Physik*, 37, pp. 863–867, 1926.
90. Born, Max: Das Adiabatenprinzip in der Quantenmechanik. In: *Zeitschrift für Physik*, 40, pp. 167–191, 1926.
91. Bose, Amar and Kenneth N. Stevens: *Introductory Network Theory.* Massachusetts Institute of Technology, Department of Electrical Engineering. New York: Harper & Row Publishers, 1965.
92. Bossel, H., N. Klaczko and N. Müller (Eds.): *System Theory in the Social Sciences.* Basel: Birkhäuser, 1976.
93. Boulaye, G. (Ed.): *Architecture and Design of Digital Computers.* Paris: Dunod, 1971.
94. Bourbaki, Nicolas.: *Élements des mathématiques.* Paris: Hermann, 1960.
95. Braizer, M. A. B. and J. S. Barlow: Some Applications of Correlation Analysis to Clinical Problems in Electroencephalography. In: *Electroencephalography and Clinical Neurophysiology*, 8, p. 325, 1956.

96. Braverman, E. M.: On the Method of Potential Functions. In: *Automation and Remote Control*, Vol. 25, No. 12, pp. 2130–2138, December 1965. (English translation of *Avtomatika i Telemekhanika*, Vol. 26, No. 12, pp. 2205–2213, December 1965).
97. Braverman, David: *Theories of Pattern Recognition*. In: [31], pp. 33–55, 1965.
98. Breidbach, Olaf: *Die Materialisierung des Ichs. Zur Geschichte der Hirnforschung im 19. und 20. Jahrhundert*. Frankfurt am Main: Suhrkamp, 1997.
99. Bremermann, Hans J.: Complexity of Automata, Brains, and Behavior. In: [131], pp. 304–331, 1973.
100. Bremermann, Hans J.: Cybernetic Functionals and Fuzzy Sets. In: *IEEE Symposium on Systems, Man, and Cybernetics*, 71C465MC, 1971.
101. Bremermann, Hans J.: What Mathematics Can and Cannot Do for Pattern Recognition. In: [207], pp. 31–45, 1970.
102. Brentano, Peter von, Karl-Achim Czemper, Bruno Fritzsch, Werner Kunz, Walter Müller and Horst Rittel: *Anhang: Beschreibung amerikanischer Forschungsinstitute*. In: [298], pp. 177–283, 1966.
103. Brix, Emil and Patrick Werkner (Ed.): *Die Wiener Moderne. Ergebnisse eines Forschungsgesprächs der Arbeitsgemeinschaft Wien um 1900 zum Thema Aktualität and Moderne*. Munich: Oldenbourg, 1990.
104. Broch, Hermann: *Hoffmannsthal und seine Zeit*. Frankfurt am Main: Suhrkamp, 1974.
105. *Brockhaus Enzyklopädie* in vierandzwanzig Bänden. (19th edition). Mannheim: Brockhaus, 1986–1994.
106. Brodman, K., A. J. Erdman, I. Lorge and H. G. Wolff: Cornell Medical Index-Health Questionnaire: As Diagnostic Instrument. In: *Journal of the American Medical Association*, 145, pp. 152–157, 20. January 1951.
107. Bundesministerium für Wissenschaft und Forschung. *Medizinisches Rechenzentrum der Universität Wien. 15 Jahre angewandte medizinische Informatik*. Vienna: Druckkunst Wien, 1982.
108. Burks, Arthur W.: Electronic Computing Circuits of the ENIAC. In: *Proceedings of the IRE*, pp. 756–768, August 1947.
109. Bush, Vannevar: The Differential Analyzer. In: *Journal of the Franklin Institute. Engineering and Applied*, Vol. 212, pp. 447–448, October 1931.
110. Cantor, Georg: Über eine Eigenschaft des Inbegriffes aller reellen algebraischen Zahlen. In: *Journal für reine und angewandte Mathematik*, 77, 1874, pp. 258–262.
111. Cantor, Georg: Beiträge zur transfiniten Mengenlehre. In: *Mathematische Annalen*, 46, 1895, pp. 481–512; 49, 1897, pp. 207–246.
112. Carnap, Rudolf: *Der logische Aufbau der Welt*. Berlin-Schlachtensee: Weltkreis Verlag, 1928.
113. Carnap, Rudolf: Die physikalische Sprache als Universalsprache der Wissenschaft. In: *Erkenntnis*, 2, 1931.
114. Carnap, Rudolf: *Logische Syntax der Sprache*. (= *Schriften zur wissenschaftlichen Weltauffassung*, Band 8). Vienna: Julius Springer, 1934.
115. Carnap, Rudolf: Testability and Meaning. In: *Philosophy of Science*, Vol. 4, No. 1, January 1937.
116. Carnap, Rudolf: *The Unity of Science*. London, 1934.
117. Carnap, Rudolf: *Mein Weg in die Philosophie*. Stuttgart: Philipp Reclam jun., 1993. (Original English edition: Carnap, Rudolf: *Intellectual Autobiography*. In: [488], pp. 1–84).

118. Cauer, Emil, Wolfgang Mathis and Rainer Pauli: Life and Work of Wilhelm Cauer (1900–1945). In: *Proceedings of the 14th International Symposium on Mathematical Theory of Networks and Systems*, June 19–23, 2000.
119. Cauer, Emil: Die Siebschaltungen der Fernmeldetechnik. In: *Zeitschrift für angewandte Mathematik und Mechanik*, 10, 5, pp. 425–433, 1930.
120. Chang, Chin-Liang and Richard T. Lee: Some Properties of Fuzzy Logic. In: *Information and Control*, 19, pp. 417–431, 1971.
121. Chang, Chin-Liang: *Fuzzy Sets and Pattern Recognition*. Ph.D. Thesis. Department of Electrical Engineering, University of California, Berkeley, December 1967.
122. Chang, Chin-Liang: Fuzzy Topological Spaces. In: *Journal of Mathematical Analysis and Applications*, 24, pp. 182–190, 1968.
123. Chang, Sheldon S. L. and Lotfi A. Zadeh: On Fuzzy Mapping and Control. In: *IEEE Transactions on Systems, Man, and Cybernetics*, SMC-2, No. 1, January 1972.
124. Chang, Sheldon S. L.: Fuzzy Dynamic Programming and the Decision Making Process. In: *Proceedings of the Third Annual Princeton Conference on Information Sciences and Systems 1979*, (papers presented March 27–28, 1969), pp. 200–203.
125. Cherry, Colin: *On Human Communication. A Review, a Survey and a Criticism*. New York: The Technology Press of Massachusetts Institute of Technology and John Wiley & Sons Inc., London: Chapman & Hall, 1957.
126. Cherry, Colin: On Communication Before the Days of Radio. In: *Proceedings of the IEEE*, pp. 1143–1145, May 1963.
127. Cherry, Colin: A History of the Theory of Information. In: *The Proceedings of the Institution of Electrical Engineers, Part III: Radio and Communication Engineering*, Vol. 98, No. 55, pp. 383–393, September 1951.
128. Chung, K. L.: Rezension von Ithaca, N. Y. In: *Mathematical Reviews*, 13, p.51, 1952.
129. Clark, W. A.: Digital Techniques in Neuroelectric Data Processing (Computer techniques in EEG analysis). In: *Electroencephalography and Clinical Neurophysiology*, Suppl. 20, 1961.
130. Clendening, Logan and Edward H. Hashinger: *Methods of Diagnosis*. St. Louis: C. V. Mosby Co., 1947.
131. Conrad, Michael, Werner Güttinger and Mario Dal Cin (Eds.): Physics and Mathematics of the Nervous System. In: *Proceedings of a Summer School*, organized by the International Centre for Theoretical Physics, Trieste, and the Institute for Information Sciences, University of Tübingen, held at Trieste, August 21–31, 1973. (= Lecture Notes in *Biomathematics*, Vol. 4). Berlin, Heidelberg, New York: Springer, 1973.
132. Cutler, P.: *Problem Solving in Clinical Medicine*. Baltimore: Williams and Wilkins Co., 1979.
133. Cutliffe, Stephen H.: *Ideas, Machines, and Values. An Introduction to Science, Technology, and Society Studies*. New York [et al.]: Rowman & Littlefield, 2000.
134. Dahms, Hans-Joachim (Ed.): *Philosophie, Wissenschaft, Aufklärung: Beiträge zur Geschichte und Wirkung des Wiener Kreises*. Berlin, New York: de Gruyter, 1985.
135. Dahms, Hans-Joachim: *Vertreibung und Emigration des Wiener Kreises zwischen 1931 und 1940*. In: [135], pp. 307–365, 1985.

136. Darlington, Sidney: A History of Network Synthesis and Filter Theory for Circuits Composed of Resistors, Inductors and Capacitors. In: *IEEE Transactions on Circuits and Systems*, 31, pp. 3–13, 1984.
137. Davidson, Mark: *Uncommon Sense. The Life and Thought of Ludwig von Bertalanffy*. Los Angeles: J. P. Tarcher 1983.
138. *Deutsche Biographische Enzyklopädie* (DBE), Band 7 (10). Killy, Walther and Rudolf Vierhaus (Ed.). Darmstadt: Wissenschaftliche Buchgesellschaft, 1998 (1999).
139. Deards, S. R.: Matrix Theory Applied to Thermionic Valve Circuits. In: *Electronic Engineering*, Vol. 24, pp. 264–277, June 1952.
140. Deiner, Reinhard: *Unscharfe Clusteranalysemethoden. Eine problemorientierte Darstellung zur unscharfen Klassifikation gemischter Daten* (= Wissenschaftliche Schriften: Reihe 2: Betriebswirtschaftliche Beiträge, Band 105). Idstein: Schulz-Kirchner, 1986.
141. Desoer, Charles A.: The Optimum Formula for the Gain of a Flow Graph or Simple Derivation of Coate's Formula. In: *Proceedings of the IRE*, p. 883, May 1960.
142. Dietzold, R. L.: Network Theory Comes of Age. In: *Electrical Engineering*, 67, pp. 895–899, September 1948.
143. Dorf, Richard C.: *Introduction to Computers and Computer Science*. San Francisco: Boyd and Fraser, 1972 (3rd edition 1981).
144. Dreyfus, Hubert: *Was Computer nicht können. Die Grenzen künstlicher Intelligenz*. Frankfurt am Main: Athenäum, 1989.
145. Driankov, Dimiter, Hans Hellendoorn and Michael Reinfrank: *An Introduction to Fuzzy Control*. Berlin [et al.]: Springer, 1993.
146. Drieschner, Michael: *Voraussage – Wahrscheinlichkeit – Objekt. Über die begrifflichen Grundlagen der Quantenmechanik*. Berlin [et al.]: Springer 1979.
147. Drösser, Christoph: *Fuzzy logic. Methodische Einführung in krauses Denken*. Reinbek bei Hamburg: Rowohlt, 1994.
148. Dunn, Michael J. and George Epstein (Eds.): Modern Uses of Multiple-Valued Logic. In: Invited Papers from *the Fifth International Symposium on Multiple-Valued Logic*. Held at Indiana University, Bloomington, Indiana, May 13–16, 1975. Dordrecht, Boston: D. Reidel, 1977.
149. Dunn, J. C.: A Fuzzy Relative of the ISODATA Process and Its Use in Detecting Compact Well-Separated Clusters. In: *Journal of Cybernetics*, 3, pp. 32–57, 1974.
150. Dupuy, Jean-Pierre: *Aux origines des sciences cognitives*. Paris: La Découverte, 1994.
151. Eckert, Michael: *Die Atomphysiker. Eine Geschichte der theoretischen Physik am Beispiel der Sommerfeldschule*. Braunschweig, Wiesbaden: Vieweg, 1993.
152. Eden, M. (Ed.): Proceedings of Conference on Diagnostic Data Processing. In: *IRE Transactions on Medical Electronics*, ME-7, p. 232, 1960.
153. Edwards, P.: *Computers and Politics of Discourse in Cold War America*. Cambridge: MIT Press, 1996.
154. Egglestone, H. G.: *Convexity Cambridge*. Cambridge: University Press, 1958.
155. Ehlers, Carl Th., N. Hollberg and A. Poppe (Ed.): *Werkzeug der Medizin*. Berlin: Springer, 1970.
156. Einhorn, Rudolf.: *Vertreter der Mathematik und der Geometrie an den Wiener Hochschulen 1900–40*. Dissertation. Verband der wissenschaftlichen Gesellschaften Österreichs, Vienna, 1985.

157. Einstein, Albert: Zur Elektrodynamik bewegter Körper. In: *Annalen der Physik*, Bd. 17, pp. 891–921, 1905.
158. Einstein, Albert: Über die von der molekularkinetischen Theorie der Wärme geforderte Bewegung von in ruhenden Flüssigkeiten suspendierten Teilchen. In: *Annalen der Physik*, Bd. 17, pp. 549–560, 1905.
159. Einstein, Albert: Über einen die Erzeugung und Verwandlung des Lichtes betreffenden heuristischen Gesichtspunkt. In: *Annalen der Physik*, Bd. 17, pp. 132–148, 1905.
160. Einstein, Albert: *Geometrie und Erfahrung*. Sitzungsberichte der Preussischen Akademie der Wissenschaften. Berlin: Springer, 1921.
161. Elder, R. C.: *Fuzzy Systems theory and Medical Decision Making*. M. S. Thesis. School of Industrial and Systems Engineering, Georgia Institute of Technology, Atlanta, Georgia, 1976.
162. Elias, P., A. Gill, R. Price, N. Abramson, P. Swerling and Lotfi A. Zadeh: Report on Progress in Information Theory in the U.S.A. 1960–1963. In: *IEEE Transactions on Information Theory*, pp. 128–144, July 1961.
163. Elias, P.: Cybernetics: Past and Present, East and West. In: *The Study of Information*. New York: Machlup & U. Mansfield, John Wiley & Sons, pp. 441–444, 1983.
164. Ellis, Clarence A.: Probabilistic Tree Automata. In: *Information and Control*, 19, pp. 401–416, 1971.
165. Esogbue, Augustine O. and Robert C. Elder: Fuzzy Sets and the Modelling of Physician Decision Processes, Part I: The Initial Interview-Information Gathering Session. In: *Fuzzy Sets and Systems*, 2, pp. 279–291, 1979.
166. Esogbue, Augustine O. and Robert C. Elder: Fuzzy Sets and the Modelling of Physician Decision Processes, Part II: Fuzzy Diagnosis Decision Models. In: *Fuzzy Sets and Systems*, 3, pp. 1–9, 1980.
167. Espenschied, Lloyd: Electric Communications, the Past and Present Illuminate the Future. A Suggestive Interpretation. In: *Proceedings of the IRE*, pp. 395–402, August 1943.
168. Everitt, William Littell and George E. Anner: *Communication Engineering*. New York, Toronto, London: McGraw Hill Book Comp., Inc., 1956.
169. Fagen, M. D. (Ed.): *A History of Engineering and Science in the Bell System. The Early Years (1875–1925)*. Bell Telephone Laboratories, Inc. New York: The Bell Laboratories, 1975.
170. Fedrowitz, Jutta, Gert Kaiser and Dirk Matejovski (Ed.): *Kultur und Technik im 21. Jahrhundert* (= Schriftenreihe des Wissenschaftszentrums Nordrhein-Westfalen, Band. 1). Frankfurt am Main: Campus Verlag, 1993.
171. Feigan, Gerry M. D.: Triple Bromides. In: *San Francisco Medical Society Bulletin*, June 1956.
172. Feldkeller, Richard: *Einführung in die Siebschaltungtheorie der elektrischen Nachrichtentechnik*. (= Monographien der elektrischen Nachrichtentechnik, Band IV). (3rd edition). Stuttgart: Hirzel Verlag, 1950.
173. Fellinger, Karl (Ed.): *Computer in der Medizin – Probleme, Erfahrungen, Projekte*. Vienna: Verlag Brüder Hollinek, 1968.
174. Fischer, Kurt Rudolf (Ed.): *Österreichische Philosophie von Brentano bis Wittgenstein*. Wien: WUV, 1999.
175. Foerster, Heinz von: *KybernEthik*. Berlin: Merve-Verlag, 1993.
176. Fölsing, Albrecht: *Heinrich Hertz. Eine Biographie*. Hamburg: Hofmann und Campe, 1997.

177. Fordon, W. A. and J. C. Bezdek: *The Application of Fuzzy Set Theory to Medical Diagnosis*. In: [218], pp. 445–461, 1979.
178. Fortner, H. and Louis Couffignal: Denkmaschinen. In: *Deutsche Zeitschrift für Philosophie*, 4, 3, pp. 371–375, 1956.
179. Foster, Ronald M.: Academic and Theoretical Aspects of Circuit Theory. In: *Proceedings of the IEEE*, pp. 866–871, May 1963.
180. Fox, Jerome (Ed.): *System Theory*. Microwave Research Institute Symposia, Series XV. Broooklyn, New York: Polytechnic Press, 1965.
181. Fox, Jerome (Ed.): *Proceedings of the Symposium on Modern Network Synthesis (Audio to Microwaves)*. New York, 16–18 April 1952.
182. Frank, Philipp: Der historische HinterGrund. Translation of *Introduction: Historical Background*. In: [183], pp. 1–52, 1949.
183. Frank, Philipp: *Modern Science and Its Philosophy*. Cambridge, Mass.: Harvard University Press, 1949.
184. Friedman, Saul: The RAND Corporation and our Policy Makers. In: *The Atlantic Monthly*, pp. 61–68, 1963.
185. Frohberg, Wolfgang: *Künstliche Intelligenz und Telekommunikation*. Berlin, Munich: Verlag Technik, 1993.
186. Fuchs-Kittowski, Klaus: *Probleme des Determinismus und der Kybernetik in der molekularen Biologie*. Jena: Gustav Fisher Verlag, 1969 (2nd edition 1976).
187. Gadol, Eugene T. (Ed.): *Rationality and Science. A Memorial Volume for Moritz Schlick in Celebration of the Centennial of his Birth*. Vienna: Springer, 1982.
188. Gaines, Brian R. and L. J. Kohout: The Fuzzy Decade: A Bibliography of Fuzzy Systems and Closely Related Topics. In: *International Journal Man-Machine Studies*, 9, pp. 1–68, 1977.
189. Gaines, Brian R.: Foundations of Fuzzy Reasoning. In: *International Journal of Man-Machine Studies*, 8, pp. 623–668, 1976.
190. Gaines, Brian R.: Precise Past – Fuzzy Future. In: *International Journal of Man-Machine Studies*, 19, pp. 117–134, 1983.
191. Galison, P.: The Ontology of the Enemy : Norbert Wiener and the Cybernetic Vision. In: *Critical Inquiry*, 21, pp. 228–66, 1994.
192. Gangl, A., Georg Grabner and Peter Bauer: Erste Erfahrungen mit einem Computerprogramm zur Differentialdiagnose der Lebererkrankungen. In: *Wiener Zeitschrift für Innere Medizin und Ihre Grenzgebiete*, 50, pp. 553–586, 1969.
193. Gardner, Howard: *Dem Denken auf der Spur. Der Weg der Kognitionswissenschaft*. Stuttgart: Klett-Cotta, 1989. Original English edition: *The Minds New Science. A History of the Cognitive Revolution*. Translation by Ebba D. Drolshagen. New York: Basic Books, 1985.
194. Geier, Manfred: *Der Wiener Kreis*. Reinbek bei Hamburg: Rowohlt, 1992.
195. Gibbs, F. A. and A. M. Grass: Frequency Analysis of Electroencephalogramms. In: *Science*, 105, 31, 1, pp. 132–134, 1947.
196. Gigch, John P. van: *Applied General Systems Theory*. New York: Harper & Row, 1974.
197. Gödel, Kurt: Über formal unentscheidbare Sätze der Principia Mathematica und verwandter Systeme I, *Monatshefte für Mathematik und Physik* 38, 1931, pp.173–198.
198. Goguen, Joseph A.: *Categories of Fuzzy Sets: Applications of a Non-Cantorian Set Theory*. Ph.D. Thesis. University of California at Berkeley, June 1968.

199. Goguen, Joseph A.: The Logic of Inexact Concepts. In: *Synthese*, 19, pp. 325–373, 1969.
200. Goguen, Joseph A.: *L*-Fuzzy Sets. In: *Journal of Mathematical Analysis and Applications*, 18, pp. 145–174, 1967.
201. Golinski, Jan: *Making Natural Knowledge. Constructivism and the History of Science*. Cambridge: University Press, 1998.
202. Golland, Louise, Brian McGuinness and Abe Sklar (Eds.): *K. Menger. Reminiscences of the Vienna Circle and the Mathematical Colloquium*. Dordrecht: Kluwer, 1994.
203. Gottwald, Siegfried: *Mehrwertige Logik und unscharfe Mengen*. In: [500], pp. 185–204, 1999.
204. Grabner, Georg (Ed.): *WAMIS. Wiener Allgemeines Medizinisches Informations-System. 10 Jahre klinische Praxis und Forschung*. (= Medizinische Informatik und Statistik, Band 59). Berlin, Heidelberg, New York, Tokyo: Springer, 1985.
205. Green, David M. and John A. Swets: *Signal Detection Theory and Psychophysics*. New York: Wiley & Sons, 1966.
206. Greniewski, Henryk: *Cybernetics without Mathematics*. Warsaw: Pergamon Press, 1960.
207. Grüsser, Otto-Joachim and Rainer Klinke (Ed.): *Zeichenerkennung durch biologische und technische Systeme*. Tagungsbericht des 4. Kongresses der Deutschen Gesellschaft für Kybernetik durchgeführt an der Technischen Universität Berlin vom 6.–9. April 1970. Berlin, Heidelberg, New York: Springer, 1971. (Original edition: *Pattern Recognition in Biological and Technical Systems*. Proceedings of the 4th Congress of the Deutsche Gesellschaft für Kybernetik, held at Berlin, Technical University, April 6–9, 1970).
208. Guillemin, Ernst A.: Guillemin Ernst A. (Autobiographical Essay). In: *McGraw-Hill Modern Men of Science*, Vol. II, pp. 197–199, 1968.
209. Guillemin, Ernst A.: Making Normal Coordinates Coincide with the Meshes of an Electrical Network. In: *Proceedings of the Institute of Radio Engineers*, Vol. 15, 11, pp. 935–944, 1927.
210. Guillemin, Ernst A.: A Recent Contribution to the Design of Electrical Filter Networks. In: *Journal of Mathematics and Physics*, 11, pp. 150–211, 1931–32.
211. Guillemin, Ernst A.: The Fourier Integral – A Basic Introduction. In: *IRE Transactions on Circuit Theory*, Vol. CT-2, No. 3, pp. 227–230, September, 1955.
212. Guillemin, Ernst A.: *Introductory Circuit Theory*. New York, London, Sydney: John Wiley & Sons, 1953.
213. Guillemin, Ernst A.: *Communication Networks*. Vol. I: The Classical Theory of Lumped Constant Networks. New York: John Wiley & Sons, 1931.
214. Guillemin, Ernst A.: *Communication Networks*. Vol. II: The Classical Theory of Long Lines, Filters and Related Networks. New York: John Wiley & Sons, 1935.
215. Guillemin, Ernst A.: *Synthesis of Passive Networks. Theory and Methods Appropriate to the Realisation and Approximation Problems*. New York, London: John Wiley & Sons, Inc., Chapman & Hall, 1957.
216. Guillemin, Ernst A.: Teaching of Circuit Theory and Its Impact on Other Disciplines. In: *Proceedings of the IRE*, pp. 872–878, May 1962.
217. Guillemin, Ernst A.: Zur Theorie der Frequenzvervielfachung durch Eisenkernkoppelung. In: *Archiv für Elektrotechnik*, XVII. Band, 1, pp. 17–51, 1926.

218. Gupta, Mandan M., Rammohan K. Ragade and Ronald R. Yager (Eds.): *Advances in Fuzzy Set Theory and Applications*. Amsterdam, New York, Oxford: North-Holland Publ. Comp., 1979.
219. Gupta, Mandan M. and Elie Sanchez (Eds.): *Fuzzy Information and Decision Processes*. Amsterdam, New York, Oxford: North-Holland Publ. Comp., 1982.
220. Gupta, Mandan M. and Elie Sanchez (Eds.): *Approximate Reasoning in Decision Analysis*. Amsterdam, New York, Oxford: North-Holland Publ. Comp., 1982.
221. Haack, Susan: *Deviant Logic – Some Philosophical Issues*. Cambridge: University Press Deviant, 1974.
222. Hagemeyer, Friedrich-Wilhelm: *Die Entstehung von Informationskonzepten in der Nachrichtentechnik*. Philosophische Dissertation. Berlin, 1979.
223. Haider, Hans: In zwei Wiener Kreisen zugleich. Presse-Gespräch mit Karl Menger über seine akademischen Anfangsjahre. In: *Die Presse*, No. 9121, p. 5, August 18, 1978.
224. Hall, Lawrence O. and Abraham Kandel: The Evolution from Expert Systems to Fuzzy Expert Systems. In: [271], pp. 3–21, 1991.
225. Haller, Rudolf: *Neopositivismus. Eine historische Einführung in die Philosophie des Wiener Kreises*. Darmstadt: Wissenschaftliche Buchgesellschaft, 1993.
226. Halmos, Paul Richard: *Naive Set Theory*. New York: Van Nostrand, 1960.
227. Hamilton, Patrick: *Künstliche Neuronale Netze. Grundprinzipien, Hintergründe, Anwendungen*. Berlin, Offenbach: vde-verlag, 1993.
228. Händle, Frank and Stefan Jensen (Ed.): *Systemtheorie and Systemtechnik. Sechzehn Aufsätze*. Munich: Nymphenburger Verlagshandlung, 1974.
229. Hartkopf, Werner and Gert Wangermann: *Dokumente zur Geschichte der Berliner Akademie der Wissenschaften von 1700 bis 1990*. Berlin, Heidelberg, New York: Spektrum Akademie Verlag, 1991.
230. Hartley, Ralph V. L.: Transmission of Information. In: *The Bell System Technical Journal*, Vol. VII, No. 3, pp. 535–563, July 1928.
231. Hartley, Ralph V. L.: A More Symmetrical Fourier Analysis Applied to Transmission Problems. In: *Proceedings of the IRE*, pp. 144–150, March 1942.
232. Haufe, Robert: Design of a Tit-Tat-Toe Machine. In: *Electrical Engineering*, Vol. 68, p. 885, October 1949.
233. Haugeland, John: *Künstliche Intelligenz – Programmierte Vernunft?* Hamburg [et al.]: McGraw-Hill Book Comp. GmbH, 1987.
234. Hebb, Donald: *The Organization of Behavior*. New York: Wiley, 1949.
235. Hegselmann, Rainer and Geo Siegwart: Zur Geschichte der ‚Erkenntnis'. In: *Erkenntnis*, 35, pp. 461–471, 1991.
236. Heims, Steve Joshua: *The Cybernetics Group*. Cambridge, Massachusetts, London: The MIT Press, 1991.
237. Heims, Steve Joshua: *John von Neumann and Norbert Wiener: From Mathematics to the Technologies of Life and Death*. Cambridge, Massachusetts: The MIT Press, 1980.
238. Heisenberg, Werner: *Der Teil und das Ganze. Gespräche im Umkreis der Atomphysik*. Munich: Piper 1969. Paperback edition, Munich: Deutscher Taschenbuch Verlag, 1973, (7th edition: 1983).
239. Heske, Franz, Jordan Pascual and Adolf Meyer-Abich: *Organik. Beiträge zur Kultur unserer Zeit*. Berlin: Haller-Verlag, 1954.
240. Hillis, Daniel: *Computerlogik. So arbeiten Computer*. Munich: Bertelsmann, 1999.

241. Hobsbawm, Eric J.: *Das imperiale Zeitalter. 1875–1914.* Frankfurt am Main: Fischer Taschenbuch Verlag, 2004.
242. Hofestädt, Ralf, Fritz Krückeberg and Thomas Lengauer (Ed.): *Informatik in den Biowissenschaften.* (1. Fachtagung der GI-FG 4.0.2 "Informatik in den Biowissenschaften", Bonn, 15./16. Februar 1993). Berlin [et al.]: Springer, 1993.
243. Hoffmann, Norbert: *Kleines Handbuch neuronale Netze – anwendungsorientiertes Wissen zum Lernen und Nachschlagen.* Braunschweig, Wiesbaden: Vieweg, 1993.
244. Holloway, David: Innovation in Science – The Case of Cybernetics in the Soviet Union. In: *Science Studies,* 4, pp. 299–337, 1974.
245. Holmblad, Lauritz P. and Jens-Jørgen Østergaard: *Control of a Cement Kiln by Fuzzy Logic.* In: [218], 1982.
246. Höpken, A.: *Künstliche Intelligenz, Expertensysteme und die Folgen.* Alsbach: Leuchtturm-Verlag, 1992.
247. Horak, W., P. Michalek, H. Richter and N. Thumb: *Computerdiagnostik in der Rheumatologie.* In: [173], pp. 232–237, 1968.
248. Houle, A.: *Searching for I. J. Good's 46, 656 kinds of Bayesians: A worthwhile endeavor?* Faculté des sciences de l'administration, Université Laval. Quebec, Canada, 1979.
249. Hucklenbroich, Peter and Richard Toellner (Ed.): *Künstliche Intelligenz in der Medizin. Klinisch-methodologische Aspekte medizinischer Expertensysteme.* (Symposium in der Werner-Reimers-Stiftung, Bad Homburg in Verbindung mit der Akademie der Wissenschaften und der Literatur, Mainz). (= Akademie der Wissenschaften und der Literatur. Mathematisch-naturwissenschaftliche Klasse. Medizinische Forschung 5, 1993). Stuttgart, Jena, New York: Gustav Fischer Verlag, 1993.
250. Humphrey, Hubert H.: *Memorandum from Subcomittee on Reorganization and International Organization: An Action Program for Strenghtening Medical Information and Communication.* SD. 10, 14. May 1962.
251. *Proceedings of the 1975 International Symposium on Multiple-Valued Logic,* Indiana University, Blomington, Indiana, May 13–16, 1975.
252. Ilgauds, Hans Joachim: *Norbert Wiener* (= Biographien hervorragender Naturwissenschaftler, Techniker und Mediziner). Band 45. Leipzig: Teubner Verlagsgesellschaft, 1980.
253. Immich, Herbert: Ereignisstatistik und Symptomenkunde. In: *Medizinische Monatsschrift,* 6, 12, 1952, I. Teil: p. 699–702, II. Teil: pp. 784–787.
254. ISI – International Statistical Institute, *Newsletter,* Vol. 25, no. 3 (75) 2001: $http://www.cbs.nl/isi/memoriam01--3.htm.$
255. Jahrbuch der Akademie der Wissenschaften der DDR. Berlin: Akademie-Verlag. Bände: (1971/72 published 1973) – (1988, published 1989), (1990/91 published 1994).
256. Janik, Alan: *Kreative Milieus. Der Fall Wien.* Vienna: Berner, pp. 45–55, 1986.
257. Janik, Alan: *Vienna 1900: Reflections on Problems and Methods.* In: [103], p. 156, 1990.
258. Jeffress, L. A. (Ed.): *Cerebral Mechanisms in Behavior. The Hixon Symposium.* New York: John Wiley & Sons, 1951. (German: Warum der Geist im Kopf ist. In: [364], pp. 93–158, 2000).
259. Kalaba, Robert: Rezension: Linear System Theory – The State Space Approach, by Lotfi A. Zadeh and Charles A. Desoer. In: *Proceedings of the IEEE,* Vol. 53, No. 9, pp. 1282f, September 1965.

260. Kalman, Rudolf E., Y. C. Ho and K. S. Narendra: Controlability of Linear Dynamical Systems. In: *Cont, Differential Equations*, 1, pp. 189–213, 1962.
261. Kalman, Rudolf E. and Nicholas DeClaris (Eds.): *Aspects of Network and System Theory*. New York [et al.]: Holt, Rinehart and Winston, 1971.
262. Kalman, Rudolf E.: *Round-Table-Discussion of 1970*. In: [351], p. 169, 1972.
263. Kalman, Rudolf E.: Book Review: Lotfi A. Zadeh, Charles A. Desoer. *System Theory – The State Space Approach*. New York: McGraw Hill, 1963, in: *Proceedings of the IEEE*, vol. 53, pp. 1282–1283, September 1965.
264. Kalman, Rudolf E.: Canonical Structure of Linear Dynamical Systems. In: *Proceedings of the National Academy of Sciences of the United States of America*, 48, pp. 596–600, 1962.
265. Kalman, Rudolf E.: *On the General Theory of Control Systems*. In: *Proceedings of the First International Congress, IFAC*, Moscow, 1960. Automatic and Remote Control. London: Butterworths and Comp. Ltd., pp. 481–92, 1961.
266. Kalman, Rudolf E., Peter L. Falb and Michael A. Arbib: *Topics in Mathematical System Theory*. New York [et al.]: McGraw-Hill, 1969.
267. Kandel, Abraham and William Byatt: Fuzzy Sets, Fuzzy Algebra, and Fuzzy Statistics. In: *Proceedings of the IEEE*, Vol. 66, No. 12, pp. 1619–1639, December 1978.
268. Kandel, Abraham and Samuel C. Lee: *Fuzzy Switching and Automata*. New York: Crane Russak, 1979.
269. Kandel, Abraham: *Fuzzy Techniques in Pattern Recognition*. New York, Chichester, Brisbane, Toronto, Singapore: John Wiley & Sons, 1982.
270. Kandel, Abraham: Reply by Abraham Kandel. In: *Proceedings of the IEEE*, Vol. 67, No. 8, p. 1169, August 1979.
271. Abraham Kandel (Ed.): *Fuzzy Expert Systems*. Boca Raton, Ann Arbor, London: CRC Press, 1991.
272. Karakash, John J.: *Transmission Lines and Filter Networks*. New York: The MacMillan Comp., 1950.
273. Kass, Seymour: Karl Menger. In: *Notices of the American Mathematical Society*, Vol. 43, No. 5, pp. 558–561, 1996.
274. Kaufmann, Arnold.: *Introduction to the Theory of Fuzzy Subsets*. Vol. I: Fundamental Theoretical Elements. New York: Academic, 1975.
275. Kay, Lily E.: *Das Buch des Lebens. Wer schrieb den genetischen Code?* Munich, Vienna: Hanser, 2000.
276. Khinchin, Aleksandr Iakovlevic: *Mathematical Foundations of Information Theory*. New York: Dover Pub. Inc. 1957 (Russian edition 1953, published in the GDR in 1957.)
277. Killmer, William L.: Designated Discussion (about [594]). In: [451], pp. 207–212, 1969.
278. Kindler, H.: Mathematische und physikalisch-technische Probleme der Kybernetik. In: *Universitätszeitung TU-Dresden*, 1, p. 3, Aprilnummer 1962.
279. Kirchenmann, P.-P.: *Kybernetik, Information, Widerspiegelung*. Munich, 1959.
280. Kittler, Friedrich, Peter Berz, David Hauptmann and Axel Roch (Ed.): *Claude E. Shannon. Ein/Aus. Ausgewählte Schriften zur Kommunikations- und Nachrichtentheorie*. Berlin: Verlag Brinkmann und Bose, 2000.
281. Kleene, Stephen C.: Analysis of Lengthening of Modulated Repetive Pulses. In: *Proceedings of the IRE*, pp. 1049–1053, October 1947.
282. Kleene, Stephen C.: *Representation of Events in Nerve Nets and Finite Automata*. In: [504], pp. 3–41, 1956.

283. Klement, Erich Peter, Radko Mesiar and Endre Pap: *Bausteine der Fuzzy Logic: t-Normen – Eigenschaften and Darstellungssätze.* In: [500], pp. 205–225, 1999.
284. Klir, George J.: *An Approach to General Systems Theory.* New York [et al.]: Van Nostrand Reinhold Comp., 1969.
285. Klir, George J.: *Facets of Systems Science.* New York, London: Plenum Press, 1991.
286. Klir, George J.: *Principles of Uncertainty: What are they? Why do we need them?* In: *Fuzzy Sets and Systems,* 74, pp. 15–31, 1995.
287. Köhler, Eckehart: *Gödel und der Wiener Kreis.* In: [299], pp. 127–158.
288. Köhler, Eckehart: *Die Metaphysik des Wiener Kreises.* In: [134], pp. 190–204.
289. Körner, Stephan: *The Philosophy of Mathematics. An Introductory Essay.* London: Hutchinson & Co. Ltd. 1960. German: *Philosophie der Mathematik. Eine Einführung.* Munich: Nymphenburger Verlagsbuchhandlung 1968.
290. Kolman, E.: *La cybernétique vue par un philosophe soviétique.* In: *La Pensée,* 68, pp. 14–34, 1956.
291. Kolman E.: *Was ist Kybernetik.* In: *Forum,* 9, No. 23, 1955.
292. Kolman, E.: *Was ist Kybernetik.* In: *Sowjetwissenschaft – Naturwissenschaftliche Beiträge,* 4, 4, pp. 309–326, 1956.
293. Kolman, E.: *What is cybernetics?* In: *Behavioral Science,* 4, pp. 132–146, 1959.
294. Kolmogorov, Alexander N.: *Interpolation and Extrapolation.* In: *Bulletin de l'Académie des Sciences de l'URSS, Serie Mathématique,* Vol. 5, pp. 3–14, 1941.
295. Kosko, Bart: *Die Zukunft ist fuzzy. Unscharfe Logik verändert die Welt.* Munich, Zürich: Piper, 2001.
296. Kosko, Bart: *Fuzzy Thinking. The New Science of Fuzzy Logic.* London: Flamingo, 1994.
297. Kraft, Viktor: *Der Wiener Kreis. Der Ursprung des Neopositivismus.* Vienna, New York: Springer, 1968.
298. Krauch, Helmut, Werner Kunz and Horst Rittel (Ed.): *Forschungsplanung. Eine Studie über Ziele und Strukturen Amerikanischer Forschungsinstitute.* Munich, Vienna: Oldenbourg, 1966.
299. Kruntorad, Paul: *Jour Fixe der Vernunft. Der Wiener Kreis und die Folgen.* Vienna: Verlag Hölder-Pichler-Tempsky, 1991.
300. Kruse, Rudolf and Klaus Dieter Meyer: *Statistics with Vague Data.* (Theory and Decision Library, Series B, Mathematical and Statistical Methods, Vol 6). Dordrecht: D. Reidel Publishing Comp., 1987.
301. Kruse, Rudolf, Detlef Nauk and Frank Klawonn: *Neuronale Fuzzy-Systeme.* In: *Spektrum der Wissenschaft,* 6, pp. 34–41, 1995.
302. Küpfmüller, K.: *Informationstheorie.* In: *Jahrbuch des elektrischen Fernmeldwesens,* 8, pp. 25–48, 1954/55.
303. Kurzrock, Ruprecht (Ed.): *Systemtheorie (= Forschung and Information,* Band 12). Berlin: Colloquium Verlag, 1972.
304. Lakoff, George.: *A Study in Meaning Criteria and the Logic of Fuzzy Concepts.* In: *Proceedings of the 8th Regional Meeting of the Chicago Linguistic Society.* University of Chicago, Linguistic Department, April 1972.
305. Lakoff, George: *Hedges: A Study in Meaning Criteria and the Logic of Fuzzy Concepts.* In: *Journal of Philosophical Logic,* 2, pp. 458–508, 1973.
306. Laplace, Pierre Simon de: *Théorie analytique des probabilités – Introduction (= Essai philosophique sur les probabilités).* Paris: Courcier, 1814.

307. Latil, Pierre de: *La Pensée artificielle – Introduction à la cybernétique. (Artificial thought – Introduction to cybernetics.)*. Paris: Gallimard, 1953.
308. Ledley, Robert Steven and Lee B. Lusted: Reasoning Foundations of Medical Diagnosis. In: *Science*, Vol. 130, No. 3366, pp. 9–21, 3. July 1959.
309. Ledley, Robert Steven: Computer Aids to Medical Diagnosis. In: *Journal of the American Medical Association*, 196, p. 933, 1966.
310. Ledley, Robert Steven: *Fortran VI Programming*. New York McGraw-Hill, 1966.
311. Ledley, Robert Steven: Digital Computational Methods in Symbolic Logic with Examples in Biochemistry. In: *Proceedings of the National Academy of Sciences of the United States of America*, 41, p. 7, 1955.
312. Ledley, Robert Steven: *Digital Computer and Control Engineering*. New York, Toronto, London: McGraw-Hill Book Comp. Inc., 1960.
313. Ledley, Robert Steven: Logical Aid to Symptomatic Medical Diagnosis. In: *Operations Research*, 3, August 1955.
314. Ledley, Robert Steven: Mathematical Foundations and Computational Methods for a Digital Logic Machine. In: *Journal of the Operations Research Society of America*, 2, p. 3, 1954.
315. Ledley, Robert Steven: Medical Informatics: A Personal View of Sowing the Seeds. In: *The Association for Computing Machinery: ACM Conference on the History of Medical Informatics*. (= Conference Proceedings. Papers Presented at the Conference, National Library of Medicine Bethesda, Maryland, November 5–6, 1987), pp. 31–41, 1987.
316. Ledley, Robert Steven: *Use of Computers in Biology and Medicine*. New York: McGraw-Hill Book Comp., 1965.
317. Lee, Samuel C. and Edward T. Lee: Fuzzy Neural Networks. In: *Mathematical Biosciences*, 23, pp. 151–177, 1975.
318. Lee, Samuel C. and Edward T. Lee: Fuzzy Neurons and Automata. In: *Proceedings of the Fourth Annual Princeton Conference on Information Sciences and Systems*. (Papers presented March 26–27, 1970), pp. 381–385, 1970.
319. Lenk, Hans and Günter Ropohl (Ed.): *Systemtheorie als Wissenschaftsprogramm*. Königstein: Athenäum Verlag, 1978.
320. Lentin, André: La cybernétique: problémes réels et mystifications. In: *La Pensée*, No. 47, pp. 47–61, mars-avril 1953.
321. Leonhard, Robert J.: Ethics and the Excluded Middle. Karl Menger and Social Science in Interwar Vienna. In: *ISIS*, p. 89, 1998.
322. Linvill W. K., R. E. Scott and E. A. Guillemin: *Evaluation of Fourier Transforms*. In: [211], pp. 243–250.
323. Lipkin, M. and J. D. Hardy: Mechanical Corellation of Data in Differential Diagnosis of Hematological Diseases. In: *Journal of the American Medical Association*, 166, pp. 113–125, 1958.
324. Loginov, V. I.: Probability Treatment of Zadeh Membership Functions and their Use in Pattern Recognition. In: *Engineering Cybernetics IEEE-Veröffentlichung*, No. 2, pp. 68–69, March-April 1966.
325. Lorenz, Dagmar: *Wiener Moderne*. Stuttgart, Weimar: Metzler, 1998.
326. Lowenschuss, Oscar: A Comment on Pattern Redundancy. In: *IRE Transactions on Informations Theory*, p. 127, December 1958.
327. Lowenschuss, Oscar: *Multi-Valued Logic in Sequential Machines*. Ph.D. Thesis. School of Engineering, Columbia University, New York, 1958.

328. Lowenschuss, Oscar: Restoring Organs in Redundant Automata. In: *Information and Control*, 2, pp. 113–136, 1959.
329. Lowenschuss, Oscar: Non-Binary Switching Theory. In: *IRE National Convention Record*, 6, Part 4, pp. 305–317, August 1958.
330. Luce, Duncan: The Decomposition Theorems for a Class of Finite Oriented Graphs. In: *American Journal of Mathematics*, Vol. 74, pp. 701–722, 1952.
331. Łukasiewicz, Jan: *Philosophical Remarks on Many-Valued Systems of Propositional Logic*, 1930. Reprinted in: Borkowski, L. (Ed.) *Selected Works* (Studies in Logic and the Foundations of Mathematics), Amsterdam: North Holland, pp. 153–179, 1970.
332. Luria, Alexander Romanovich: *The Role of Speech in the Regulation of Normal and Abnormal Behavior*. New York: Irvington, 1960.
333. Lusted, Lee B. and R. S. Ledley: Mathematical Models in Medical Diagnosis. In: *Journal of Medical Education*, Vol. 35, No. 3, pp. 214–222, March 1960.
334. Lusted, L. B.: Computer Techniques in Medical Diagnosis. In: [518], pp. 319–338, 1965.
335. Lusted, Lee B.: Twenty Years of Medical Decision Making Studies. In: *Proceedings of the Third Annual Symposium on Computer Application in Medical Care*, Washington D.C., pp. 4–8, Oct. 14–17, 1979.
336. Lusted, Lee B.: Design for Decisions – A 25 Year Perspective. In: [58], p. XI–XV, 1983.
337. Lusted, L. B.: Medical Electronics. In: *The New England Journal of Medicine*, pp. 580–585, April 7, 1955.
338. Lusted, Lee B.: The Clearing "Haze". A View from My Window. In: *Medical Decision Making*, 11, No. 2, pp. 76–87, April-June, 1991.
339. Lusted, Lee B.: *Introduction to Medical Decision Making*. Springfield: Charles C. Thomas, 1968.
340. Lusted, Lee B.: Computers in Medicine – A Personal Perspective. In: *Journal of Chronic Diseases and Therapeutic Research*, Vol. 19, pp. 365–372, 1966.
341. Lusted, Lee B.: Computer Programming of Diagnostic Tests. In: *IRE Transactions on Medical Electronics*, pp. 255–258, October 1960.
342. Lusted, Lee B.: Quantification in the Life Sciences. In: *IRE Transactions on Bio-Medical Electronics*, pp. 1–3, January 1962.
343. Lusted, Lee B.: Some Roots of Clinical Decision Making. In: *The Association for Computing Machinery (= ACM Conference on the History of Medical Informatics*. Conference Proceedings. Papers presented at the Conference, National Library of Medicine Bethesda, Maryland, November 5–6, 1987), pp. 165–193, 1987.
344. Mach, Ernst: *Die Principien der Wärmelehre. Historisch-kritisch entwickelt von Dr. Ernst Mach*. (4th edition). Leipzig: Verlag von Johann Ambrosius Barth, 1923.
345. Maiers, Jerald: Fuzzy Set Theorie and Medicine: The First Twenty Years and Beyond. In: *IEEE*, pp. 325–329, 1985.
346. Mainzer, Klaus: *Computer – Neue Flügel des Geistes? Die Evolution computergestützter Technik, Wissenschaft, Kultur und Philosophie*. Berlin, New York: de Gruyter, 1995.
347. Maletzke, Gerhard: *Kommunikationswissenschaft im Überblick. Grundlagen, Probleme, Perspektiven*. Opladen, Wiesbaden: Westdeutscher Verlag, 1998.

348. Mamdani, Ebrahim H.: Twenty Years of Fuzzy Control: Experiences Gained and Lessons Learnt. In: *IEEE Transactions on Fuzzy Systems.* 1, pp. 19–24, 1993.
349. Mamdani, Ebrahim H. and S. Assilian: An Experiment in Linguistic Synthesis with a Fuzzy Logic Controller. In: *International Journal of Man-Machine Studies*, Vol. 7, No. 1, pp. 1–13, 1975.
350. Mamdani, Ebrahim H.: Advances in the Linguistic Synthesis of Fuzzy Controllers. In: *International Journal of Man-Machine Studies*, 8, pp. 669–678, 1976.
351. Marois, M. (Ed.): Man and Computer. In: *Proceedings of the First International Conference on Man and Computer*, Bordeaux, June 22–26, 1970. Basel: S. Karger, 1972.
352. Martin, C.: Dianne: The Myth of the Awesome Thinking Machine. In: *Communications of the ACM*, Vol. 36, No. 4, pp. 120–133, 1993.
353. Masani, Pesi Rustom: *Norbert Wiener 1894–1964.* Basel, Boston, Berlin: Birkhäuser Verlag, 1990.
354. Mason, Samuel J. and Henry J. Zimmermann: *Electronic Circuits, Signals, and Systems.* Cambridge, Massachusetts, London: The MIT Press, 1970.
355. Mason, Samuel J.: Signal Flow Graphs. In: *Proceedings of the IRE*, 39, p. 297, 1951.
356. Mason, Samuel J.: Feedback Theory – Some Properties of Signal Flow Graphs. In: *Proceedings of the IRE*, Vol. 41, pp. 1144–1156, 1953.
357. Mason, Warren P.: *Electromechanical Transducers and Wave Filters.* Toronto, New York, London: D. Van Nostrand Comp., Inc., 1942. (2nd edition 1948).
358. Maturana, Humberto R., Jerome Y. Lettvin, Warren S. McCulloch and Walter H. Pitts: Anatomy and Physiology of Vision in the Frog. In: *Journal of General Physiology*, 43,1960.
359. Mayr, Ernst: *Das ist Biologie. Die Wissenschaft des Lebens.* Heidelberg, Berlin: Spektrum Akademischer Verlag 1997. (Translation from English of Jorum Wissmann: Mayr, Ernst: *This is Biology.* Harvard: University Press, 1997.)
360. McCartney, Scott: *ENIAC. The Triumphs and Tragedies of the World's First Computer.* New York: Walker and Comp., 1999.
361. McCulloch, Warren S. and Walter H. Pitts: A Logical Calculus of the Ideas Immanent in Nervous Activity. In: *Bulletin of Mathematical Biophysics*, 5, pp. 115–133, 1943.
362. McCulloch, Warren S. and Walter H. Pitts: Ein Logikkalkül für die Nerventätigkeit immanenter Gedanken. In: [364], pp. 24–40, 2000. (English original: [361].)
363. McCulloch, Warren S.: The Brain as a Computing Machine. In: *Electrical Engineering. The Journal of the AIEE*, Vol. 68, pp. 492–497, 1949.
364. McCulloch, Warren S.: *Verkörperungen des Geistes* (= Computerkultur Band VII). Vienna, New York: Springer, 2000.
365. McCulloch, Warren S.: Why the Mind is in the Head. In: [258], pp. 93–158, 1951.
366. McCulloch, Warren S.: What Is a Number, that a Man May Know It, and a Man, that He May Know a Number? In: *Ninth Alfred Korybski Memorial Lecture, General Semantics Bulletin*, No. 26 and 27. Lakeville, Conn.: Institute of General Semantics, pp. 7–18, 1961.
367. McGuinness, Brian (Ed.): *Hans Hahn, Empirismus, Logik, Mathematik.* Frankfurt am Main: Suhrkamp, 1988.

368. McKeen, Cattell J. and Jaques Cattell (Eds.): *American Men of Science. A Biographical Directory*. New York: The Science Press, 1938.
369. McMillian, B. and D. Slepian: Information Theory. In: *Proceedings of the IEEE*, pp. 1151–1157, May 1963.
370. McNeill, Daniel and Paul Freiberger: *Fuzzy Logic*. New York: Simon and Schuster, 1993. (German edition: *Fuzzy Logic: die „unscharfe" Logik erobert die Technik*. Munich: Drömer Knaur, 1994.)
371. Meadows, Dennis [et al.]: *Die Grenzen des Wachstums. Bericht des Club of Rome zur Lage der Menschheit*. Stuttgart: Deutsche Verlags Anstalt, 1972.
372. Mehrtens, Herbert: *Moderne, Sprache, Mathematik*. Frankfurt am Main: Suhrkamp 1990.
373. Mendel, Jerry M. and King Sun Fu: *Adaptive Learning and Pattern Recognition Systems: Theory and Applications*. New York: Academic Press, 1970.
374. Menger, Karl: *Calculus – A Modern Approach*. Boston [et al.]: Ginn and Comp., 1953.
375. Menger, Karl: *Dimensionstheorie*. Leipzig, Berlin: B. G. Teubner, 1928.
376. Menger, Karl: *Einleitung*. In: [367], pp. 9–18, 1988.
377. Menger, Karl: *Erinnerungen an Kurt Gödel* (Manuscript from 1981). Quoted from [520], p. 237, 1997.
378. Menger, Karl: Ensembles flous et fonctions aléatoires In: *Comptes Rendus Académie des Sciences*, 37, pp. 226–229, 1951.
379. Menger, Karl: Bericht über Metrische Geometrie. In: *Jahresbericht der Deutschen Mathematiker-Vereinigung*, 40, pp. 201–219, 1931.
380. Menger, Karl: Geometry and Positivism. A Probabilistic Microgeometry (1970). In: [389], pp. 225–234, 1979.
381. Menger, Karl: Bemerkungen zu Grundlagenfragen. In: *Jahresberichte der Deutschen Mathematiker-Vereinigung*, 37, Teil I: pp. 213–226, Teil II: pp. 298–302, Teil III: pp. 303–308, Teil IV: pp. 309–325, 1928.
382. Menger, Karl: Grundzüge einer Theorie der Kurven. In: *Proceedings, Royal Academy of Science, Amsterdam*, 28, pp. 67–71, 1925, sowie *Mathematische Annalen*, 95, pp. 277–306, 1925.
383. Menger, Karl: The Ideas of Variable and Function. In: *Proceedings of the National Academy of Sciences of the United States of America*, Vol. 39, pp. 956–961, 1953.
384. Menger, Karl: Der Intuitionismus. In: *Blätter für Deutsche Philosophie*, 4, pp. 311–325, 1930.
385. Menger, Karl: Zur allgemeinen Kurventheorie. In: *Fundamenta Mathematicae*, 10, pp. 96–115, 1926.
386. Menger, Karl: Zur Dimensions- und Kurventheorie. Unveröffentlichte Aufsätze aus den Jahren 1921–1923. Published in *Monatshefte für Mathematik und Physik*, 36, p. 411–432, 1929: 1) Zur Theorie der Punktmengen (Herbst 1921); 2) Über die Dimensionalität von Punktmengen (Januar 1922), 3) Ergänzung zu 2), 4) Über die Dimensionalität von Punktmengen (Herbst 1922), 5) Zur Theorie der Punktmengen II (Herbst 1923).
387. Menger, Karl: Memories of Moritz Schlick. In: [187], pp. 83–103, 1982.
388. Menger, Karl: Untersuchungen über allgemeine Metrik I-III. In: *Mathematische Annalen*, 100, 1928: 1) Theorie der Konvexität, p. 75–113; 2) Die Euklidische Metrik, p. 113–141; 3) Eine n-dimensionale Metrik, pp. 142–163, 1928.

389. Menger, Karl: *Selected Papers in Logic and Foundations, Didactics, Economics.* Vienna Circle Collection. Band 10. Dordrecht, Holland: D. Reidel Publ. Comp., 1979.
390. Menger, Karl: Über die Dimensionalität von Punktmengen I. In: *Monatshefte für Mathematik und Physik*, 33, pp. 148–160, 1923.
391. Menger, Karl: Probabilistic Theories of Relations. In: *Proceedings of the National Academy of Sciences of the United States of America*, Vol. 37, pp. 178–180, 1951.
392. Menger, Karl: Statistical Metrics. In: *Proceedings of the National Academy of Sciences of the United States of America*, Vol. 28, pp. 535–537, 1942.
393. Menger, Karl: Das Hauptproblem über die dimensionelle Struktur der Räume. In: *Proceedings, Royal Academy of Science, Amsterdam*, 29, pp. 138–144, 1926.
394. Menger, Karl: Zur Theorie der Punktmengen II (Herbst 1923). Published in [386].
395. Menger, Karl: Logical Tolerance in the Vienna Circle. In: [389], pp. 11–16, 1979.
396. Menger, Karl: Variable de diverses natures. In: *Bulletin des sciences mathématiques*, 78, 2, pp. 229–234, 1954.
397. Menger, Karl: On Variables in Mathematics and in Natural Sciences. In: *The British Journal for the Philosophy of Sciences*, Vol. V, pp. 134–142, 1955.
398. Mesarovic, Mihajlo D. and Yasuhiko Takahara: *General Systems Theory: Mathematical Foundations.* New York, San Francisco, London: Academic Press, 1975.
399. Mesarovic, Mihajlo D.: Views on General Systems Theory. In: *Proceedings of The Second Systems Symposium at Case Institute of Technology.* Huntington, New York: Robert E. Krieger Publ. Comp., 1964.
400. Michels, K., F. Klawonn, R. Kruse, and A. Nürnberger: *Fuzzy Control. Fundamentals, Stability and Design of Fuzzy Controllers.* Heidelberg: Springer (Studies in Fuzziness and Soft Computing, Vol. 200), 2006.
401. Miller, Kenneth S. and Lotfi A. Zadeh: Generalization of the Fourier Integrals. In: *IRE Transactions on Circuit Theory*, Vol. CT-2, No. 3, pp. 256–260, September, 1955.
402. Miller, Kenneth S. and R. J. Schwarz: Analysis of a Sampling Servomechanism. In: *Journal of Applied Physics*, Vol. 21, pp. 290–294, April 1950.
403. Minsky, Marvin and Seymour Papert: *Perceptrons.* Cambridge, Mass.: MIT Press, 1969.
404. Minsky, Marvin: *Computation: Finite and Infinite Machines.* Englewood Cliffs: Prentice Hall, 1967.
405. Minsky, Marvin (Ed.): *Semantic Information Processing.* Cambridge, London: The MIT Press, 1968.
406. Moon, R. E., S. Jordanov, A. Perez and I. N. Türksen: Medical Diagnostic System with Human-Like Reasoning Capability. In: [512], pp. 115–119, 1977.
407. Moore, Edward F.: *Gedanken Experiments on Sequential Machines.* (= Automata Studies, Annals of Mathematical Studies, No. 34). Princeton: University Press, 1956.
408. Mulder, Hank L.: Wissenschaftliche Weltauffassung. Der Wiener Kreis. In: *Journal of the History of Philosophy*, 6, pp. 368–390, 1968.
409. Murakami, T. and Murlan S. Corrington: Applications of the Fourier Integral in the Analysis of Color Television Systems. In: *IRE Transactionson Circuit Theory*, Vol. CT-2, 3, pp. 250–255, September 1955.

410. Musil, Robert: *Der Mann ohne Eigenschaften*. Reinbek bei Hamburg: Rowohlt, 1970.
411. Nagy, George: State of the Art in Pattern Recognition. In: *Proceedings of the IEEE*, Vol. 56, No. 5, pp. 836–863, May 1968.
412. Nash, F. A.: Differential Diagnosis: An Apparatus to Assist the Logical Faculties. In: *Lancet*, 1, pp. 874–875, 1954.
413. Nautz, Jürgen and Richard Vahrenkamp (Ed.): *Die Wiener Jahrhundertwende. Einflüsse, Umwelt, Wirkungen*. Vienna, Cologn, Graz: Böhlau, 1993.
414. *Neue Deutsche Biographie*, published by the Historical Commission at the Bavarian Academy of Sciences, Vol. 17. Berlin: Duncker and Humblot, 1994.
415. Negoita, Constantin B. and Dan A. Ralescu: *Applications of Fuzzy Sets to System Analysis*. Basel, Stuttgart: Birkhäuser Verlag, 1975.
416. Neidhardt, Peter: *Einführung in die Kybernetik*. Berlin, Stuttgart: Verlag Technik, 1957.
417. Neumann, John (Johann) von: *Mathematische Grundlagen der Quantenmechanik*. Berlin: Verlag Julius Springer 1932.
418. Neumann, John von: *Probabilistic Logic*. The Institute for Advanced Study, January 1952.
419. Neumann, John von: The General and Logical Theory of Automata. Delivered at the Hixon Symposium, September 1948. In: [258], pp. 1–41, 1951.
420. Neumann, John von: *The Computer and the Brain*. New Haven: Yale University Press, 1958.
421. Neumann, John von: *First Draft of a Report on the EDVAC*. The document can be accessed online at: $http://wwwalt.ldv.ei.tum.de/lehre/pent-/skript/VonNeumann.pdf$. (April 29, 2003).
422. Neumann, John von: *Probabilistic Logics and the Synthesis of Reliable Organisms from Unreliable Components*. In: [504], pp. 43–98, 1956.
423. Neumann, John von: *Allgemeine und logische Theorie der Automaten*. In: *Kursbuch*, 8, pp. 139–175, March 1967.
424. Neurath, Otto: *Gesammelte philosophische und methodologische Schriften*, published by Rudolf Haller and Heiner Rutte, 2 volumes. Vienna, 1981.
425. Nierhaus, Gerhard: Ludwig von Bertalanffy 1901–1972. In: *Sudhoffs Archiv*, 65, p. 144–172, 1981.
426. Noll, P.: Nachrichtentechnik an der TH/TU Berlin – Geschichte, Stand und Ausblick. In: $http://www.nue.tu-berlin.de/geschichte/th\_nachr.pdf$ (June 21, 2001).
427. Novikoff, A.: On Convergence Proofs for Perceptions. In: *Proceedings of the Symposium on Mathematical Theory of Automata*. Published by the Polytechnic Institute of Brooklyn, Vol. XII, pp. 615–622, April 1962.
428. Nyquist, H.: Certain Factors Affecting Telegraph Speed. In: *Journal of the AIEE*, Vol. 43, p. 124, 1924. (Reprint in: *The Bell System Technical Journal*, Vol. III, pp. 324–346, 1924.)
429. Nyquist, H.: Certain Topics in Telegraph Transmission Theory. In: *Transactions of the AIEE*, pp. 617–644, 1928.
430. Ogata, Katshiko: *State Space Analysis of Control Systems*. Englewood Cliffs, N. J.: Prentice Hall, 1967.
431. Owens, Larry: Vannevar Bush and the Differential Analyzer: The Text and Context of an Early Computer. In: *Technology and Culture*, 27, 1, pp. 63–95, 1986.

432. Page, Arthur W., John E. Otterson, Ralph H. D. Brown, Harvey Arnold, Frank B. Fletcher and Herbert Jewett: *Modern Communication*. Boston, New York: Houghton Mifflin Comp., The Riverside Press Cambridge, 1932.
433. Page, Arthur W.: *Social Aspects of Communication Development*. In: [432], pp. 1–24, 1932.
434. Page, C. H.: Applications of the Fourier Integral in Physical Science. In: *IRE Transactions on Circuit Theory*, Vol. CT-2, No. 3, pp. 231–237, September 1955.
435. Parsi, R. A., U. Stürmer, R. Seidel and B. Wolf: Ein Modell zur Einschätzung des Schweregrades der Koronarsklerose mittels unblutiger Untersuchungsmethoden unter Anwendung der Diskriminanzanalyse. In: *Das deutsche Gesundheitswesen*, 51, 1974.
436. Parzen, Philip: On the Resonant Frequencies of n-meshed Tuned Circuits. In: *Proceedings of the IRE*, pp. 284–285, March 1947.
437. Pask, G.: *Learning Machines*. In: Proceedings of the 2nd IFAC International Congress. Basel, 1963.
438. Pask, G.: A Cybernetic Experimental Method and its anderlying Philosophy. In: *International Journal of Man-Machine Studies*, 3, pp. 279–337, 1971.
439. Paycha, C. F.: Diagnosis by Slide Rule. In: *What's New (Abbott Laboratories)*, No. 189, 1955.
440. Paycha, C. F.: Mémoire diagnostique. In: *Montpellier médical*, 47, 588, 1955.
441. Peray, K. E. and J. J. Wadell: *The Rotary Cement Kiln*. New York: Chemical Publishing Comp., 1972.
442. Perez-Ojeda, A.: *Medical Knowledge Network. A Database for Computer Aided Diagnosis*. Master Thesis. Department of Industrial Engineering, University of Toronto, 1976.
443. Philips, D. C.: System Theory – A Discredited Philosophy. In: *Abacus*, September 1969. (Nachdruck: [494], pp. 55–65, 1971.)
444. Pichler, Franz: *Mathematische Systemtheorie. Dynamische Konstruktionen*. Berlin, New York: Walter de Gruyter, 1975.
445. Pierce, John R.: *An Introduction into Information Theory. Symbols, Signals and Noise*. New York: Dover Publications, 1961.
446. Plail, Michael: *Die Entwicklung der optimalen Steuerungen*. Göttingen: Vandenhoeck und Ruprecht, 1998.
447. Poincaré, Henri J.: *La science et l'hypothèse. La valeur de la science. Des fondements de la géométrie*. Paris: Flammarion, 1902.
448. Poincaré, Henri J.: Du rôle de l'intuition et de la logique en mathématiques. In: *Compte rendu du deuxième Congrès International de mathématique*. Paris: Gauthiers-Villars 1902, pp. 115–130.
449. Poincaré, Henri J.: *Science et Méthode*. Paris: Flammarion (1908). German: *Wissenschaft und Methode*. Leipzig: Teubner 1914.
450. Proceedings of the Institution of Electrical Engineering, London, 98, Part III, September 1951.
451. Proctor, Lorne D. (Ed.): *The Proceedings of an International Symposium on Biocybernetics of the Central Nervous System*. London: Little, Brown and Comp., 1969.
452. Puckett, T. H.: A Note on the Admittance and Impedance Matrices of an n-terminal Network. In: *IRE Transactions on Circuit Theory*, Vol. CT-3, pp. 70–75, March 1956.

453. Pullen, Roscol L.: *Medical Diagnosis.* Philadelphia: W. B. Saanders Co., 1944.
454. Quillian, M. Ross: The Teachable Language Comprehender: A Simulation Program and Theory of Language. In: *Communications of the ACM*, Vol. 12, No. 8, pp. 459–476, August 1969.
455. Quillian, M. Ross: *Semantic Memory.* In: [405], pp. 227–270, 1968.
456. Ragazzini, John R. and Lotfi A. Zadeh: The Analysis of Sampled-Data Systems. In: *Transactions of the American Institute of Electrical Engineers*, Vol. 71, Part II: Applications and Industry, pp. 225–232, Discussion, pp. 232–234, November 1952.
457. Ragazzini, John R. and S. S. L. Chang: Noise and Random Processes. In: *Proceedings of the IEEE*, pp. 1146–1151, May 1963.
458. Ragazzini, John R., Robert H. Randall and Frederick A. Russell: Analysis of Problems in Dynamics by Electronic Circuits. In: *Proceedings of the IRE*, pp. 444–452, May 1947.
459. Ragazzini, John R. and A. R. Bergen: A Mathematical Technique for the Analysis of Linear Systems. In: *Proceedings of the IRE*, pp. 1645–1651, November 1954.
460. Rider, Jacques le: *Das Ende der Illusion. Die Wiener Moderne und die Krisen der Identität.* Vienna 1990.
461. Riemer, R. (Ed.): *Die DDR – Bildung, Wissenschaft and Forschung.* Munich: Kopernikus Verlag, 1970.
462. Rine, David C. (Ed.): *Computer Science and Multiple-Valued Logic.* Amsterdam: North Holland, 1984.
463. Robbins, Herbert M.: An Empirical Bayes Approach to Statistics. In: *Proceedings of the Third Berkeley Symposium on Mathematical Statistics and Probability*, 1955.
464. Robbins, Herbert M.: An Extension of Wiener Filter Theory to Partly Sampled Systems. In: *IRE Transactions on Circuit Theory*, Vol. CT-6, pp. 362–370, December 1959.
465. Roch, Axel: Die Maus. Von der elektrischen zur taktischen Feuerleitung. In: $http://mikro.org/Events/19991006/roch.htm$ (May 7, 2003).
466. Rogers, Everett M. and Lawrence D. Kincaid: *Communication Networks. Toward a New Paradigm for Research.* New York, London: The Free Press (Macmillian Publ. Comp.), 1981.
467. Ropohl, Günter: Einführung in die allgemeine Systemtheorie. In: [319], pp. 9–49, 1978.
468. Rosch, Eleanor: Natural Categories. In: *Cognitive Psychology*, 4, pp. 328–350, 1973.
469. Rosen, Judah Ben: *Pattern Separation by Convex Programming.* Technical Report No. 30, June 28, 1963. Applied Mathematics and Statistics Laboratories, Stanford University, Stanford, California.
470. Rosen, Judah Ben: Pattern Separation by Convex Programming. In: *Journal of Mathematical Analysis and Applications*, Vol. 10, pp. 123–134, 1965.
471. Rosenblatt, Frank: The Perceptron: A Probabilistic Model for Information Storage and Organization in the Brain. In: *Psychological Review*, Vol. 65, No. 6, pp. 386–408, 1958.
472. Rosenblueth, Arturo, Norbert Wiener and J. Bigelow: Behavior, Purpose and Teleology. In: *Philosophy of Science*, 10, pp.18–24, 1943.
473. Ruspini, Enrique Hector: A New Approach to Clustering. In: *Information and Control*, 15, pp. 22–32, 1969.

474. Ruspini, Enrique Hector: Numerical Methods for Fuzzy Clustering. In: *Information Sciences*, 2, pp. 319–350, 1970.
475. Ruspini, Enrique Hector: *A Theory of Mathematical Classification*. Ph.D. Thesis. University of California, Los Angeles, 1977.
476. Russell, Bertrand: Vagueness. In: *Australian Journal of Philosophy*, Vol. 1, 1923.
477. Rutherford D. A. and G. C. Bloore: The Implementation of Fuzzy Algorithms for Control. In: *Proceedings of the IEEE*, Proceedings Letters, pp. 572–573, April 1976.
478. Salleron, L.: *L'automation*. Paris: Presses Universitaires de France, 1956.
479. Sanchez, Elie: *Compositions of Fuzzy Relations*. In: [218], pp.421–433, 1979.
480. Sanchez, Elie: *Medical Diagnosis and Composite Fuzzy Relations*. In: [218], pp. 437–444, 1979.
481. Sanchez, Elie: *Equations de Relations Floues*. Thése Biologie Humaine. Faculté de Médecine de Marseille, 1974.
482. Sanchez, Elie (Ed.): Fuzzy Information, Knowledge Representation and Decision Analysis. In: *Proceedings of the IFAC Symposium*, Marseille, France 19–21 July. Oxford, New York, Toronto, Sydney, Paris, Frankfurt am Main: Pergamon Press, 1983.
483. Sanchez, Elie: Resolution of Composite Fuzzy Relation Equations. In: *Information and Control*, Vol. 30, pp. 38–48, 1976.
484. Santos, Eugene S.: Fuzzy Algorithms. In: *Information and Control*, 17, pp. 326–339, 1970.
485. Santos, Eugene S.: Maximin Sequential Chains. In: *Journal of Mathematical Analysis and Applications*, 26, pp. 28–38, 1969.
486. Santos, Eugene S. and William G. Wee: General Formulation of Sequential Machines. In: *Information and Control*, Vol 12, pp. 5–10, 1968.
487. Santos, Eugene S.: Maximin Automata. In: *Information and Control*, Vol 13, pp. 363–377, 1968.
488. Schilpp, Paul Arthur (Ed.): *The Philosophy of Rudolf Carnap* (= The Library of Living Philosophers. 11). London: Open Court, 1963.
489. Schmidt, H.: Regelungstechnik. Die technische Aufgabe und ihre wirtschaftliche, sozialpolitische und kulturpolitische Auswirkung. In: *Zeitschrift des Vereines Deutscher Ingenieure*, 85, 4 (1941),
490. Schneider, B.: Mathematische Grundlagen der medizinischen Diagnostik. In: [155], pp. 160–182, 1970.
491. Schneider, Ivo: *Johannes Faulhaber: 1580–1635. Rechenmeister in einer Welt des Umbruchs*. (= Vita mathematica, Band 7). Basel [et al.]: Birkhäuser, 1993.
492. Schneider, Ivo: Der Mathematiker Abraham de Moivre (1667–1754). In: *Archive for History of Exact Sciences*, 5, pp. 177–317, 1968.
493. Schneider, Ivo: *Isaac Newton*. (= Beck'sche Reihe 514: Grosse Denker). Munich: Beck, 1988.
494. Schoderbek, Peter P. (Ed.): *Management Systems*. (2nd edition). New York: Wiley, 1971.
495. Schorske, Carl E.: *Fin-de-Siècle Vienna: Politics and Culture*. German edition: Wien. Geist und Gesellschaft im Fin de siècle. Frankfurt am Main: Fischer, 1982.
496. Schreiber, F. and Nielsen, A.: Punch Card Code for Classification of Craniocerebral Injuries. In: *Journal of the Michigan Mathematical Society*, 37, pp. 909–912, 1938.

497. Schrödinger, Erwin: Grundlinien einer Theorie der Farbenmetrik im Tagessehen. In: *Annalen der Physik*, 63, 4, pp. 397–456, 481–520, 1920.
498. Schrödinger, Erwin: Quantisierung als Eigenwertproblem (Erste Mitteilung). In: *Annalen der Physik*, 79, 4, pp. 361–363, 1926.
499. Seising, Rudolf: *Probabilistische Strukturen der Quantenmechanik.* Frankfurt am Main [et al.]: Peter Lang (Europäische Hochschulschriften, Reihe XX, Philosophie, Bd./Vol. 500), 1996.
500. Seising, Rudolf (Ed.): *Fuzzy Theorie und Stochastik – Modelle und Anwendungen in der Diskussion. Computational Intelligence.* Braunschweig, Wiesbaden: Vieweg, 1999.
501. Seitz, Frederick and G. Norman: *Einspruch: Electronic Genie. The Tangled History of Silicon.* Urbana and Chicago: University of Illinois Press, 1998.
502. Selfridge, Oliver G.: Pandemonium: A Paradigm for Learning Mechanization of Thought Processes. In: *Proceedings of a Symposium.* Held at the National Physical Laboratory, London, November 1958.
503. Seshu, Sanduran: *Norman Balabanian: Linear Network Analysis.* New York: John Wiley & Sons, London: Chapman & Hall, 1959.
504. Shannon, Claude E. and J. McCarthy (Eds.): *Automata Studies.* Princeton, New Jersey: University Press, 1956.
505. Shannon, Claude E. and Warren, Weaver: *Mathematische Grundlagen der Informationstheorie.* Munich: Oldenbourg, 1976.
506. Shannon, Claude E.: The Bandwagon. Editorial. In: *IRE Transactions on Information Theory*, March 1956.
507. Shannon, Claude E.: The Mathematical Theory of Communication. In: *Bell System Technical Journal*, 27, No. 3 and No. 4, pp. 379–423 and pp. 623–656, 1948.
508. Shannon, Claude E.: A Symbolic Analysis of Relay and Switching Circuits. In: *Transactions of the AIEE*, Vol. 57, pp. 713–723, 1938.
509. Shekel, J.: The Gyrator as a 3-terminal Element. In: *Proceedings of the IRE*, Vol. 41, pp. 1493–1497, November 1952.
510. Shekel, J.: Indefinite Admittance Representation of Linear Network Elements. In: *Bulletin of the Research Council of Israel*, Vol. 3, pp. 390–394, June 1954.
511. Shekel, J.: Voltage Reference Node. In: *Wireless Engineering*, Vol. 31, pp. 6–10, January 1954.
512. Shires, David B. and Wolf, Hermann (Eds.): *MEDINFO 77. Proceedings of the Second World Conference on Medical Informatics.* Toronto, August 8–12, 1977 (= IFIP World Conferences Series on Medical Informatics, Vol. 2). Amsterdam, New York, Oxford: North-Holland Publishing Comp., 1977.
513. Siegmund-Schultze, Reinhard: *Mathematiker auf der Flucht vor Hitler. Quellen und Studien zur Emigration einer Wissenschaft* (= Dokumente zur Geschichte der Mathematik. Band 10. Deutsche Mathematiker-Vereinigung). Braunschweig, Wiesbaden: Vieweg, 1998.
514. Sloane, Neil J. A. and Aaron D. Wyner (Eds.): *Claude Elwood Shannon.* Collected Papers. New York: IEEE-Press, 1993.
515. Smets, Ph., H. Vainsel, R. Bernard and F. Kornreich: *Bayesian Probability of Fuzzy Diagnosis* In: [512], pp. 121–122, 1977.
516. Spindelberger W. and Georg Grabner: *Ein Computerverfahren zur diagnostischen Hilfestellung.* In: [173], pp. 189–221, 1968.
517. Stach, Heike: *Als Rechner zu abstrakten Maschinen wurden.* In: $http://tal.cs.tu-berlin.de/ifp/fiff/Stach.html$ (April 21, 2003).

518. Stacy, Ralph W. and Bruce D. Waxmann (Eds.): *Computers in Biomedical Research*. Vol. 1. New York, London: Academic Press, 1965.
519. Stadler, Friedrich (Ed.): *Kontinuität und Bruch 1938-1945-1955. Beiträge zur österreichischen Kultur- und Wissenschaftsgeschichte*. Vienna-Munich: Jugend & Volk, 1988.
520. Stadler, Friedrich: *Studien zum Wiener Kreis. Ursprung, Entwicklung und Wirkung des Logischen Empirismus im Kontext*. Frankfurt am Main: Suhrkamp, 1997.
521. Stadler, Friedrich: *Vertriebene Vernunft. Emigration und Exil österreichischer Wissenschaft*. Vol. 2. Vienna, Munich: Jugend & Volk, 1988.
522. Steinbuch, Karl: *Automat und Mensch*. (2nd edition). Berlin: Springer, 1961.
523. Steinbuch, Karl: *Die informierte Gesellschaft. Geschichte und Zukunft der Nachrichtentechnik*. Stuttgart: Deutsche Verlags Anstalt, 1966.
524. Steinbuch, Karl: *Die Lernmatrix. Kybernetik*. Vol. 1. Berlin: Springer Verlag, 1961.
525. Stevenson, Robert L.: *The Strange Case of Dr. Jekyll and Mr. Hyde*. Edinburgh: Edinburgh University Press, 2004.
526. Stewart, D. J.: *An Essay on the Origins of Cybernetics. Human Factors Research*. (1959, online 2000).
    URL:$http://www.hfr.org.uk/cybernetics - pages/origins.htm$ (June 14, 2005).
527. Stout, T. M.: Block Diagram Transformations for Systems with One Nonlinear Element. In: *American Institute of Electrical Engineering, Applications and Industry*, pp. 130–139, Discussion, pp. 139–141, July 1956.
528. Strejc, Vladimir: *State Space Theory of Discrete Control*. Chichester, New York, Brisbane, Toronto: John Wiley & Sons, 1981.
529. Strubecker, Karl (Ed.): *Geometrie* (= Wege der Forschung, Band CLXXVII). Darmstadt: Wissenschaftliche Buchgesellschaft, 1972.
530. Szaniawski, Klemens (Ed.): *The Vienna Circle and the Lvov-Warsaw School*. Dordrecht: Kluwer Academic Publishers, 1989.
531. Takagi, T. and M. Sugeno: *Derivation of Fuzzy Control Rules from Human Operator's Control Actions*. In: [482], pp. 55–60, 1983.
532. Tautu, Petre and Gustav Wagner: *The Process of Medical Diagnosis: Routes of Mathematical Investigations*. In: *Methods of Information in Medicine*, Vol. 17, No. 1, pp. 1–10, 1978.
533. Terman, Frederick E.: *Network Theory, Filters, and Equalizers*. Part III. In: *Proceedings of the IRE*, pp. 288–302, June 1943.
534. Thiel, Christian: *Philosophie und Mathematik. Eine Einführung in ihre Wechselwirkungen und in die Philosophie der Mathematik*. Darmstadt: Wissenschaftliche Buchgesellschaft, 1995.
535. Timms, Edward: *Die Wiener Kreise. Schöpferische Interaktionen in der Wiener Moderne*. In: [413], pp. 128–143, 1993.
536. Tribus, Myron: Comment on " Fuzzy Sets, Fuzzy Algebra, and Fuzzy Statistics". In: *Proceedings of the IEEE*, Vol. 67, No. 8, p. 1168, August 1979.
537. Trillas, Enric: Menger's Trace in Fuzzy Logic. In: *Theoria – Seganda Época*, Vol. 11, No. 27, pp. 89–96, 1996.
538. Truxal, J. G.: Adaptive Control. In: *Proceedings of the 2nd IFAC International Congress*. Basel, 1963.

539. Turing, Alan Mathison: On Computable Numbers, with an Application to the Entscheidungsproblem. *Proceedings of the London Mathematical Society*, Series 2, Vol. 42, pp. 230–265, 1936–37, with corrections from *Proceedings of the London Mathematical Society*, Series 2, Vol.43, pp. 544–546, 1937.
540. Turing, Alan Mathison: Computing Machinery and Intelligence. In: *Mind*, Vol. 49, No. 236, pp. 433–460, 1950. Also appeared under the title „Can a Machine Think?" in: James R. Neuman (Ed.): *The World of Mathematics*, Vol. 4, New York: Simon & Schuster, pp. 2099–2123, 1956.
541. Turing, Alan Mathison: Kann eine Maschine denken? In: *Kursbuch*, 8, pp. 106–138, March 1967.
542. Ulrich, W.: Nonbinary Error Correction Codes. In: *Bell Systems Technical Journal*, pp. 1341–1142, November 1957.
543. Valkenburg, M. E. van: *Introduction to Modern Network Synthesis*. New York, London: John Wiley & Sons, 1960.
544. Varela, Francisco J: *Kognitionswissenschaft und Kognitionstechnik. Eine Skizze aktueller Perspektiven*. Frankfurt am Main: Surkamp, 1990.
545. Wagner, Richard: *Probleme und Beispiele biologischer Regelung*. Stuttgart: Thieme, 1954.
546. Waismann, Friedrich: *Wittgenstein und der Wiener Kreis*. Werkausgabe Band 3. Frankfurt am Main: Surkamp, 1984.
547. Wald, Abraham: *Statistical Decision Functions*. New York: John Wiley and Sons Inc., 1950.
548. Weaver, Warren: *Ein aktueller Beitrag zur mathematischen Theorie der Kommunikation*. In: [505], 1976.
549. Wechsler, H.: Applications of Fuzzy Logic to Medical Diagnosis. In: *Proceedings of the 1975 International Symposium on Multiple-Valued Logic, IEEE*, 75, CH 0959-7C, 1975.
550. Wee, William Go. and K. S. Fu: A Formulation of Fuzzy Automata and its Application as a Model of Learning Systems. In: *IEEE Transactions on Systems Science and Cybernetics*. Vol. SSC-5, No. 3, pp. 215–223, 1969.
551. Wee, W. G.: On a Generalization of Adaptive Algorithms and Applications of the Fuzzy Set Concept to Pattern Classification. In: *Technical Report*, 67, 7, 1967.
552. Weizsäcker, Carl Friedrich von: *Aufbau der Physik*. Munich, Vienna: Carl Hanser, 1985.
553. Wheeler, Harvey (Ed.): *Beyond the Punitive Society. Operant Conditioning: Social and Political Aspects*. San Francisco: W. H. Freeman and Comp., 1973.
554. Whiteman, R. A.: A Contribution to the Theory of Network Synthesis. In: *Proceedings of the IRE*, pp. 244–246, May 1942.
555. Whitman, Walt: Of the Terrible Doupt of Appearances. In: *Leaves of Grass*, 1891. Quoted from: [167], pp. 395–402, 1943. See e. g.: $http://www.iath.virginia.edu/whitman/works/leaves/1891/text/fulltext.html$.
556. Widrow, Bernard and Marcian E. Hoff: Adaptive Switching Circuits. In: *IRE Wescon Convention Record*. New York: IRE, pp. 96–104, 1960.
557. Wiener, Norbert: A New Concept of Communication Engineering. In: *Electronics*, pp. 74–76, January 1949.
558. Wiener, Norbert: *Cybernetics or Control and Communications in the Animal and the Machine*. Cambridge, Massachusetts: MIT Press, 1948.
559. Wiener, Norbert: *Ex-prodigy: My Childhood and Youth*. New York: Simon and Schuster, 1953.

560. Wiener, Norbert: *Extrapolation, Interpolation, and Smoothing of Stationary Time Series. With Enginnering Applications.* New York: The Technology Press of The Massachusetts Institute of Technolgy and John Wiley & Sons Inc., London: Chapman & Hall, 1949.
561. Wiener, Norbert: *The Human Use of Human Beings.* Boston: Houghton Mifflin, 1950. (German edition: Norbert Wiener: *Mensch und Menschmaschine. Kybernetik und Gesellschaft.* Frankfurt am Main: Metzner Verlag 1958.)
562. Wiener, Norbert: *The Human Use of Human Beings.* (2nd edition). Boston: Houghton Mifflin, 1950.
563. Wiener, Norbert: What is Information Theory? In: *IRE Transactions on Information Theory*, Editorial, June 1956.
564. Wiener, Norbert: *Über Informationstheorie.* Lecture for the 101st meeting of the „Gesellschaft Deutscher Naturforscher und Ärzte" on September 26, 1960 in Hannover. In: [568], pp. 212–214.
565. Wiener, Norbert: *Invention. The Care and Feeding of Ideas.* London: The MIT Press, 1993.
566. Wiener, Norbert: *Kybernetik. Regelung und Nachrichtenübertragung in Lebewesen und Maschine.* (2nd edition). Düsseldorf and Vienna: Econ, 1963. Licensed edition: Rohwolts Deutsche Enzyklopädie (Published by Prof. Ernesto Grossi), Band 294/295. Reinbek bei Hamburg: Rohwolt, 1968.
567. Wiener, Norbert: *Mathematik – Mein Leben.* Frankfurt am Main, Hamburg: Fischer (Lizenzausgabe, Econ 1962), 1965.
568. Wiener, Norbert: *Collected Works with Commentaries.* Vol. 4, edited by P. Masani, Cambridge, Massachusetts: The MIT Press, 1985.
569. Winograd, Terry: *Understanding Natural Language.* New York: Academic Press, 1972. (Also: *Cognitive Psychology*, 3, 1, pp. 1–191, 1972.)
570. Wittgenstein, Ludwig: *Logisch-philosophische Abhandlung. Wilhelm Ostwalds „Annalen der Naturphilosophie" 1921.* Frankfurt am Main: Suhrkamp, 1980.
571. Wolenski, Jan: *The Lvov-Warshaw School and the Vienna Circle.* In: [530], pp. 443–453, 1989.
572. Wong, Priscilla Dik-Chin: *Mathematical Techniques of Computer-Aided Medical Diagnosis.* Master Thesis in Mathematics. Graduate Division of the University of California, Berkeley, December 20, 1975.
573. Yovits, M. C., G. T. Jacobi and G. D. Goldstein (Eds.): *Self Organising Systems.* Washington: Spartan Books, 1962.
574. Zadeh, Fay: *My Life and Travels with the Father of Fuzzy Logic.* Albuquerque, New Mexico: TSI Press, 1998.
575. Zadeh, Lotfi A. and John R. Ragazzini: The Analysis of Sampled-Data Systems. In: *AIEE Transactions*, Vol. 71, Part II, pp. 225–232, Discussion, pp. 232–234, November 1952.
576. Zadeh, Lotfi A. and John R. Ragazzini: An Extension of Wiener's Theory of Prediction. In: *Journal of Applied Physics*, Vol 21, pp. 645–655, 1950.
577. Zadeh, Lotfi A. and Kenneth S. Miller: Generalized Ideal Filters. In: *Bulletin of the American Mathematical Society*, Vol. 58, p. 48, January 1952.
578. Zadeh, Lotfi A. and Kenneth S. Miller: Generalized Ideal Filters. In: *Journal of Applied Physics*, Vol. 23, No. 2, pp. 223–228, February 1952.
579. Zadeh, Lotfi A. and Kenneth S. Miller: On an Integral Equation Occuring in the Theory of Prediction. In: *Bulletin of the American Mathematical Society*, Vol. 60, p. 69, January 1954.

580. Zadeh, Lotfi A. and Kenneth S. Miller: Fundamental Aspects of Linear Multiplexing. In: *Proceedings of the IRE*, pp. 1091–1097, September 1952.
581. Zadeh, Lotfi A. and John R. Ragazzini: Optimum Filters for the Detection of Signals in Noise. In: *Proceedings of the IRE*, pp. 1223–1231, October 1952.
582. Zadeh, Lotfi A. and Kenneth S. Miller: Signal Representation in Terms of Periodic Functions. In: *Bulletin of the American Mathematical Society*, Vol. 61, p. 300, July 1955.
583. Zadeh, Lotfi A. and J. B. Thomas: A Simple Proof of an Inequality of McMillan. In: *IEEE Transactions on Information Theory*, pp. 118–120, April 1961.
584. Zadeh, Lotfi A. and J. H. Eaton: Optimal Pursuit Strategies in Discrete State Probabilistic Systems. In: *Journal of Basic Engineering*, pp. 23–29, March 1962.
585. Zadeh, Lotfi A. and Kenneth S. Miller: Solution of an Integral Equation Occuring in the Theories of Prediction and Detection. In: *IRE Transactions on Information Theory*, Vol. CT-2, pp. 72–75, June 1956.
586. Zadeh, Lotfi A. and Charles A. Desoer: *Linear System Theory: The State Space Approach*. New York, San Francisco, Toronto, London: McGraw-Hill Book Comp., 1963.
587. Zadeh, Lotfi A. and E. Polak (Eds.): *System Theory*. New York: McGraw-Hill, 1969.
588. Zadeh, Lotfi A.: On the Extended Definition on Linearity. In: *Proceedings of the IRE*, pp. 200–201, February 1962.
589. Zadeh, Lotfi A.: Toward Fuzziness in Computer Systems: Fuzzy Algorithms and Languages. In: [93], pp. 9–18, 1971.
590. Zadeh, Lotfi A.: On the Definition of Adaptivity. In: *Proceedings of the IEEE*, Correspondence, pp. 469–470, March 1963.
591. Zadeh, Lotfi A.: Fuzzy Algorithms. In: *Information and Control*, Vol. 12, pp. 99–102, February 1968.
592. Zadeh, Lotfi A.: An Application of Fourier Transformation to the Solution of Linear Differential Equations with Variable Coefficients. In: *Bulletin of the American Mathematical Society*, Vol. 56, p. 52, January 1950.
593. Zadeh, Lotfi A.: In Memoriam Richard Bellman (1920–1984). In: *IEEE Transactions on Automatic Control*, Vol. AC-29, No. 11, November 1984.
594. Zadeh, Lotfi A.: Biological Application of the Theory of Fuzzy Sets and Systems. In: [451]), pp. 199–206, 1969.
595. Zadeh, Lotfi A.: An Operational Calculus for Time-Dependent Heaviside Operators. In: *Bulletin of the American Mathematical Society*, Vol. 57, pp. 61f, January 1951.
596. Zadeh, Lotfi A.: Circuit Analysis of Linear Varying-Parameter Networks. In: *Journal of Applied Physics*, Vol. 21, pp. 1171–1177, November 1950.
597. Zadeh, Lotfi A.: Resonant Frequencies of n-meshed Tuned Circuits. In: *Proceedings of the IRE*, p. 1335, Discussion, November 1947.
598. Zadeh, Lotfi A.: A Fuzzy-Algorithmic Approach to the Definition of Complex or Imprecise Concepts. In: *International Journal of Man-Machine Studies*, Vol. 8, No. 3, pp. 249–291, 1976.
599. Zadeh, Lotfi A.: From Computing with Numbers to Computing with Words – from Manipulation of Measurements to Manipulation of Perceptions. In: *IEEE Transactions of Circuits and Systems*, Vol. 45, No.1, pp. 105–119, January 1999.

600. Zadeh, Lotfi A.: The Concepts of System, Aggregate, and State in System Theory. In: [663], pp. 3–42, 1969.
601. Zadeh, Lotfi A.: Initial Conditions in Linear Varying-Parameter Systems. In: *Bulletin of the American Mathematical Society*, Vol. 56, p. 540, November 1950.
602. Zadeh, Lotfi A.: Correlation Functions and Power Spectra in Variable Networks. In: *Proceedings of the IRE*, pp. 1342–1345, November 1950.
603. Zadeh, Lotfi A.: A Critical View of Our Research in Automatic Control. In: *IRE Transactions on Automatic Control*, AC-7, No. 3, pp. 74–75, April 1962.
604. Zadeh, Lotfi A.: Linguistic Cybernetics. In: *Proceedings of the International Symposium on Systems Sciences and Cybernetics*, Oxford 1972.
605. Zadeh, Lotfi A.: A New Direction in AI. Toward a Computational Theory of Perceptions. In: *AI-Magazine*, Vol. 22, No. 1, pp. 73–84, 2001.
606. Zadeh, Lotfi A.: The Birth and Evolution of Fuzzy Logic – A Personal Perspective. Part 1 and 2. In: *Journal of Japan Society for Fuzzy Theory and Systems*, Vol 11, No. 6, pp. 891–905, 1999.
607. Zadeh, Lotfi A.: A Note on the Initial Excitation of Linear Systems. In: *Journal of Applied Physics*, Vol. 22, p. 1216, September 1951.
608. Zadeh, Lotfi A.: Can Expert Systems be Designed Without Using Fuzzy Logic? In: *Proceedings of the Seventeenth Annual Conference on Information Sciences and Systems*, Department of Electrical Engineering and Computer Sciences. The Johns Hopkins University, Baltimore, Maryland 21218, p. 490.
609. Zadeh, Lotfi A.: Theory of Filtering of Signals in a λ Domain. Presented at the New York Meeting of the American Mathematical Society, April 27, 1951. In: *Bulletin of the American Mathematical Society*, Vol. 57, p. 278, 1951.
610. Zadeh, Lotfi A.: On the Theory of Filtration of Signals. In: *Bulletin of the American Mathematical Society*, Vol 57, p. 483, November 1951.
611. Zadeh, Lotfi A.: On the Theory of Filtration of Signals. In: *Zeitschrift für Angewandte Mathematik und Physik*, Vol 3, pp. 149–156, 1952.
612. Zadeh, Lotfi A.: Fuzzy Sets. In: *Information and Control*, 8, pp. 338–353, 1965.
613. Zadeh, Lotfi A.: A General Theory of Linear Signal Transmission Systems. In: *Journal of The Franklin Institute*, pp. 293–312, April 1952.
614. Zadeh, Lotfi A.: Generalization of the Fourier Integrals. In: *IRE Transactions On Circuit Theory*, Vol. CT-2, No. 3, pp. 256–260, September 1955.
615. Zadeh, Lotfi A.: Fuzzy Sets and Information Granularity. In: [218], pp. 3–18, 1979.
616. Zadeh, Lotfi A.: On Stochastic Heaviside Operators. In: *Bulletin of the American Mathematical Society*, Vol. 57, pp. 279f, July 1951.
617. Zadeh, Lotfi A.: A Fuzzy-Set-Theoretic Interpretation of Linguistic Hedges. In: *Journal of Cybernetics*, 2, pp. 4–34, 1972.
618. Zadeh, Lotfi A.: On the Identification Problem. In: *IRE Transactions on Circuit Theory*, pp. 277–281, December 1956.
619. Zadeh, Lotfi A.: The General Identification Problem. In: *Proceedings of the Conference on Identification Problems in Communication and Control Systems*, Princeton, New Jersey, Princeton University, 1963.
620. Zadeh, Lotfi A.: The Determination of the Impulsive Response of Variable Networks. In: *Journal of Applied Physics*, Vol. 21, pp. 642–645, July 1950.
621. Zadeh, Lotfi A.: Initial Conditions in Linear Varying-Parameter Systems. In: *Journal of Applied Physics*, Vol 22, pp. 782–786, June 1951.

622. Zadeh, Lotfi A. and Edward T. Lee: Fuzzy Languages and their Acceptance by Automata. In: *Proceedings of the Third Annual Princeton Conference on Information Sciences and Systems*, (Papers presented March 27–28, 1969), p. 399, 1979.
623. Zadeh, Lotfi A.: Fuzzy Languages and their Relation to Human and Machine Intelligence. In: [351], pp. 130–165, Round Table Discussions, pp. 166–178, 1972.
624. Zadeh, Lotfi A.: An Extended Definition of Linearity. In: *Proceedings of the IRE*, Correspondence, pp. 1452–1453, September 1961.
625. Zadeh, Lotfi A.: On the Extended Definition of Linearity. In: *Proceedings of the IRE*, Correspondence, pp. 1452f, February 1962.
626. Zadeh, Lotfi A.: Making Computers Think Like People. In: *IEEE Spectrum*, pp. 26–32, August 1984.
627. Zadeh, Lotfi A.: Mathematics – A Call for Reorientation. In: *Newsletter, Conference Board of the Mathematical Sciences*, Vol. 7, No. 3, pp. 1–3, May 1972.
628. Zadeh, Lotfi A.: Multipole Analysis of Active Networks. In: *IRE Transactions on Circuit Theory*, pp. 97–105, September 1957.
629. Zadeh, Lotfi A.: Nonlinear Multipoles. In: *Proceedings of the National Academy of Sciences of the United States of America*, Washington D.C., Vol. 39, pp. 274–280, April 1953.
630. Zadeh, Lotfi A.: Frequency Analysis of Variable Networks. In: *Proceedings of the IRE*, Vol. 38, pp. 291–299, March 1950.
631. Zadeh, Lotfi A.: Constant-Resistance Networks of the Linear Varying-Parameter Type. In: *Proceedings of the IRE*, pp. 688–691, June 1951.
632. Zadeh, Lotfi A.: Time-Varying Networks. I. In: *Proceedings of the IRE*, pp. 1488–1503, October 1961.
633. Zadeh, Lotfi A.: A Contribution on the Theory of Nonlinear Systems. In: *Journal of The Franklin Institute*, pp. 387–408, May 1953.
634. Zadeh, Lotfi A.: An Operational Calculus for Time-Dependent Heaviside Operators. In: *Bulletin of the American Mathematical Society*, Vol. 57, p. 61, January 1951.
635. Zadeh, Lotfi A.: Operational Analysis of Variable-Delay Systems. In: *Proceedings of the IRE*, pp. 564–568, May 1952.
636. Zadeh, Lotfi A.: Time-Dependent Heaviside Operators. In: *Journal of Mathematics and Physics*, Vol. 30, No. 2, pp. 73–78, July 1951.
637. Zadeh, Lotfi A.: What is optimal? In: *IRE Transactions on Information Theory*, IT-4, No. 1, Editorial, March 1958.
638. Zadeh, Lotfi A.: On Optimal Control and Linear Programming. In: *IRE Transactions on Automatic Control*, AC-7, No. 4, pp. 45–46, July 1962.
639. Zadeh, Lotfi A.: Optimality and Non-Scalar-Valued Performance Criteria. In: *IEEE Transactions on Automatic Control*, Correspondence, pp. 59f, January 1963.
640. Zadeh, Lotfi A.: Optimum Nonlinear Filters. In: *Journal of Applied Physics*, Vol 24, pp. 396–404, April 1953.
641. Zadeh, Lotfi A.: Outline a New Approach to the Analysis of Complex Systems and Decision Processes. In: *IEEE Transactions on Systems, Man and Cybernetics*, Vol. SMC-3, No. 1, pp. 28–44, January 1973.
642. Zadeh, Lotfi A.: Part 5: Prediction and Filtering. In: [162], pp. 128–144, 1961.
643. Zadeh, Lotfi A.: Probability Measures of Fuzy Events. In: *Journal of Mathematical Analysis and Applications*, 10, pp. 421–427, 1968.

644. Zadeh, Lotfi A.: Fuzzy Sets versus Probability. In: *Proceedings of the IEEE*, Vol 68, No. 3, p. 421, March 1980.
645. Zadeh, Lotfi A.: Some Basic Problems in Communication of Information. In: *The New York Academy of Sciences*, Ser. II, Vol. 14, No. 5, pp. 201–204, March 1952.
646. Zadeh, Lotfi A.: Report on Progress in Information Theory in the U.S.A. 1960–1963. In: *IEEE Transactions on Information Theory*, pp. 221ff, 1963.
647. Zadeh, Lotfi A.: A Rationale for Fuzzy Control. In: *Transactions of the ASME-Journal of Dynamic Systems, Measurement, and Control*, pp. 3–4, March 1972.
648. Zadeh, Lotfi A.: General Input-Output Relations for Linear Networks. In: *Proceedings of the IRE*, Correspondence, p. 103, 1952.
649. Zadeh, Lotfi A.: Remark on the Paper by Bellman and Kalaba. In: *Information and Control*, 4, pp. 350–352, 1961.
650. Zadeh, Lotfi A.: Toward an Institute for Research in Communication Sciences. In: *IRE Transactions on Information Theory*, Editorial, p. 3, March 1960.
651. Zadeh, Lotfi A.: The Correlation Function of the Response of Variable System. In: *Bulletin of the American Mathematical Society*, Vol. 56, p. 347, January 1950.
652. Zadeh, Lotfi A.: Quantitative Fuzzy Semantics. In: *Information Sciences*, 3, pp. 159–176, 1971.
653. Zadeh, Lotfi A.: Shadows of Fuzzy Sets. In: *Problemy peredachi informatsii*, Akademija Nauk SSSR Moskva, Vol. 2, No. 1, pp. 37–44, 1966. (An English translation was published in: *Problems of Information Transmission: A Publication of the Academy of Sciences of the USSR*; the Faraday Press cover-to-cover transl., New York, NY, Vol. 2, 1966.)
654. Zadeh, Lotfi A.: Similarity Relations and Fuzzy Orderings. In: *Information Sciences*, 3, pp. 177–200, 1971.
655. Zadeh, Lotfi A.: Fuzzy Logic, Neural Networks, and Soft Computing. In: *Communications of the ACM*, Vol. 37, No. 3, pp. 77–84, 1994.
656. Zadeh, Lotfi A.: Correlation Functions and Spectra of Phase- and Delay-Modulated Signals. In: *Proceedings of the IRE*, pp. 425–428, 1951.
657. Zadeh, Lotfi A.: On Stability of Linear Varying-Parameter Systems. In: *Journal of Applied Physics*, Vol. 22, pp. 402–405, April 1951.
658. Zadeh, Lotfi A.: The Concept of State in System Theory. In: [399], pp. 39–50, 1964.
659. Zadeh, Lotfi A.: The Concept of State in System Theory. In: [79], pp. 21–26, 1968.
660. Zadeh, Lotfi A.: On a Class of Stochastic Operators. In: *Journal of Mathematics and Physics*, Vol. 32, pp. 48–53, April 1953.
661. Zadeh, Lotfi A.: System Theory. In: *Columbia Engineering Quarterly*, Vol. 8, pp. 16–19, 34, November 1954.
662. Zadeh, Lotfi A.: From Circuit Theory to System Theory. In: *Proceedings of the IRE*, Vol. 50, No. 5, pp. 856–865, May 1962.
663. Zadeh, Lotfi A. and E. Polak (Eds.): *System Theory*. New York: McGraw-Hill, 1969.
664. Zadeh, Lotfi A.: Fuzzy Sets and Systems. In: [180], pp. 29–37, 1965.
665. Zadeh, Lotfi A.: On Passive and Active Networks and Generalized Norton's and Thevenin's Theorems. In: *Proceedings of the IRE*, p. 378, March 1956.
666. Zadeh, Lotfi A.: Theory of Filtering. In: *Journal of the Society for Industrial and Applied Mathematics*, 1, 1, pp. 35–51, 1953.

667. Zadeh, Lotfi A.: Thinking Machines. A New Field in Electrical Engineering. In: *Columbia Engineering Quarterly*, pp. 12–30, January 1950.
668. Zadeh, Lotfi A.: Towards a Theory of Fuzzy Systems. In: [261], pp. 469–490, 1971.
669. Zadeh, Lotfi A.: Band-Pass Low-Pass Transformation in Variable Networks. In: *Proceedings of the IRE*, pp. 1339–1341, November 1950.
670. Zadeh, Lotfi A.: A Note on the Analysis of Vacuum Tube and Transistor Circuits. In: *Proceedings of the IRE*, pp. 989–992, August 1953.
671. Zadeh, Lotfi A.: A System-Theoretic View of Behavior Modification. In: [553], pp. 160–169.
672. Zadeh, Lotfi A.: Fuzzy Logic = Computing with Words. In: *IEEE Transactions of Circuits and Systems*, Vol. 4, No. 2, pp. 103–111, May 1996.
673. Zeilinger, Anton: *Einsteins Schleier. Die neue Welt der Quantenphysik*. Munich, Vienna: Verlag C. H. Beck, 2003. Paperback edition: Munich: Wilhelm Goldmann Verlag, 2005.
674. Zimmermann, Hans-Jürgen: *Fuzzy Technologien. Prinzipien, Werkzeuge, Potentiale*. Düsseldorf: VDI Verlag, 1993.

Reports and Memoranda

675. Bellman, Richard E., Robert Kalaba and Lotfi A. Zadeh: Abstraction and Pattern Classification. In: Memorandum RM-4307-PR, Santa Monica, California: The RAND Corporation, October 1964.
676. Bellman, Richard E.: An Introduction to the Theory of Dynamic Programming. In: *RAND Report*, 245, 1953.
677. Zadeh, Lotfi A.: A Theory of Commonsense Knowledge. In: Memorandum No. UCB/ERL M 83/26, April 17, 1983. Electronic Research Laboratory, University of California, Berkeley, California.
678. Zadeh, Lotfi A.: Fuzzy Sets. In: ERL Report No. 64–44, University of California, Berkeley, November 16, 1964.
679. Zadeh, Lotfi A.: On the Analysis of Large Scale Systems. In: Memorandum No. ERL-M 418, January 8, 1974. Electronic Research Laboratory, College of Engineering, University of California, Berkeley, California 14720.
680. Zadeh, Lotfi A.: Test-Score Semantics as a Basis for a Computational Approach to the Representation of Meaning. In: Memorandum No. UCB/ERL M 84/8, 24, January 1984. Electronic Research Laboratory, University of California Berkeley, California.
681. Zadeh, Lotfi A.: PRUF – A Meaning Representation Language for Natural Language. In: Memorandum No. UCB/ERL M 77/61, October 17, 1977. Electronic Research Laboratory, University of California, Berkeley, California.
682. Zadeh, Lotfi A.: A Computational Approach to Fuzzy Quantifiers in Natural Languages. In: Memorandum No. UCB/ERL M 82/36, May 17, 1982. Electronic Research Laboratory, University of California, Berkeley, California.
683. Zadeh, Lotfi A.: A Theory of Approximate Reasoning (AR). In: Memorandum No. UCB/ERL M 77/58, 30, August 1977. Electronic Research Laboratory, University of California, Berkeley, California.
684. Zadeh, Lotfi A.: Shadows of Fuzzy Sets. In: Notes of System Theory, Vol. VII, pp. 165–170, May 1965. Report No. 65–14, Electronic Research Laboratory, University of California, Berkeley, California.

685. Zadeh, Lotfi A.: Syllogistic Reasoning in Fuzzy Logic and its Application to Reasoning with Dispositions. In: Memorandum No. UCB/ERL M 84/17, February 2, 1984. Electronic Research Laboratory, University of California, Berkeley, California.
686. Zadeh, Lotfi A.: Theory of Fuzzy Sets. In: Memorandum No. UCB/ERL M 77/1, January 4, 1977. Electronic Research Laboratory, University of California, Berkeley, California.
687. Zadeh, Lotfi A.: Toward a Theory of Fuzzy Systems. Electronic Research Laboratory, College of Engineering, University of California, Berkeley 94720, Report No. ERL-69-2, June 1969.
688. Zadeh, Lotfi A.: A Fuzzy-Algorithmic Approach to the Definition of Complex or Imprecise Concepts. In: Memorandum No. ERL-M 474, October 11, 1974. Electronic Research Laboratory, University of California, Berkeley, California.

Unpublished Material.

Archive of the Deutsches Museum in Munich

689. Letter from Arnold Sommerfeld to Walter Rogowski, February 18, 1926, Archiv NL 89, 003; facsimile, Sommerfeld-Project: $http : //www.lrz-muenchen.de/ Sommerfeld/gif$100/02177_01.$gif$.
690. Letter from Walter Rogowski to Arnold Sommerfeld, Oktober 20, 1912, Archiv NL 89, 012; summary, Sommerfeld-Project: $http : //www.lrz-muenchen.de/ Sommerfeld/KurzFass$/01208.$html$.
691. Letter from Arnold Sommerfeld to Frederick Norton, Januar 27, 1926, Archiv NL 89, 003.
692. Photographs of Arnold Sommerfeld.

Archive of the Core Unit for Medical Statistics and Informatics, Medical University of Vienna, formerly: Institute for Medical Computer Sciences (IMC), Medical Faculty of the University of Vienna.

693. Photographs from the former Institute for Medical Computer Sciences (IMC), Medical Faculty of the University of Vienna.

Archive of the Institute Vienna Circle

694. Photographs of the Vienna Circle, also in [520], Stadler, Kreis (1997).

University Archive of the Ludwig-Maximilians University, Munich

695. Arnold Sommerfeld, Votum informativum for the dissertation of Ernst A. Guillemin, Archive of the Ludwig-Maximilians University, Munich (OCI 52p).

Documents and Photographs from the Private Archive of Lotfi A. Zadeh

696. Letter from Richard Bellman to Lotfi Zadeh dated September 9, 1964. Private archive of Lotfi Zadeh.
697. Zadeh, Lotfi A.: "Autobiographical Note 1" – andated two-page typewritten manuscript by Lotfi Zadeh, written after 1978.

698. Discussion of "Fuzzy Sets and Concepts", Berkeley, California on November 18, 1966 (Mailed to all those who attended; others on request). Discussion notes, Interdisziplinary Colloquium on Mathematics in the Behavioral Sciences, No. 3 (1966/67); organized by Jacob Marschak at the University of California, Los Angeles (UCLA).
699. Zadeh, Lotfi A.: The Search for Metrics of Intelligence – A Critical View. Draft dated July 18, 2000 of a lecture for the Washington Conference of the National Institute of Science and Standards, August 14, 2000.
700. Images of people photographed by Lotfi A. Zadeh.

Photographs from other Private Archives

701. Photographs from the private archive of Ebrahim Mamdani.
702. Photographs from the private archive of Rosemary Menger Gilmore.
703. Photographs from the private archive of Elie Sanchez.

Interviews

704. R. S. Interview with L. A. Zadeh on September 8, 1998 in Aachen, RTWH Aachen, at the margin of the European Congress on Intelligent Techniques and Soft Computing EUFIT 1998.
705. R. S. Interview with L. A. Zadeh on September 8, 1999 in Zittau, at the margin of the 7th Zittau Fuzzy Colloquium at the University Zittau/Görlitz.
706. R. S. Interview with L. A. Zadeh on July, 26, 2000, University of California, Berkeley, Soda Hall.
707. R. S. Interview with L. A. Zadeh on June 15, 2001, University of California, Berkeley, Soda Hall.
708. R. S. Interview with L. A. Zadeh on June 16, 2001, University of California, Berkeley, Soda Hall.
709. R. S. Interview with L. A. Zadeh on June 19, 2001, University of California, Berkeley, Soda Hall.
710. R. S. Interview with L. A. Zadeh on July 5, 2002, University of California, Berkeley, Soda Hall.
711. R. S. Interview with L. A. Zadeh on July 9, 2002, University of California, Berkeley, Soda Hall.
712. R. S. Interview with L. A. Zadeh on February 13, 2003 in Vienna, during the International Conference on Computational Intelligence for Modelling, Control & Automation – CIMCA 2003.
713. R. S. Interview with Fay Zadeh on Juli 15, 2002 in Berkeley.
714. R. S. Interview with Prof. Dr. Klaus-Peter Adlassnig (University of Vienna) on July 26, 2000 in Berkeley, University of California, Berkeley.
715. R. S. Interview with Prof. Dr. Hans Bandemer on October 25, 1999 in Halle, Germany.
716. R. S. Interview with Prof. Dr. James C. Bezdek on September 8, in Aachen, RWTH Aachen, at the margin of the European Congress on Intelligent Techniques and Soft Computing EUFIT 1998.
717. R. S. Interview with Prof. Dr. Dimiter Driankov on June 26, 1998 in Munich, Siemens, Munich-Neuperlach.

718. R. S. Interview with Prof. Dr. Didier Dubois on September 10, in Aachen, RWTH Aachen, at the margin of the European Congress on Intelligent Techniques and Soft Computing EUFIT 1998.
719. R. S. Interview with Dr. Rudolf Felix on Juli 2, 1998 in Dortmand, Fuzzy Demonstrationszentrum Dortmand.
720. R. S. Interview with Dr. Erwin Gerstorfer on April 6, 1998 in Munich, Siemens, Munich-Neuperlach.
721. R. S. Interview with Dr. Erwin Gerstorfer on Oktober 10, 1998 in Munich, Siemens, Munich-Neuperlach.
722. R. S. Interview with Prof. Dr. Siegfried Gottwald on August 10 in Aachen, RWTH Aachen, at the margin of the European Congress on Intelligent Techniques and Soft Computing EUFIT 1998.
723. R. S. Interview with Prof. Dr. Siegfried Gottwald on March 17, 1999 in Leipzig, University of Leipzig.
724. R. S. Interview with Prof. Dr. Joseph Goguen on September 13, 2002 in Munich.
725. R. S. Interview with Prof. Dr. Peter Hajek on March 17, 1999 in Leipzig, at the margin of the 6th International Workshop Fuzzy-Neuro Systems '99 at the University of Leipzig.
726. R. S. Interview with Prof. Dr. Harro Kiendl on September 19, 1998 in Aachen, RWTH Aachen, at the margin of the European Congress on Intelligent Techniques and Soft Computing EUFIT 1998.
727. R. S. Interview with Prof. Dr. Ernst-Peter Klement on March 25, 1998 in Munich, Munich Stochastik Days 1998, University of the Armed Forces Munich in Neubiberg.
728. R. S. Interview with Prof. Dr. Rudolf Kruse on September 19, 1998 in Aachen, RWTH Aachen, at the margin of the European Congress on Intelligent Techniques and Soft Computing EUFIT 1998.
729. R. S. Interview with Prof. Dr. George Lakoff on August 6, 2002, University of California, Berkeley, Dwinell Hall.
730. R. S. Interview with Dr. Karl Lieven on July 1, 1998 in Aachen.
731. R. S. Interview with Prof. Dr. Ebrahim Mamdani on September 9, 1998 in Aachen, RWTH Aachen, at the margin of the European Congress on Intelligent Techniques and Soft Computing EUFIT 1998.
732. R. S. Interview with Dr. Rainer Palm on April 6, 1998 in Munich, Siemens, Munich-Neuperlach.
733. R. S. Interview with Dr. Rainer Palm on February 12, 1999 in Munich, Siemens, Munich-Neuperlach.
734. R. S. Interview with Prof. Zdzislaw Pawlak on March 19, 1998 in Munich at the margin of the 5th International GI-Workshop Fuzzy-Neuro-Systeme '98 (FNS'98) at the Technical University of Munich.
735. R. S. Interview with Prof. Dr. Manfred Peschel on September 9, 1999 in Grossschönau.
736. R. S. Interview with Dr. Paul-Theo Pilgram on October 8, 1998 in Munich, Siemens, Munich-Neuperlach.
737. R. S. Interview with Dr. Michael Reinfrank on June 30, 1998 (by telephone).
738. R. S. Interview with Prof. Dr. Bernd Reusch on July 2, 1998 in Dortmand, University of Dortmund.
739. R. S. Interview with Dr. E. H. Ruspini on Juli 30, 2002, Stanford Research Institute, Stanford, California.

740. R. S. Interview with Prof. Dr. Kokichi Tanaka on September 9, 1998 in Aachen, RWTH Aachen, at the margin of the European Congress on Intelligent Techniques and Soft Computing EUFIT 1998.
741. R. S. Interview with Dr. Richard Weber on Juli 1, 1998 in Aachen, RWTH Aachen, at the margin of the European Congress on Intelligent Techniques and Soft Computing EUFIT 1998.
742. R. S. Interview with Prof. Dr. Hans-Jürgen Zimmermann on Juli 1, 1998 in Aachen, RWTH Aachen, at the margin of the European Congress on Intelligent Techniques and Soft Computing EUFIT 1998.
743. R. S. Interview with Prof. Dr. Hans-Jürgen Zimmermann on November 26, 1999 in Aachen.

List of Figures

1.1 Lloyd Espenschied. Photograph in: Information about the authors in *Proceedings of the I.R.E.*, August 1943, p. 461. 8
1.2 Ernst A. Guillemin. Fig. in [208] Guillemin, Autobiography (1968), p. 198. 9
1.3 George A. Campbell. Photograph in: $http://www.ieee.org/organi-zations/history-center/legacies/campbell.htm$. (May 3, 2005) 16
1.4 Wilhelm Cauer. Photograph in the private archive of E. Cauer, reprint courtesy of Emil Cauer. 18
1.5 Various filter types. Fig. in [172] Feldkeller, Siebschaltungtheorie (1950), p. 2. 19
1.6 Arnold Sommerfeld. Photograph: German Museum (Deutsches Museum) Munich [692]. 20
1.7 Arnold Sommerfeld with physicists at the California Institute of Technology in Pasadena. Photograph: German Museum (Deutsches Museum) Munich [692]. 24
1.8 Ernst A. Guillemin in conversation. Photograph in: *Proceedings of the I.R.E.*, May 1962, p. 44A. 27
1.9 Lotfi A. Zadeh at his desk in Tehran in 1937. Photograph in the private archive of L. A. Zadeh [700]. 37
1.10 John R. Ragazzini. Photograph in the private archive of L. A. Zadeh [700]. 38
1.11 Schematic representation of a signal transmission system. Fig. in [613] Zadeh, General (1952), p. 294. 40
1.12 Filter combinations (Zadeh's functional symbolism). Fig. in [578] Zadeh et al., Ideal (1952), p. 225. 41
1.13 Recovery of the input signal by comparing the distances between the obtained signals and all possible transmitted signals. Fig. in [645] Zadeh, Problems (1952), p. 202. 42
1.14 Geometric representation of linear (left) and nonlinear (right) filtering. Fig. in [645] Zadeh, Problems (1952), p. 203. 44
1.15 Representation of a multipolemultipole ($m + n$-pole). Fig. in [629] Zadeh, Multipoles (1953), p. 274. 44

392   List of Figures

1.16 Illustration accompanying Zadeh's article *Thinking Machines. A New Field in Electrical Engineering*. Fig. in [667] Zadeh, Thinking (1950), p. 12. 50
1.17 Title page of the *Columbia Engineering Quarterly*, January 1950. Fig. in [667] Zadeh, Thinking (1950). 51
1.18 Zadeh's table showing the similarity between statements and circuits. Fig. in [667] Zadeh, Thinking (1950), p. 30. 52
1.19 Shannon's representation of Boolean statements using electrical circuits. Fig. in [508] Shannon, Relay (1938), p. 713. 53
1.20 Shannon's table showing the similarity between statements and circuits. Fig. in [508] Shannon, Relay (1938), p. 713. 54
1.21 The two units comprising Robert Haufe's Tit-Tat-Toe machine. Fig. in [667] Zadeh, Thinking (1950), p. 30. 55
1.22 Zadeh's chart for the basic elements of a "Thinking Machine". Fig. in [667] Zadeh, Thinking (1950), p. 13. 56
1.23 Illustration of a system as a black box for Zadeh's article *System Theory*. Fig. in [661] Zadeh, System (1954), p. 16. 58
1.24 Block diagram with input-output relationships. Fig. in [661] Zadeh, System (1954), p. 17. 60
1.25 Oriented linear graph. Fig. in [661] Zadeh, System (1954), p. 18. 60
1.26 Matrix representation of a system. Fig. in [661] Zadeh, System (1954), p. 19. 61
1.27 Graphic representation of the toggle switch system. Fig. in [661] Zadeh, System (1954), p. 19. 63

2.1 Otto Neurath. Photograph Archive Institute Vienna Circle [694]. Photograph reprint courtesy of Friedrich Stadler. 70
2.2 Moritz Schlick. Photograph Archive Institute Vienna Circle [694]. Photograph reprint courtesy of Friedrich Stadler. 72
2.3 Hans Hahn. Photograph Archive Institute Vienna Circle [694]. Photograph reprint courtesy of Friedrich Stadler. 73
2.4 Kurt Gödel. Photograph Archive Institute Vienna Circle [694]. Photograph reprint courtesy of Friedrich Stadler. 75
2.5 Illustration depicting the murder of Moritz Schlick in Vienna's *Illustrierten Kronenzeitung* newspaper on June 22, 1936. Photograph Archive Institute Vienna Circle [694]. Photograph reprint courtesy of Friedrich Stadler. 76
2.6 Karl Menger. Photograph $http://www.iit.edu/ \sim am/Menger-/menger.html$ (May 3, 2005). Photograph in the private archive of R. Menger Gilmore, reprint courtesy of R. Menger Gilmore [702]. 78
2.7 Ludwig Wittgenstein. Photograph Archive Institute Vienna Circle [694]. Photograph reprint courtesy of Friedrich Stadler. 82
2.8 Rudolf Carnap. Photograph Archive Institute Vienna Circle [694]. Photograp reprint courtesy of Friedrich Stadler. 84
2.9 Alfred Tarski. Photograph Archive Institute Vienna Circle [694]. Photograph reprint courtesy of Friedrich Stadler. 90
2.10 Alfred Tarski and Kurt Gödel in Vienna, 1935. Photograph: $http://www2-data.informatik.unibw-muenchen.de/relmics-/html/pictures/history/Friend/Tarski.html$ (May 3, 2005). 92

List of Figures   393

2.11 Computer-generated likeness of Karl Menger. Photograph: $http://www.iit.edu/\sim am/Menger/menger.html$ (May 3, 2005). Private archive of R. Menger Gilmore [702]. Reprint courtesy of R. Menger Gilmore. 104
2.12 Color Spectrum. Figure in [497] Schrödinger, Farbenmetrik (1920), p. 429. 107
2.13 Schrödinger's Color Cone. Figure in [497] Schrödinger, Farbenmetrik (1920), p. 428. 108

3.1 Ludwig von Bertalanffy. Photograph on the cover of [71] Bertalanffy, Robots (1976). 114
3.2 Illustration of Bertalanffy's General Systems Theory. Figure in [72] Bertalanffy, System (1954), p. 8. 114
3.3 Norbert Wiener. Figure in [353] Masani, Wiener (1990), p. 44. 122
3.4 Julian Bigelow. Figure in [353] Masani, Wiener (1990), p. 183. 123
3.5 Arturo Rosenblueth. Figure in [353] Masani, Wiener (1990), p. 199... 125
3.6 Andrei N. Kolmogorov. Photograph in the private archive of L. A. Zadeh [700]. 130
3.7 Claude E. Shannon. Photograph in the private archive of L. A. Zadeh [700]. 133
3.8 Harry Nyquist. Figure in [40] Beaulieu, Topics (2002), p. 276. 133
3.9 Ralph Hartley. Figure in: Information on the authors in *Proceedings of the I.R.E.*, March 1942, p. 155. 135
3.10 Shannon's Communication Model. Figure in [507] Shannon, Communication (1948), p. 381. 136
3.11 Bertalanffy's Stimulus-Reaction Diagram. Figure in [72] Bertalanffy, System (1954), p. 5. 139
3.12 Members of the "Cybernetic Group". Figure in [307] Latil de, Pensée (1953). 144
3.13 Warren S. McCulloch. Photograph: $http://www.csulb.edu/\sim cwallis/artificialn/warren\_mcculloch.html.$ (May 3, 2005) 145
3.14 Walter H. Pitts. Figure in [353] Masani, Wiener (1990), p. 224. 147
3.15 John von Neumann. Photograph: $http://scidiv.bcc.ctc.edu/Math-/vonNeumann.html$ (May 3, 2005) 149
3.16 Problem of linear separability. Figure in [346] Mainzer, Computer (1995), p. 307. Reprint courtesy of Klaus Mainzer. 157
3.17 Drawing of a Turing machine by Klaus Mainzer. Figure in [346] Mainzer, Computer (1995), p. 80. Reprint courtesy of Klaus Mainzer. 159
3.18 Marvin L. Minsky and John McCarthy. Photograph in the private archive of L. A. Zadeh [700]. 160
3.19 Automaton A with inputs and outputs. 163

4.1 Claude E. Shannon, picture in the editorial [506] Shannon, Bandwagon (1956). 165
4.2 Norbert Wiener, picture in the editorial [563] Wiener, Information (1956). 166
4.3 Communication channel as a black box. Figure in [46] Bellman et al., Communication (1957), p. 198. 168

4.4 Richard E. Bellman. Photograph in the private archive of L. A. Zadeh [700]. ... 168
4.5 Herbert Robbins. Photograph in the private archive of L. A. Zadeh [700]. ... 172
4.6 Steven Kleene. Figure in: Information about the authors in *Proceedings of the I.R.E.*, October 1947, p. 1102. ... 173
4.7 Lotfi A. Zadeh, Picture in the editorial [637] Zadeh, Optimal (1958). ... 177
4.8 Illustration of the definitions of $\mathcal{C}$ and $\Sigma_>(x)$. Figure in [639] Zadeh, Optimality (1963), p. 59. ... 182
4.9 The set of non-inferior points at the edge of $\mathcal{C}$. Figure in [639] Zadeh, Optimality (1963), p. 59. ... 182
4.10 Charles Desoer. Figure in the information about the authors in *Automatica*, Vol. 12, 1976, p. 543. ... 187
4.11 Electrical and mechanical realization of a system. Fig. in [586] Zadeh et al., System (1963), p. 14. ... 188
4.12 Lev Pontryagin. Photograph in the private archive of L. A. Zadeh [700]. ... 189
4.13 Two consecutive segments. Figure in [600] Zadeh, Concepts (1969), p. 6. ... 197

5.1 Richard E. Bellman. Photograph in the private archive of L. A. Zadeh [700]. ... 201
5.2 Lotfi A. Zadeh with his parents. Photograph in the private archive of L. A. Zadeh [700]. ... 202
5.3 Richard E. Bellman and his wife Nina. Photograph in the private archive of L. A. Zadeh [700]. ... 207
5.4 Robert Kalaba. Figure in: Information about the authors in *IRE Transactions on Information Theory*, September 1956, p. 209. ... 208
5.5 Letter from R. E. Bellman to L. A. Zadeh dated September 9, 1964. [696]. ... 209
5.6 Title page of the RAND memorandum by Bellman, Kalaba and Zadeh. [675] Bellman et al., Abstraction (1964). ... 210
5.7 Illustrations of the union (1, 2) and intersection (3, 4) of two fuzzy sets. Fig. in [612] Zadeh, Fuzzy (1965), p. 342. ... 211
5.8 First page (excerpt) of Zadeh's contribution to the Symposium on System Theory in Brooklyn in April 1965. [664] Zadeh, Systems (1965), p. 29. ... 213
5.9 Membership functions of a convex and a non-convex fuzzy set. Fig. in [664] Zadeh, Systems (1965), p. 33. ... 215
5.10 First page (excerpt) of Zadeh's first article about fuzzy sets. Fig. in [612] Zadeh, Fuzzy (1965), p. 338. ... 220
5.11 Convex and non-convex fuzzy sets in $\mathbf{E}^1$. Fig. in [612] Zadeh, Fuzzy (1965), p. 347. ... 222
5.12 Illustration of the separation theorem for fuzzy sets in $\mathbf{E}^1$. Fig. in [612] Zadeh, Fuzzy (1965), p. 352. ... 223
5.13 Illustration of the parallel and serial connection of sieves. Fig. in [612] Zadeh, Fuzzy (1965), p. 343. ... 224
5.14 Combination of filter circuits. Fig. in [612] Zadeh, Fuzzy (1965), p. 344. ... 228

5.15 Elijah Polak. Fig. in: Informations about the authors in *Automatica*, Vol. 12, 1976, p. 389. 228
5.16 Tandem connection of two systems to an overall system. Fig. in [600] Zadeh, Concepts (1969), p. 18. 230
5.17 Example of a system as a combination of three subsystems. Fig. in [600] Zadeh, Concepts (1969), p. 21. 231

6.1 Rudolf Kalman. Photograph in the private archive of L. A. Zadeh [700]. 238
6.2 Mark Aizerman. Photograph in the private archive of L. A. Zadeh [700]. 242
6.3 Hierarchical structure of the linguistic variable "Age". Figure in: [679] Zadeh, ERL-Large (1974), p. 23. 248
6.4 Recipe for a chocolate sauce Zadeh adopted from R. S. Ledley [310] Ledley, Fortran (1966). Figure in [641] Zadeh, Outline (1973), p. 41. . 252
6.5 Richard E. Bellman. Photograph in the private archive of L. A. Zadeh [700]. 255
6.6 Joseph Goguen. Figure in: $http : //www - cse.ucsd.edu/-users/goguen/$. Photograph reprint courtesy of J. Goguen. 262
6.7 Enrique H. Ruspini. Photograph in $http : //www.ai.svi.com/people/-ruspini/$. Photograph reprint courtesy of E. Ruspini. 265
6.8 Attendees at the 1st NAFIPS Meeting in Logan, Utah, 1982. Photograph reprint courtesy of Elie Sanchez. 265
6.9 Sheldon L. Chang. Figure in: Information about the authors in *IEEE Transactions on Systems, Man, and Cybernetics*, Vol. SMC-2, No. 1, January 1972, p. 117. 269
6.10 Ebrahim H. Mamdani. Photograph in: $http : //www.iis.ee.ic.ac.uk/-\sim e.mamdani$. Photograph reprint courtesy of E. Mamdani. 271
6.11 The fuzzy-controlled steam engine. Photograph in: [26] Assilian, Control (1974), p. 20. 272
6.12 Diagram of the system consisting of a steam engine and a boiler. Figure in [26] Assilian, Control (1974), p. 18. 273
6.13 The system of the fuzzy steam engine. Figure in [26] Assilian, Control (1974), p. 41. 273
6.14 The process variables of the fuzzy steam engine. Figure in [26] Assilian, Control (1974), p. 31. 274
6.15 Fuzzy sets for the variables PE (Pressure Error) and SE (Speed Error). Figure in: [348] Mamdani, Fuzzy (1975), p. 8. 275
6.16 Fuzzy sets for the variables CPE (Change in Pressure Error) and CSE (Change in Speed Error). Figure in: [348] Mamdani, Fuzzy (1975), p. 8. 275
6.17 Fuzzy sets for the variable HC (Heat Change). Figure in: [348] Mamdani, Fuzzy (1975), p. 8. 275
6.18 Fuzzy sets for the variable TC (Throttle Change). Figure in: [348] Mamdani, Fuzzy (1975), p. 9. 276
6.19 Illustration of the application of the min-max rule. Figure based on [370] McNeill et al., Fuzzy (1993), p. 161. 277
6.20 Fuzzy control commands for the steam engine designed by Assilian and Mamdani. Figure in: [350] Mamdani, Advances (1976), p. 327. ... 278

List of Figures

6.21 Illustration of the selection of the centroid as a defuzzification method by devised Assilian and Mamdani. Figure in: [26] Assilian, Control (1974), p. 5. 279

6.22 The result of the Assilian-Mamdani Fuzzy Controller (○) compared to a conventional controller. (Dynamic Divergence Caching (DDC) algorithm damped (□) and undamped (x).) Figure in: [348] Mamdani, Fuzzy (1975), p. 6. 279

6.23 Ebrahim Mamdani, Lotfi Zadeh and several of Mamdani's students. Photograph in the private archive of E. Mamdani [701]. Photograph reprint courtesy of E. Mamdani. 280

7.1 Mechanical table to assist in diagnosis, developed by F. A. Nash, 1954. Figure in [412] Nash, Diagnosis (1954), p. 874. 285

7.2 Notch card system developed by Lipkin and Hardy, 1958. Figure in [323] Lipkin et al., Corellation (1958), p. 119. 288

7.3 Robert S. Ledley. Photograph in $http://www.amia.org/acmi$-$/fellows/fellows/ledley.htm$ (May 3, 2005). 289

7.4 Lee B. Lusted. Photograph in $http://www.amia.org/acmi$-$/fellows/fellows/lusted.htm$ (May 3, 2005). 291

7.5 Members of the NIH Advisory Committee on Computers in Research, May 1962. Figure in [343] Lusted, Roots (1987), p. 169. ... 293

7.6 Vladimir K. Zworykin. Photograph in $http://www.earlytelevision.org/yanczer\_history.html$. (May 3, 2005). 294

7.7 Logical combinations of attributes in a patient population as devised by Ledley and Lusted, 1959. Figure in [308] Ledley et al., Foundations (1959), p. 10. 300

7.8 Ledley's vision of a computer network in the health care system of the 1960s. Figure in [312] Ledley, Digital (1960), p. 22. 305

7.9 Karl Fellinger, Peter Bauer, Georg Grabner (left to right). Photograph: Archive of the Core Unit for Medical Statistics and Informatics, Medical University of Vienna [693]. 307

7.10 Karl Fellinger, Peter Bauer, Georg Grabner (left to right). Photograph: Archive of the Core Unit for Medical Statistics and Informatics, Medical University of Vienna [693]. 308

7.11 IBM/2260 computer screen unit and IBM/1053 printer for the WAMIS system. Figure in [35] Bauer et al., Computer (1972), p. 3. ... 309

7.12 Diagram of the computer for the WAMIS system. Figure in [35] Bauer et al., Computer (1972), p. 3. 311

7.13 Punch cards No. 1 and No. 2 for the Vienna diagnosis system. Figure in [516] Spindelberger et al., Computerverfahren (1968), p. 196. 312

7.14 Punch card No. 3 for the Vienna diagnosis system. Figure in [516] Spindelberger et al., Computerverfahren (1968), p. 198. 313

7.15 K.-P. Adlassnig. Photograph reprint courtesy of K.-P. Adlassnig. 319

7.16 Classical indicator function for "fever". Figure in [406] Moon et al., Diagnostic (1977), p. 116. 323

7.17 Membership function for "fever". Figure in [406] Moon et al., Diagnostic (1977), p. 116. 323

7.18 Modifiers for "low blood pressure". Figure in [406] Moon et al., Diagnostic (1977), p. 116. 324

7.19 Attendees of the IIZUKA-88 International Video-Session, KIT-NASA at the Kyushu Institute of Technology in Iizuka, Japan, 1988. Photograph in the private archive of E. Sanchez [703], reprint courtesy of E. Sanchez... 325

7.20 Relationships between symptoms and diagnoses according to Perez-Ojeda. Figure in [442] Perez-Ojeda, Knowledge (1976), p. 3.3. . 326

7.21 Semantic network of symptoms and diseases, disease complexes and statements according to Perez-Ojeda. Figure in [442] Perez-Ojeda, Knowledge (1976), p. 3.11.. 327

7.22 Elie Sanchez. Photograph in the private archive of E. Sanchez [703], reprint courtesy of E. Sanchez. 328

7.23 L. A. Zadeh and E. Sanchez at a dinner at the IEEE Conference on Decision and Control, New Orleans, Louisiana, December 7–9, 1977. Photograph in the private archive of E. Sanchez [703], reprint courtesy of E. Sanchez... 330

7.24 Membership functions for the fuzzy sets "occurrence" and "strength of confirmation". Figure in [9] Adlassnig, Fuzzy (1980), p. 145. 335

7.25 Structure of the computer-assisted diagnosis system (dashed lines: components that take effect before the diagnosis process commences), partially relabeled for legibility. Figure in [5] Adlassnig, Diagnosesystem (1983). 336

8.1 Honda Prize presentation ceremony, awarded to Lotfi Zadeh in 1989. Center: Lotfi Zadeh; to his right, his wife Fay and company founder Soichiro Honda (1906–1991). Photograph in the private archive of L. A. Zadeh [700].. 344

8.2 Lotfi Zadeh and the author of this book in the lecture hall of the Medical University of Vienna, Core Unit for Medical Statistics and Informatics, after a lecture Zadeh delivered there on March 2, 2005. Photograph in the private archive of DI Andrea Rappelsberger, photograph reprint courtesy of Andrea Rappelsberger. 349

Index of Names

Adlassnig, Klaus-Peter, 317–320, 331–334
Aizerman, Mark A., 190, 241
Albin, Merle A., 322
Alt, Franz, 89
Ampère, André-Marie, 121
Arco, Georg von, 13
Arnold, Harold De Forest, 10–12
Ashby, William R., 162
Assilian, Sedrak, 6, 272, 274, 275, 278, 279, 345

Babbage, Charles, 53
Bahr, Hermann, 66
Banach, Stefan, 91
Bartlett, A. C., 17
Bateson, Gregory, 141
Bauer, Peter, 317
Bellman, Richard E., 109, 167–170, 183–185, 190, 201, 202, 208, 209, 254, 255
Berbard, R., 318
Bergmann, Gustav, 73, 98
Berkeley, Edmund C., 49
Berlekamp, Elwyn, 222
Bertalanffy, Ludwig von, 4, 64, 111, 113–120, 138, 139, 189, 192, 198
Bigelow, Julian, 30–32, 120, 124, 126–128, 141
Birkhoff, Garrett, 220
Birkhoff, George D., 193

Black, Max, XI, 113, 237
Bode, Hendrik W., 17
Bohr, Niels H. D., 21, 340
Boltzmann, Ludwig, 68
Boole, George, 91
Born, Max, 21, 341
Boulding, Kenneth E., 119, 192–194
Bowles, Edward L., 26, 27
Brandt, W., 17
Braun, Carl F., 14, 24, 294
Braverman, Emmanuel, 241
Breidbach, Olaf, 147, 148, 157
Bremermann, Hans-Joachim, 260, 261, 322
Broch, Hermann, 67
Broglie, Louis V. P. R., Duc de, 21
Brouwer, Luitzen E., 79, 84–86, 93, 97, 101, 178
Brune, Otto W. H. O., 17, 27, 28
Bruner, Jerome S., 237
Brunswik, Egon, 112
Bühler, Charlotte, 82
Bühler, Karl, 70, 82
Bush, Vannevar, 15, 18, 26, 28, 52, 124, 286

Campbell, George, 14–16, 18, 28
Cantor, Georg, 1, 84, 338
Caratheodory, Constantin, 24
Carnap, Rudolf, 4, 72–75, 80–82, 84, 86–88, 92, 93, 95–99, 109–113, 143, 146

Carson, John R., 17
Cauer, Wilhelm, 17, 18, 20
Chang, Chin-Liang, 241, 264, 266
Chang, Sheldon S. L., 5, 266–268, 280
Chernoff, Herman, 178
Cherry, Colin, 34
Chwistek, Leon, 90
Clendening, Logan, 283, 296
Culbertson, James T., 162
Curry, Walter A., 33

Darlington, Sidney, 17, 18
Davis, M. D., 162
de Morgan, Augustus, 220
de No, Rafael L., 142
Desoer, Charles A., 172, 186, 199
Dirac, Paul, 21
Dirichlet, Peter G. L., 338
Dreyfus, Alfred, 66
Duhem, Pierre, 72

Edmundson, H. P., 237
Ehrenhaft, Felix, 85
Einstein, Albert, XV–XVII, 20–22, 68, 100, 341
Elder, Richard C., 322–324
Esogbue, Augustine O., 322–324
Espenschied, Lloyd, 8–10

Fano, Robert M., 33
Feigan, Gerry, 295
Feigl, Herbert, 72, 82, 85, 87
Fleming, Ambrose, 295
Foerster, Heinz von, 142, 198
Forest, Lee de, 295
Foster, Ronald M., 17, 28
Fréchet, Maurice René, 105, 106
Frank, Philipp, 71, 113
Frege, Gottlob, X, 74
Freiberger, Paul, X
Fremont-Smith, Frank, 141
Frenkel-Brunswik, Else, 112
Freud, Sigmund, 66, 69
Fried, Alfred H., 69
Fu, King S., 241

Gödel, Kurt, 101, 339
Gabor, Denis, 34
Gauß, Carl F., XV, 338

Geier, Manfred, 68
Geoffrion, Arthur M., 237
Gerard, Ralph W., 119
Gibbs, Josiah W., 166
Gödel, Kurt, 73, 74, 77, 87, 88, 92, 93, 101, 102, 158, 173
Goguen, Joseph, 261, 262
Goldscheid, Rudolf, 69
Goldstine, Herman, 148
Gomperz, Heinrich, 70
Gomperz, Theodor, 65
Grabner, Georg, 306, 310, 314, 317, 318
Guillemin, Ernst A., 4, 10, 18–20, 23–28, 31–36, 39, 63, 199, 343
Guillemin, Victor, 23
Gupta, Madan M., 328

Hahn, Hans, 70–72, 74, 75, 77–81, 85–88, 92, 99, 101
Hahn, Olga, 71, 92
Halmos, Paul R., 221
Hamming, Richard W, 264
Hardy, James D., 286, 299, 307
Hartley, Ralph V. L., 134, 135
Hashinger, Edward H., 296
Haufe, Robert, 53, 54
Hausdorff, Felix, 105
Hayek, Friedrich August von, 69
Heaviside, Oliver, 15, 45
Hebb, Donald O., 152, 157
Heims, Steve J., 132
Heisenberg, Werner K., XVI, 22, 340, 341
Hertz, Heinrich R., 12
Herzl, Theodor, 66
Hilbert, David, 84, 91, 120, 339
Hitler, Adolf, 99
Hodges, Richard, 260
Hönigschmid, Otto, 24
Hoff, Marcian Edward, 157
Hoffmann, Josef, 66
Hoffmannstahl, Hugo von, 65
Holmblad, Lauritz P., 281
Horvath, William J., 289
Hubel, David H., 147
Hurewicz, Witold, 79, 178

Husserl, Edmund, 120
Huxley, Aldous, 114

Janik, Allan, 67
Jodl, Friedrich, 69
Johnson, Lyndon B., 140
Jordanov, S., 318, 322–325

Kalaba, Robert, 109, 167, 168, 183–185
Kalman, Rudolf E., 26, 202, 239, 240
Kasper, Maria, 82
Kass, Seymour, 100
Kaufmann, Felix, 73, 87, 95, 96
Kelvin, William T., Lord, 302, 303
Kilmer, William L., 260
Kleene, Steven C., 162, 173, 211, 235, 315
Klement, Ernst-Peter, 3
Knaster, Bronislaw, 90
Kokoschka, Oskar, 65
Kolmogorov, Andrei N., 38, 121, 128
Kornreich, F., 318
Kosko, Bart, 2
Kraft, Viktor, 87
Krasovskii, Nikolai N., 190
Kraus, Karl, 66
Kreisel, Georg, 173, 174
Kronecker, Leopold, 84
Kuhn, Thomas, 292
Kuratowski, Kazimierz, 90

Lakoff, George, 262, 263
Landau, Edmund, 120
Langevin, Paul, 294
Ledley, Robert S., 251, 283, 289, 290, 296–299, 302–304, 306, 310, 314
Lee, Edward T., 264
Lee, Samuel C., 264
Leeuw, Karel de, 162
Lesniewski, Stanislaw, 91
Letov, Alexandr M., 190
Lettwin, Jerome Y., 147
Lewis, Clarence I., 98
Lindenbaum, Adolf, 90
Lipkin, Martin, 286, 299, 307
Loginov, Vasilij I., 235
Loos, Adolf, 66

Lowenschuss, Oscar, 174
Luce, Robert D., 46
Łukasiewicz, Jan, XI, 91
Luria, Alexander R., 271
Lurie, Anatolii I., 190
Lusted, Lee B., 283, 289–293, 295–299, 302, 303, 305, 306, 310, 314

Mach, Ernst, 65, 68, 72, 74, 75, 82, 102, 107, 109
MacKay, Donald M., 162
Malkin, Ioel G., 190
Mamdani, Ebrahim H., 6, 270–272, 274, 275, 278, 279, 345
Marconi, Guglielmo, 12–14
Marinelli, Wilhelm, 111
Mark, Hermann, 89
Markov, Andrei A., 193
Mason, Samuel J., 46
Maturana, Humberto, 147
Maxwell, James C., 12, 121
McCarthy, John, 160–162
McCulloch, Warren S., 141–148, 150, 151, 158
McNeill, Daniel, X
Mead, Margaret, 141
Menger, Karl, 3, 4, 73, 74, 76, 77, 79, 82, 83, 85–93, 95, 96, 98–103, 105–109
Mesarovic, Mihajlo D., 192, 197
Milhaud, Gaston, 72
Miller, Kenneth S., 39
Minkowski, Hermann, 340
Minsky, Marvin L., 153, 155, 156, 160–162, 182, 203, 204, 305, 326
Mises, Ludwig von, 69
Mises, Richard von, 71, 73
Montgomery, Deane, 172
Moon, Richard, 318, 322–325
Moore, Edward F., 162
More, Trenchard, 161
Morgenstern, Oscar, 77, 169
Morris, Charles W., 113, 117
Murray, Francis J., 49
Musil, Robert, 68

Nagy, George, 204
Nash, F. A., 284
Nemytskii, Victor, 193
Neumann, John von, 101, 142, 143, 148–152, 158–160, 163, 168, 190, 298, 341
Neurath, Otto, 71, 72, 74, 75, 82, 87, 88, 113
Newell, Allen, 161
Norton, Frederick J., 26
Novikoff, Albert B. I., 241
Nyquist, Harry, 134

Østergaard, Jens-Jorgen, 281

Page, Arthur W., 10
Papert, Seymour, 155, 156, 182, 203, 204, 305
Pask, Andrew G. S., 271
Pauli, Wolfgang E., 22
Paycha, François, 285, 286
Perez-Ojeda, Alonso, 318, 322–326, 328
Pierce, John R., 121
Piloty, Hans, 17
Pitts, Walter H., 143, 144, 146–148, 150, 151, 158
Plail, Michael, 169, 170, 178
Planck, Max K. E. L., 20, 21
Plato, 121
Poincaré, Jules Henri, 71, 72, 83, 103, 105, 107, 193, 338
Polak, Elijah, 228
Polanyi, Karl, 111
Pontryagin, Lev S., 188, 190, 193
Popper, Karl R., 76, 83
Popper-Lynkeus, Josef, 68

Quillian, M. Ross, 326
Quine, Willard van Orman, 173, 174

Ragade, Rommohan K., 328
Ragazzini, John R., 36, 38, 39
Ramsey, Frank P., 88
Rapoport, Anatol, 119, 120, 198
Rashevsky, Nicolas, 117, 142, 143
Reich, Wilhelm, 112
Reichenbach, Hans, 75, 88, 113
Reidemeister, Kurt W. F., 73, 80, 81

Reininger, Robert, 70
Rey, Abel, 72
Rider, Jacques Le, 67
Riemann, Bernhard, 338
Righi, Augusto, 13
Robbins, Herbert E., 172, 185, 186
Rochester, Nathaniel, 160, 161
Röntgen, Wilhelm K., 23
Rogowski, Walter, 25
Rosch, Eleanor, 262
Rosen, Judah B., 204–206
Rosenblatt, Frank, 153, 155, 156, 162, 203–205
Rosenblueth, Arturo, 30, 31, 120, 122, 125–128, 141
Roy, Édouard Le, 72
Rozonoér, Lev I., 241
Russell, Bertrand, 74, 80, 83, 120, 146

Samuel, Arthur, 161
Sanchez, Elie, 325, 327–331, 333
Santos, Eugene S., 243–246
Schönberg, Arnold, 65
Schlesinger Kubie, Lawrence, 141
Schlick, Moritz F. A., 70, 72–75, 77, 79–81, 84–89, 96, 98, 99
Schmid, Josef, 314
Schnitzler, Arthur, 65, 66
Schorske, Carl E., 66
Schröder, Ernst F. W. K., 91
Schrödinger, Erwin, 21, 107
Schurz, Carl, 22
Schuschnigg, Kurt von, 99
Schwarz, R. J., 39
Selfridge, Oliver, 157, 161
Shannon, Claude E., 4, 32, 33, 41, 48–50, 52, 131–136, 138, 144, 146, 160–162, 165–167, 177, 190, 218, 224
Shapiro, Neil, 162
Sierpinski, Waclaw, 91
Siforov, Vladimir I., 218
Simon, Herbert A., 161
Skolem, Thoralf A., 91
Slaby, Adolf, 13
Smets, Philippe, 318, 319, 324
Solomonoff, Raymond J., 161
Sommerfeld, Arnold J. W., 19, 22–26

Spindelberger, Walter, 306, 310, 314, 317
Stach, Heike, 150, 151
Stadler, Friedrich, 67
Stalin, Josef W., 236
Steinbuch, Karl, 157
Stevenson, Robert L., 34
Strig, Richard, 111
Suttner, Berta von, 69

Türksen, I. Burhan, 318, 322–325
Tarski, Alfred, 4, 90–93, 110, 145
Tautu, Petre, 319, 322
Thirring, Hans, 77, 89
Thirring, Walter, 77
Todesco, Sophie, 65
Turing, Alan M., 158, 159, 161, 254, 339

Ulrich, Werner, 174
Uttley, Albert M., 162

Vainel, H., 318
Varela, Francesco J., 152
Veblen, Oswald, 122

Wagner, Gustav, 319, 322
Wagner, Karl W., 14
Wagner, Otto, 66
Waismann, Friedrich, 72, 76, 82, 87, 88
Wald, Abraham, 77, 169, 183
Watt, James, 121
Weaver, Warren, 136, 137
Wechsler, Harry, 322
Wee, William G., 241–244, 246
Weiss, Lionel, 178

Wertheimstein, Josefine von, 65
Whitehead, A. N., 146
Whitman, Walt, 8, 9
Widrow, Bernard, 157
Wien, Wilhelm (Gulielmo) C. W. O. F. F., 23
Wiener, Norbert, 4, 28–32, 38, 39, 44, 46, 48–50, 52, 119–138, 141–143, 152, 165–167, 175, 177, 198
Wiesel, Torsten N., 147
Wightman, Charles, 153
Winograd, Terry, 271
Wittgenstein, Ludwig, 65, 74, 76, 80–83, 86, 88, 92, 95, 96, 98
Wolber, U., 319

Yager, Ronald R., 328

Zadeh, Lotfi A., IX, X, XVIII, 1–6, 32–34, 36–50, 52–55, 57–59, 61–64, 109, 110, 170–196, 198–204, 206, 208–212, 214–219, 221–233, 235–243, 245–247, 249–251, 253–257, 259–264, 266–268, 270, 272, 276, 280, 303, 319–324, 327–329, 332–334, 342–345, 347–349
Zeilinger, Anton, 342
Zeisel, Hans, 111
Zenneck, Jonathan, 24
Zilsel, Edgar, 73, 88
Zobel, Otto J., 17, 18, 28
Zuckerkandl, Berta, 65
Zworykin, Vladimir K., 294, 295, 297

Index of Subjects

algebra Boolean, 52, 57
algebraic
    product of fuzzy sets, 221
    sum of fuzzy sets, 221
algorithm, 5, 153, 158, 241, 246, 254, 266, 270, 272, 277, 279
all-or-none law of nervous activity, 144–146
anti-aircraft weaponry, 121–126
antinomy, 338
arithmetic, 83, 84, 95, 99, 102, 150, 271
artificial
    intelligence, 143, 160, 161, 204, 270, 278, 305
    neural networks, 5
automata theory, 5, 49, 61, 109, 173–175, 188, 190, 192, 194, 204, 241–245, 253
automaton
    maximin, 245
    probabilistic, 244
    pseudo-, 244

BINAC (Binary Automatic Computer), 54
biomathematics, 328
biomedicine, 290, 292, 303
BISC (Berkeley Intitiative in Soft Computing), 343
black box, 58, 167, 170, 171, 185, 190, 191

Boolean algebra, 52, 57, 131, 146, 301, 310
brain and computer, 50, 51, 55, 148, 151, 152, 157

CADIAG (Computer-Assisted Diagnosis), 306, 309, 311, 312, 320
CADIAG-I, 6, 314, 315, 317, 331
CADIAG-II, 6, 317, 331, 335
card and needle system, 287
central nervous system, 295
circuit, electrical, 4, 13, 121, 144, 147, 148, 171, 174, 175, 188, 193, 204, 224, 226, 227
cognitive science, 152, 155, 157
colorimetry (color metrics), 107
Columbia University, New York, 17, 33, 36, 39, 49, 50, 56, 172, 174, 201, 202, 239
communication, 8, 10, 320
    engineering, 4, 29, 121, 128, 135, 188
    model, 132, 136
    network, 9, 10, 18, 27–29, 188, 304
    system, 8, 13, 15, 28, 40, 42, 44–46, 109, 134, 168
    technology, 4, 7–13, 109, 129, 134, 163, 227, 292
    theory, X, 13, 41, 63, 120, 131, 132, 135, 166, 167, 170, 175, 184, 206, 320
    via machine, 9

complement of a fuzzy set, 214
complexity, IX, 2, 11, 176, 194, 198, 199, 246, 260, 269, 270, 272, 281, 292, 303, 321
computational intelligence, 3
Computational Theory of Perceptions, 347, 349
computer, 4, 53, 206, 228, 239, 246, 261, 276, 283, 286, 288, 292, 295, 296, 299, 303–305, 308, 315, 337
    analog, 52
    and brain, 148, 151, 152, 157, 347
    digital, 47, 49, 54, 58, 149–151
computer sciences, 198, 245, 246
    medical, 298, 307, 308, 317, 319, 328
computing and sorting machines, 286
Computing with Words, 347, 349
constructivity, mathematical, 86, 93, 94, 98
continuum, 102–104, 107
control
    engineering, IX, XI, 28, 128, 178, 185, 269–272, 280, 304
    theory, 202, 204, 206, 242, 266, 268–270, 279, 280
conventionalism, 71, 72, 85, 86
convex combination
    of fuzzy sets, 221
    of sets, 221
cybernetics, IX, 4, 5, 28–32, 48, 50, 63, 109, 118–121, 127, 131, 133, 136–140, 142, 143, 166, 192, 198, 218, 235, 236, 259, 260

Dartmouth Conference, 160, 161
decision theory, 5, 168–170, 174, 178, 183, 184, 204, 264, 305, 306
diagnosis
    medical, 6, 261, 283–286, 288–291, 294–299, 301–307, 309–311, 314, 315, 317–331, 333, 335, 336, 345
    computer-assisted, 6, 261, 283, 289, 290, 297, 305, 306, 309–312, 314, 315, 317, 318, 320, 335, 345
Differential Analyzer, 50, 52, 54
digital computer, 4, 239, 246, 261, 276, 288, 304, 305, 308, 315
disease, 287, 310, 312, 314–317, 319, 325–327, 330–332
distance function, 42, 45, 105–107, 109, 154, 205, 206, 212
dynamic
    equilibrium, 118
    programming, 167–170, 183, 184, 188, 201, 266

EDVAC (Electronic Discrete Variable Automatic Computer), 149, 150
electrical
    engineering, X, 4, 9, 79, 119, 128, 179, 188, 199, 236, 290, 295
    and mathematics, 11, 36, 57
    and ROC-analysis, 305
electrical circuit, 4, 13, 121, 144, 147, 148, 171, 174, 175, 188, 193, 204, 224, 226, 227
electronic brain, 303
electronics, 8, 291
Empiricism, Logical, 4
energy information, matter, 137
ENIAC(Electronic Numerical Integrator And Computer), 148–150, 158
ensemble flou, 4, 98, 108, 109
entropy, 130, 133, 165
equality
    mathematical, 104, 105, 109, 147
    of fuzzy sets, 214, 220
    physical, 103
Erkenntnis (philosophical journal), 75, 112, 113
Ernst Mach Society, 74, 75, 111
ether, 11
expert system, 345

feedback, 5, 6, 30, 118, 119, 124–128, 138, 139, 267, 268
filter
    band-pass, 16

electrical (Siebschaltung), 4, 5, 13, 17, 129, 137, 170, 177, 186, 224, 226, 227, 343
  high-pass, 16
  ideal, 4, 40, 41, 43–45, 226, 343
  low-pass, 16
  optimum, 4, 44, 45, 49, 177, 226, 343
  theory, 30, 43, 343
fineness of a fuzzy set, 267
Ford Foundation, 141
Foundations of the Theory of Probability (A. Kolmogorov), 128
Fourier
  analysis, 12, 25, 48
  integral, 34
function space, 40–42, 45
fuzzification, XI, 3, 5, 6, 233, 235, 240, 241, 246, 247, 254, 256, 257, 259, 263, 266, 276, 283, 310, 323
fuzziness, IX, X, 1, 2
  of sensations, 104
fuzzy
  -IF-THEN rule, 6, 249, 250, 253, 275, 329
  algebra, 266
  algorithm, XI, 5, 6, 232, 240, 242, 245–251, 253, 254, 268, 270, 272, 320, 330
    behavioral, 253
    decisional, 253
    definitional, 250
    generational, 251
    relational, 253
  automaton, XI, 5, 241–243, 253
  cement kiln, 281
  control, 6, 232, 266, 273, 278–281, 330, 345
  inference rule, 256, 330
  instruction, 249
  intersection, X
  logic, IX, 4, 5, 255, 256, 263, 266, 320
    controller, XI
  mapping, 267, 268
  mathematics, 64, 189, 259, 268, 303, 320, 321, 343
  methods, IX, XII, 6, 318

  in medical diagnosis, 322–325, 327–334, 336
  operator, X, 5, 256
  ordering relation, 256, 257
  output function, 243
  probability theory, 5, 266
  relation, X, 1, 6, 221, 232, 242, 243, 253, 256, 257, 276, 327, 329–331, 333, 334, 344
  semantics, 257
  set, X, XI, 1, 2, 5, 6, 109, 202–204, 208–228, 232, 236, 237, 240, 242, 245, 246, 250, 251, 254–257, 261–264, 266, 267, 274, 276, 277, 318, 320, 323–325, 344
    bounded, 222, 223
    convex, X, 214, 215, 221–223
    cylindrical, 215, 218
  similarity relation, 256
  state, 254
  statistics, 266
  steam engine, 270, 272, 278, 280
  subset, 214, 220, 267, 332
  system, IX, XI, 3, 5, 6, 176, 201, 202, 215, 216, 227, 228, 231, 246, 254, 281, 320, 321, 344
  topology, 5, 266
  transition function, 243
  Turing machine, XI, 254
Fuzzy Set Theory, IX–XI, 3–6, 34, 109, 110, 155, 176, 182, 194, 203, 210, 213, 218, 224, 227, 231–233, 235, 236, 238, 239, 248, 259–263, 268, 276, 318, 319, 321, 322, 328, 332, 334, 336, 343

game theory, 168, 174, 178, 184, 188, 298
General Systems Theory (Allgemeine Systemtheorie), 4, 5, 63, 109, 110, 116–119, 138, 139, 245, 321
geometry, 73, 79, 80, 93, 106–108
  Euclidean, 79, 95
  micro, 107, 108
  non-Euclidean, 95

of lumps, 108
positivistic, 107
graph theory, 46, 61

hazy set, 108
heavy current engineering
(Starkstromtechnik), 29
Hebb rule of learning, 154

IF-THEN rule, 248, 250, 253, 315
imprecise (unsharp), 170, 185, 199, 202, 203, 215, 216, 245, 246, 248, 250, 251, 263, 343, 344
incompleteness
of information, 199
theorem (K. Gödel), 93
indicator function of a set, 225, 232
indistinct, 4
inference rule, 256, 329, 345
inferiority of a system, 180, 181
information, 128–131, 133–135, 137–139, 152, 163, 286, 320, 347
incompleteness, IX
matter, energy, 137
technology, 4, 246, 292
theory, X, 4, 121, 165–168, 175, 188, 204, 320
transmitting telegraph, 12
input
of a system, 126, 127, 150, 153, 163, 168, 171, 179, 184, 185, 187, 189–191, 193–197, 215–217, 230, 246, 253
of an automaton, 163
pole, 46
input-output
analysis of a system, XI, 5, 229, 230
relationship of a system, 39, 46, 49, 57, 59, 60, 171, 185, 191, 193, 195–197, 199, 217, 229–232
Institute
for Advanced Study (IAS), Princeton, 73, 131, 173–176
for Automatics and Telemechanics in the Soviet Union, 175

of Radio Engineers (IRE), 8, 17, 34, 36, 63
Institute for Medical Computer Sciences, University of Vienna, Medical School, 307
intention tremor, 125
International Encyclopedia of Unified Science (book series), 113
intersection
of fuzzy sets, X, 1, 5, 210, 214, 220, 223–227, 232
of sets, 225
intuition, 86, 95
intuitionism, 4, 78, 79, 83–87, 93–98, 339

laboratory tests, 286–288, 298, 302, 305, 314, 323, 326
language
in science, 4
scientific, 84–88, 91, 93, 95–97, 100, 109, 110, 112, 115
language – science – world
three-way relationship, 82
lattice theory, 220
law of excluded middle
(tertium non datur), X
law of incompleteness (K. Gödel), 339
Lemberg-Warsaw school, 90, 91, 109
life sciences
and electrical engineering, 289, 292, 295, 296, 321
and fuzzy sets, 260, 321
linguistics, 204, 233, 237, 262, 263, 271, 272, 320
logic, X, XI, 4, 17, 52, 54, 57, 70–72, 74, 76, 81–87, 89–91, 93, 95, 96, 98, 100, 109, 110, 112, 132, 143, 144, 146, 147, 150–152, 155, 157, 161, 162, 172–174, 199, 211, 224, 232, 237, 314, 315
and diagnosis, 290
and mathematics, 83, 86, 89, 91–94, 100, 288, 305
and reality, 109
multi-valued, X, 57, 83, 91, 109, 173, 174, 255, 262
propositional, XI

symbolic, 298
three-valued, 6, 91, 210, 211, 315, 317, 331
two-valued, 314
Logic of Scientific Discovery (Logik der Forschung), 76, 83
Logical Empiricism, 4, 71, 75
Logical Theorist (program), 161, 203
logical tolerance principle, 4, 87, 93, 112, 113
Logicism, 83, 84, 93

machine intelligence, 238
Machine Intelligence Quotient, 349
Macy
    conference, 142, 143, 152
    Foundation, 141
Massachusetts Institute of Technology (MIT), Cambridge, 14, 15, 17, 33
Mathematical Colloquium in Vienna, 88–90, 99
mathematics, 3, 337
    and electrical engineering, XIX, 11, 36, 57
    and logic, 83, 86, 89, 91–94, 100, 288, 305, 338, 339
    and reality, XVII, XVIII, 100, 340–342
    exact, 2, 5, 11, 64, 173, 178, 199, 200, 268, 270, 303, 320, 321, 342, 343
matter information, energy, 137
max-min-composition, 6, 224, 232, 243, 257, 276, 277, 328–330, 333, 345
Max-Planck-Institutes, 175
maximum of membership functions, 225, 226, 257
McCulloch-Pitts neuron, 146, 148, 151, 152, 154
medical information, 297, 303
medical knowledge, 6, 297, 298, 302, 309, 311, 314, 315, 320, 325, 326, 328–330, 334
medicine
    and computers, 297
    and electrical engineering, 290–292

membership function of a fuzzy set, 203, 207–215, 217–221, 224–227, 229–232, 235, 237, 243, 248, 254, 256–258, 263, 267, 277, 279, 323, 324, 329, 332–336
membership function of an introduced fuzzy set, 324
membership, set theoretical, 154, 203, 276, 277, 332
metamathematics, 4, 91, 92, 110, 339
methods
    exact, IX, 198
    fuzzy, IX, XII, 6
    in medical diagnosis, 6, 260, 318
metrics, 42, 79, 105, 106, 108, 109, 168, 237
    probabilistic, 3, 4, 106
    statistical, 3, 4, 106, 107, 109
minimum of membership functions, 225, 226
Mises Circle, 69
multipole, 44, 46, 47, 58–61, 170
    linear, 47
    non-linear, 46, 47, 61

Neo-Positivism, 4
network
    analysis, 4, 17, 26, 28
    communication, 18, 28, 29
    electrical, 26, 27, 52, 170, 171, 187, 188, 224, 226, 227
    engineering, 47
    ladder (Kettenschaltung), 15
    neural
        artificial, 109, 153, 188, 246, 260, 271
        natural, 122, 128, 138, 141, 143, 144, 146, 150, 151
    semantic, 326
    synthesis, 4, 17, 26, 28
    theory, 13, 18, 27, 32, 34–36, 46
        electrical, 17
    time-varying, 179
    variable, 36, 37, 49, 59
neuron, 146, 147, 153
noise, 31, 36, 42, 44, 129, 131, 137, 167, 190

non-inferiority of a system, 180, 181, 183

observation, 267, 268
occurrence of a symptom, 315, 316, 329, 331–333
operations research, 63, 184, 255, 305
operator theory, 30, 31, 37, 39, 45, 46
optimality, 169, 170, 176–178, 180, 181, 184, 194
output
    of a system, 126, 127, 154, 163, 167, 171, 176, 179, 185, 186, 189–191, 193, 195, 196, 215–217, 230, 246, 253
    of an automaton, 159
    pole, 46

paradoxes, 338
parallel combination
    of electrical circuits, 224
    of electrical filters (sieves), 224
passband of filters, 226
pattern recognition, 1, 5, 132, 148, 153–157, 162, 174, 202–209, 212–214, 218, 219, 221, 222, 241, 242, 246, 247, 260, 261, 265, 266, 271, 320
perception, 347
perceptron, 5, 153–156, 162, 182, 203–205
performance of a system, 168, 170, 177, 180, 181, 184, 198
power engineering (Starkstromtechnik), 29
prediction
    operator, 30
    theory, 30, 38, 46, 121, 126, 128, 129, 177
principle
    of logical tolerance, 4, 87, 88, 93, 95–97, 109, 110, 112, 113
    of the excluded third (tertium non datur), X, 93
    of uncertainty (W. Heisenberg), 341, 342
Principles of Thermodynamics (Prinzipien der Wärmelehre), 102

probability, 4, 56, 64, 104–109, 129–131, 135, 163, 202, 216, 235, 262, 310, 327
    theory, XI, 106, 109, 128, 178, 186, 235, 263, 298, 310
product of sets, 217, 243, 254, 256, 334
prototype theorie (E. Rosch), 262
psychoanalysis community (Wednesday Society), 66
psychon, 145
punch cards, 286, 290, 299, 311, 312

quantum
    mechanics, XVI, 20, 40, 41, 340, 341
    theory, 20, 74, 166, 340, 342

radar, 123, 129, 138
radio, 8–13, 17, 19, 64, 134, 136–138, 291, 304
RAND Corporation, Santa Monica, 161, 201, 202, 208, 254
reality
    and logic, 109
    and mathematics, 100
relation, set theoretical, 221, 256, 315, 325, 326
relay switch, 135, 148–150
response separation property, 196, 197
rigid set, 108
ROC (Receiver Operating Characteristic) analysis, 305, 306

s-norm, 109
science – world – language three-way relationship, 82
scientific variable, 101, 102
Scientific World View (Wissenschaftliche Weltauffassung), 74
separability, linear, 5, 181, 182, 204–206, 212, 222, 223, 226
separation
    theorem for convex fuzzy sets, 241
    theorem for convex sets, 206, 223

series combination
    of electrical circuits, 224
    of electrical circuits (sieves), 227
set
    convex, 181, 182
    hazy, 108
    of the real world, 83, 268
    rigid, 108
set of the real world, 2
set theory, XI, 1, 2, 5, 84, 85, 87, 91,
        203, 204, 207, 208, 213, 225,
        232, 246, 256, 262, 338
shadow of a fuzzy set, 214, 217
similarity relation, 257
Sloan Foundation, 141
soft computing, 3, 343, 344
Stanford Research Institute (SRI),
        Menlo Park, California, 241
state
    equation, 190, 197
    of a quantum mechanical object,
        341, 342
    of a system, 62, 159, 163,
        169–171, 179, 180, 183, 185,
        189–191, 193–197, 215, 216,
        246, 253, 254, 267, 268
    of an automaton, 163
    transition function, 244
state space approach
        (L. A. Zadeh), 5, 177,
        186, 188
statistical mechanics, 130, 135
statistics, XI, XVIII, 21, 29–32, 38,
        44, 45, 59, 121, 129, 130,
        132, 134, 135, 155, 166, 167,
        170, 174, 178, 183, 184, 204,
        207, 211, 227, 235, 298, 310
steam engine, 121, 270, 272, 278, 280
stopband of filters, 226
strength of confirmation of a symptom, 331–334
symptom
    facultative, 313, 314
    medical, 283, 284, 286, 287,
        297–299, 310, 313–317, 319,
        325–327, 329–332
    obligatory, 313, 314
symptom-disease relationship, 299,
        315, 317

system
    abstract, 58, 116, 119, 129, 139,
        140, 163, 165, 169, 170,
        177–181, 183–186, 188–195,
        198, 215–217, 230–232
    adaptive, 5, 183, 184, 213
    analog, 48, 49
    biological, 50, 51, 63, 64, 116,
        118, 175, 189, 199, 260,
        303, 321
    characterization, 190, 191,
        216, 229
    classification, 191, 216
    closed, 118
    complex, 1, 2, 199, 240, 245, 251,
        260, 268, 269, 272, 303, 321,
        336, 343
    digital, 49, 50
    dynamic, 267
    ideal, 227
    identification, 171, 185, 186, 191,
        192, 247
    inanimate, 50, 63, 64
    knowldege-based, 304
    linear, 5, 47, 61, 171–173,
        180, 199
    non-linear, 39, 46, 47, 57, 61, 171
    open, 118, 119
    optimum, 5, 177, 180, 181, 194
    physical, 12, 38, 170
    real, 5, 14, 171, 177, 178, 187,
        189, 199, 200, 227, 232, 233,
        268, 342–344
    theory, 5, 57–59, 63, 174, 175,
        177, 179, 185, 188, 190–194,
        198, 199, 216, 217, 227–232
    general, 63
    technical, 13, 139, 140, 163,
        197, 246, 321

t-norm (triangular norm), 4
tautology, 82, 83, 86, 87
telegraph, 9, 10, 14, 24, 30, 129, 136,
        137, 150
    wireless, 12, 13
telemedicine, 304
telephone, 9, 10, 14, 129, 136–138, 304
teletype, 304

Index of Subjects

television, 9, 29, 129, 136, 137, 290, 294
tertium non datur
    (law of excluded middle, principle of the excluded third), X
The Limits to Growth
    (D. Meadows), 140
theory of relativity, XV, 340
thinking machine, 4, 49, 50, 53–57, 158
Tic-Tac-Toe, 53, 54
time series, 28–32, 38, 121, 128, 129
Tit-Tat-Toe, 53, 54
topology, 17, 42, 78, 106
Tractatus logico-philosophicus, 74, 80–83
triangular norm (t-norm), 107, 109
truth value, logical, 82, 83, 146, 155, 255, 256, 262, 314, 315
Turing machine, 158, 159, 161, 162, 189, 190, 254
    fuzzy, XI, 254
    nonfuzzy deterministic, 254
    nonfuzzy nondeterministic, 254
two-pole (2-pole), 46, 47, 59, 61, 171

uncertainty, IX, XVIII, 178, 183, 199, 270, 343, 344
uncertainty principle
    (W. Heisenberg), XVI, 341, 342
union
    of fuzzy sets, 1, 5, 210, 214, 220, 224–227, 232
    of sets, 225
unity of science (Einheitswissenschaft), 110, 113, 115

UNIVAC (Universal Automatic Computer), 54
University
    of Bonn, 74
    of California, Berkeley, IX, X, 174, 176, 186, 198, 201, 202, 208, 222, 235, 236
    of California, Los Angeles, 254
    of Chicago, 69
    of Munich, 19, 22, 23, 25
    of Notre Dame, Indiana, 99
    of Vienna, 66, 68, 71, 80
unsharp (imprecise), 170, 185, 199, 202, 203, 215, 216, 245, 246, 248, 250, 251, 263, 343, 344

variable, linguistic, 248, 251, 272, 274, 331, 332, 344
verifiability, 82, 96
Vienna
    avant-garde, 66, 67
    coffeehouse, 66, 71
    modern, 65–67
    workshop (Wiener Werkstätte), 66
Vienna Circle, 4, 70–72, 74–76, 80–91, 93, 97–100, 102, 103, 109–112, 141

WAMIS (Wiener Allgemeines Medizinisches Informationssystem), Vienna General Medical Information System, 308, 309, 317
wave, electromagnetic, 10–14
weak current engineering
    (Schwachstromtechnik), 29
world – language – science
    three-way relationship, 82

Printing: Krips bv, Meppel
Binding: Stürtz, Würzburg